Field-effect Transistor Biosensors for Rapid Pathogen Detection

Detection Science Series

Titles in the Series:

1: Sensor Technology in Neuroscience
2: Detection Challenges in Clinical Diagnostics
3: Advanced Synthetic Materials in Detection Science
4: Principles and Practice of Analytical Techniques in Geosciences
5: Microfluidics in Detection Science: Lab-on-a-chip Technologies
6: Electrochemical Strategies in Detection Science
7: Peroxynitrite Detection in Biological Media: Challenges and Advances
8: Biological Fluid-Surface Interactions in Detection and Medical Devices
9: Advanced Environmental Analysis: Applications of Nanomaterials, Volume 1
10: Advanced Environmental Analysis: Applications of Nanomaterials, Volume 2
11: Quenched-phosphorescence Detection of Molecular Oxygen: Applications in Life Sciences
12: Carbon-based Nanomaterials in Analytical Chemistry
13: Forensic Analytical Methods
14: Immunosensors
15: Analytical Electrogenerated Chemiluminescence: From Fundamentals to Bioassays
16: *In Situ* Analysis of Cellular Functional Molecules
17: Confining Electrochemistry to Nanopores
18: Detection Methods in Precision Medicine
19: Analytical Strategies for Cultural Heritage Materials and their Degradation
20: Challenges in Detection Approaches for Forensic Science
21: Disposable Electrochemical Sensors for Healthcare Monitoring

How to obtain future titles on publication:
A standing order plan is available for this series. A standing order will bring delivery of each new volume immediately on publication.

For further information please contact:
Book Sales Department, Royal Society of Chemistry, Thomas Graham House, Science Park, Milton Road, Cambridge, CB4 0WF, UK
Telephone: +44 (0)1223 420066, Fax: +44 (0)1223 420247
Email: booksales@rsc.org
Visit our website at www.books.rsc.org

Field-effect Transistor Biosensors for Rapid Pathogen Detection

Edited by

Naif H. M. Al-Hardan
Department of Applied Physics, Faculty of Science and Technology, Universiti Kebangsaan, Malaysia
Email: naifalhardan@ukm.edu.my

Azman Jalar
Department of Applied Physics, Faculty of Science and Technology, Universiti Kebangsaan, Malaysia
Email: azmn@ukm.edu.my

Muhammad Azmi Abdul Hamid
Department of Applied Physics, Faculty of Science and Technology, Universiti Kebangsaan, Malaysia
Email: azmi@ukm.edu.my

and

Mohd Firdaus Raih
Department of Applied Physics, Faculty of Science and Technology, Universiti Kebangsaan, Malaysia
Email: firdaus@mfrlab.org

ROYAL SOCIETY OF CHEMISTRY

Detection Science Series No. 24

Print ISBN: 978-1-83767-081-9
PDF ISBN: 978-1-83767-342-1
EPUB ISBN: 978-1-83767-343-8
Print ISSN: 2052-3068
Electronic ISSN: 2052-3076

A catalogue record for this book is available from the British Library

The Royal Society of Chemistry is a charity, registered in England and Wales, Number 207890, and a company incorporated in England by Royal Charter (Registered No. RC000524), registered office: Burlington House, Piccadilly, London W1J 0BA, UK, Telephone: +44 (0)20 7437 8656.

For further information see our website at www.rsc.org

Printed in the United Kingdom by CPI Group (UK) Ltd, Croydon, CR0 4YY, UK

Preface

Disease outbreaks caused by various pathogens that have rapidly evolved into pandemics are becoming more common and frequent. Within the span of slightly over a century, the effects of these pandemics from Spanish flu in the early 20th century to COVID-19 in the first quarter of the 21st century have been devastating. Detecting pathogens, such as viruses, bacteria, and fungi, have become a crucial task in preventing future outbreaks and to avoid the need for widespread and crippling lockdowns that have significantly impacted societies and economic activities. The absence of rapid detection methods that are able to diagnose early stages of infections has led to the rapid spread of these diseases that quickly developed into pandemics. Therefore, the capacity to rapidly detect infections and disrupt their spread are the major motivations driving the advancements in the use of field effect transistors (FETs) that we elaborate on in this book. The concept of such devices was introduced to the world by J. E. Lilienfeld *via* a patent application for a FET in 1925. However, it was only decades later that the first biosensor was developed by Clark in 1962. A few years later, Bergveld in 1970 introduced for the first time an ion-sensitive FET (ISFET), which had been used to detect several ions such as hydrogen (pH), potassium and others that are of physiological interest. This paved the way for the introduction of biologically modified field-effect transistors (BioFETs).

In this title, **Field-Effect Transistor Biosensors for Rapid Pathogen Detection**, we have invited prominent researchers in the field of BioFETs for pathogen detection to contribute chapters. This book outlines a selection of BioFET applications in this field, which has been divided in two main parts. The first part includes chapters that deal with the history of ISFETs and their future, the physics of FET devices, electrolyte gated FET biosensors, and the challenges in the detection of emerging novel pathogens and evolving

Detection Science Series No. 24
Field-effect Transistor Biosensors for Rapid Pathogen Detection
Edited by Naif H. M. Al-Hardan, Azman Jalar, Muhammad Azmi Abdul Hamid and Mohd Firdaus Raih

Published by the Royal Society of Chemistry, www.rsc.org

known pathogens. The second part includes chapters on the rapid detection of microorganisms based on FET devices, field-effect transistor biosensors based on nanomaterials for zoonotic pathogen detection, and field-effect transistor biosensors for rapid SARS-COV-2 detection, with the last chapter in this part discussing the future of commercializing FET-based biosensors.

Naif H. M. Al-Hardan
Azman Jalar
Muhammed Azmi Abdul Hamid
Mohd Firdaus Raih

Acknowledgements

We express our gratitude to the authors for their willingness to share their experiences in the field of BioFET application for the detection of various infections. We also express our gratitude to the reviewers for their valuable assistance in assessing the substance of each chapter and providing input that has greatly enhanced the quality of the material. The editors of the book would like to express their gratitude for the support received from the Faculty of Science and Technology at Universiti Kebangsaan Malaysia (UKM). We would like to express our gratitude for the support received from the Centre for Research Management and Instrumentation (CRIM-UKM) through a short-term grant, identified as GPS-DPK-2023-002.

We express our gratitude for the valuable contributions provided by the publishing team at the Royal Society of Chemistry in facilitating the editing process of the book. Additionally, we extend our appreciation to the editorial board of the RSC Detection Science Series, who accepted our proposal on the subject.

Detection Science Series No. 24
Field-effect Transistor Biosensors for Rapid Pathogen Detection
Edited by Naif H. M. Al-Hardan, Azman Jalar, Muhammad Azmi Abdul Hamid and Mohd Firdaus Raih

Published by the Royal Society of Chemistry, www.rsc.org

About the Book

Field-effect transistors (FETs) have been investigated widely as biosensors for pathogen detection, with advantages such as label-free and real-time detection capabilities. These biosensors utilize the changes in electrical properties, such as current and voltage, induced by the presence of pathogens on the channel surface of the transistor. The devices have a high level of sensitivity, a remarkable capacity for miniaturization, a molecular minimum limit of detection (LoD), and seamless integration with semiconductor technology. Infectious diseases continue to be a challenging field that necessitates increased precision in detection and integration in order to achieve accurate diagnosis at the point of care (PoC). In this title, we invited several expert scientific researchers to share their experience in this field. This book is dedicated to focusing on the application of FETs as biosensors, for rapid and real-time detection of pathogens that affect the human life. The lack of commercially available efficient devices that can be deployed for this task hampered efforts to curb the recent global spread of the SARS-CoV-19 virus. The book is an attempt to keep the interested audience up to date on this subject.

Detection Science Series No. 24
Field-effect Transistor Biosensors for Rapid Pathogen Detection
Edited by Naif H. M. Al-Hardan, Azman Jalar, Muhammad Azmi Abdul Hamid and Mohd Firdaus Raih

Published by the Royal Society of Chemistry, www.rsc.org

Contents

Detection Science Series No. 24
Field-effect Transistor Biosensors for Rapid Pathogen Detection
Edited by Naif H. M. Al-Hardan, Azman Jalar, Muhammad Azmi Abdul Hamid and Mohd Firdaus Raih

Published by the Royal Society of Chemistry, www.rsc.org

Part I: Introduction to Biologically Sensitive Field-effect Transistors (BioFETs)

CHAPTER 1

The Physics and Operating Principles of Field-effect Transistor-based Biosensors

M. F. MOHD RAZIP WEE,[a] NAIF H. AL-HARDAN,*[b] ASAD MASOOD,[c] MUHAMMED AZMI ABDUL HAMID,[b] AZMAN JALAR[b] AND NASER M. AHMED[d]

[a] Institute of Microengineering and Nanoelectronics (IMEN), Universiti Kebangsaan Malaysia 43600 UKM Bangi, Selangor, Malaysia; [b] Department of Applied Physics, Faculty of Science and Technology, Universiti Kebangsaan Malaysia, 43600 UKM Bangi, Selangor, Malaysia; [c] School of Energy and Chemical Engineering, Xiamen University Malaysia, 43900 Sepang, Selangor, Malaysia; [d] School of Physics, Universiti Sains Malaysia, 11800 Penang, Malaysia
*Email: naifalhardan@ukm.edu.my

1.1 Introduction

Biosensors were first developed by Leland Clark Jr in 1956 for oxygen detection in the blood. This was followed by the demonstration of an amperometric enzyme electrode for the detection of glucose by Clark and Lyons in 1962,[1,2] following which the concept has been widely employed in various applications such as environmental monitoring, genetic screening, sports surveillance, agriculture, food pathogens, and medical diagnostics.[3–7] The current biosensors, especially those used in healthcare, provide more precise detection, are lightweight and easy to use, and work on low energy

Detection Science Series No. 24
Field-effect Transistor Biosensors for Rapid Pathogen Detection
Edited by Naif H. M. Al-Hardan, Azman Jalar, Muhammad Azmi Abdul Hamid and Mohd Firdaus Raih

Published by the Royal Society of Chemistry, www.rsc.org

consumption to obtain information quickly compared to conventional assays. The commercialization of biosensors in the marketplace is evident from recent breakthroughs in this domain. This development aims to offer point-of-care testing (POC) capabilities, namely in the area of glucose monitoring, cholesterol, and Alzheimer's disease biosensors,[5] especially for remote areas. These commercial biosensor products offer label-free detection with better reliability, high signal-to-noise ratio (SNR), and low instrumentation. Biosensors typically consist of three components: specific bioreceptors that are able to recognize and capture a specific target molecule, such as nucleic acids, antibodies, aptamers, microorganisms, or enzymes; a transducer which will convert the reaction into an electrical signal that can be evaluated, quantified, and processed by hardware and software and displayed by the third component of the biosensor.[8,9] The primary component of a biosensor is depicted in a schematic diagram shown in Figure 1.1.

Amongst various types of biosensors, field effect transistor (FET)-based biosensors are the most promising due to their immense advantages such as high sensitivity, scalability, rapid real-time detection, inherent amplification, lower power requirements, direct electrical readout, and low-cost mass production compared to other established techniques.[9,10] FET-based biosensors are considered label-free electrochemical biosensors using electrochemical principles to detect biomolecules in a sample without the need for a labelling or tagging process. In other words, they do not require the use of a secondary molecule (such as a fluorescent or radioactive label) to detect the target molecule.[11,12] Label-free electrochemical biosensors have several advantages over traditionally labelled biosensors, including simplicity, speed, and cost-effectiveness.[13,14]

FETs are electronic devices that control the flow of electrical current through a semiconductor (acts as the channel) by varying an electric field. They consist of a gate electrode, a source electrode, and a drain electrode.

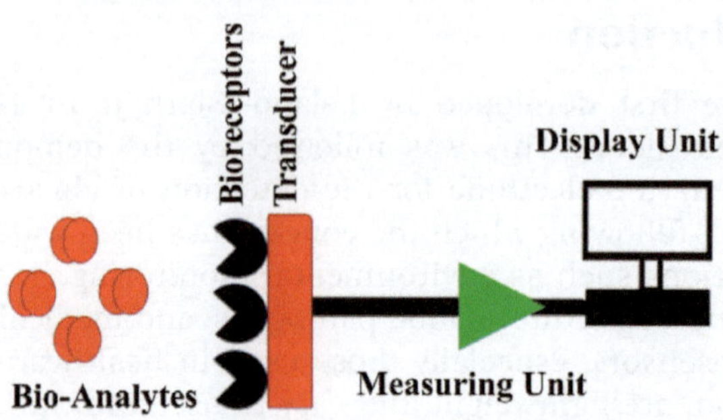

Figure 1.1 The schematic diagram illustrates the key components of a biosensor, comprising bioreceptors, a transducer, signal processing hardware (measuring unit), and a display unit. The bioreceptor is a molecule designed to selectively recognize the analyte, while the transducer converts the resulting reaction into an electrical signal. The signal is then amplified and processed by the hardware before being displayed and visualized by the display unit.

FET-based biosensors operate by utilizing the gate region of the FET to detect the presence of biological molecules, such as antibodies, DNA probes, or enzymes. The operating principle of FET biosensors is based on the modulation of the gate voltage induced by the binding of target molecules to the surface (mostly the channel), which alters the electrical properties of the FET.[15–17] The schematic diagram of a field effect transistor (FET) is depicted in Figure 1.2.

The sensitivity of FET-based biosensors depends on the ability of the surface to capture target molecules selectively. This can be achieved by modifying the surface with specific receptor molecules, such as antibodies or peptides, that recognize and bind to the target molecules with high affinity and selectivity, allowing for the detection of even trace amounts of target molecules. The binding of the target molecules to the receptor molecules on the surface induces a local charge distribution that alters the gate potential and hence the FET electrical characteristics. The term biologically sensitive field-effect transistors (BioFETs) became familiar due to the binding of the bioreceptors to the channel/gate surface.[16,18–20] Furthermore, FET biosensors could make use of mature manufacturing techniques from the complementary metal–oxide semiconductor (CMOS) process to further enhance its advantages in terms of miniaturization, parallel sensing, and the integration with other solid-state platforms and systems for the development of biosensing devices.[10,21,22]

Further enhancement could be achieved by the incorporation of various nanomaterials to enhance its sensitivity and selectivity, and organic/inorganic materials such as polymeric and metal oxide materials could contribute to the overall improvement.[23] The properties of such nanomaterials provide excellent interfacing for biological identification, which improves electronic signal transduction and is useful for designing novel bioelectronic devices.[24,25] Different nanomaterials in various forms, such as magnetic nanoparticles, silicon nanowires, and different structures of titanium dioxide, zinc oxide, gallium nitride, and various carbon-based materials (graphene, graphite, and carbon nanostructures),[26,27] have been

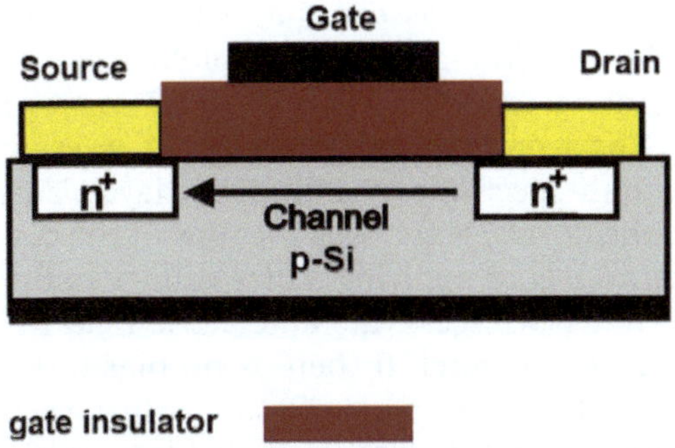

Figure 1.2 The FET has three electrodes, with an insulator material in between the channel electrodes and the gate electrode.

employed for the development of FET-based biosensors.[9,17,21,28] The incorporation of nanomaterials for the fabrication of FET-based biosensors has demonstrated a very high successful rate of binding the target-specific entities in the specific gate area for high functionality. This chapter is organized as follows: Section 1.2 will describe the structure, materials, biofunctionalization, and operating principle of FET-based biosensors. Section 1.3 will summarize the "performance criteria" of BioFET sensors. In Section 1.4, the most common configurations of FETs used as biosensors will be explained, and Section 1.5 will explore the current state of affairs in FET-based biosensor applications.

1.2 Structure and Operating Principle of FET Biosensors

In this section, we will briefly go through the structure of FETs, the material that is typically utilized in the production of FETs, as well as the operating principle of these devices.

1.2.1 The Structure of FET Biosensors

Field effect transistors are manufactured according to the standard semiconductor manufacturing process. However, due to the emergence of new applications such as wearable and flexible electronics, materials such as paper and polymers have been employed to fulfill the requirements for such applications. Given its abundance in nature and the established manufacturing process, silicon is expected to remain the predominant element in conventional semiconductor technology.[29] Silicon is used as the substrate of most semiconductor devices. Initially, silicon is covered with an insulator layer by oxidizing its surface with silicon dioxide, which will play the role of a gate dielectric to create the device known as a MOSFET (metal oxide semiconductor field effect transistor). As previously mentioned, the device has three terminals: a source, a drain, and a gate. The space between the drain and source terminals is known as the channel through which the current flows.[10,30] Usually, it is functionalized with a biological recognition element (bioreceptors) that interacts with the analyte and senses its presence and concentration.[31,32] The third electrode "the gate" is separated from both the source and the drain electrode by a dielectric layer. This electrode is the control terminal of the device. It controls the flow of the current between the drain and source terminals by applying a bias voltage to the gate terminal.[33] Biasing the gate terminal will generate an electric field that modulates the current flow through the channel. If there is no bias or low bias voltage at the gate terminal, the FET is in the 'off' state.[8] In fact, the FET operates by controlling the bias voltages over its both (drain and the gate) source terminals. There are several FET configurations according to the position of the gate terminal towards the drain–source terminals, which can be back, top, both sides (dual gate) and solution/electrolyte gated FET biosensors,[15,34] which are known as electrolyte-gated field effect transistors[35] (due to the importance of electrolyte-gated FET biosensors, further discussion will be undertaken in the subsequent chapter). However, most FET biosensors are based on an architecture known as an ion sensitive field effect transistor (ISFET), where the metal gate is replaced by a reference electrode in contact with the gate oxide *via* the electrolytic solution.[16,31,33,36] The reference electrode will maintain the performance stability of a biosensing system. Figure 1.3 shows the schematics of both devices: a MOSFET and an ISFET.

1.2.2 BioFET Sensing Materials

Silicon dioxide was the first material to be tested for biosensing applications.[37] Due to the low stability and drift issues in aqueous solutions,[38] several alternative materials were introduced to achieve better stability and lower the drifting issue. Here, we will summarize some of these materials that display better performance. Oxides such as ZnO, TiO_2, Al_2O_3, SnO_2, ZrO_2, HfO_2, GO_x and others have been heavily investigated.[39,40] There are several reasons why these oxide-based materials are preferable as sensing-based materials, such as a cost-effective growth process, the ability to grow over different types of substrates with good quality, biocompatibility, and easy integration in the conventional semiconductor CMOS process.[22,41–43] Nitride-based materials such as Si_3N_4 and TiN were tested as BioFET sensors,[44,45] and both reveal a better performance than that reported for SiO_2 due to their thermodynamic stability.[46]

Recently, there has been significant interest in advanced materials centered around carbon and its derivatives, notably carbon nanotubes (CNTs) along with single- and multi-walled CNTs. CNTs are unique one-dimensional

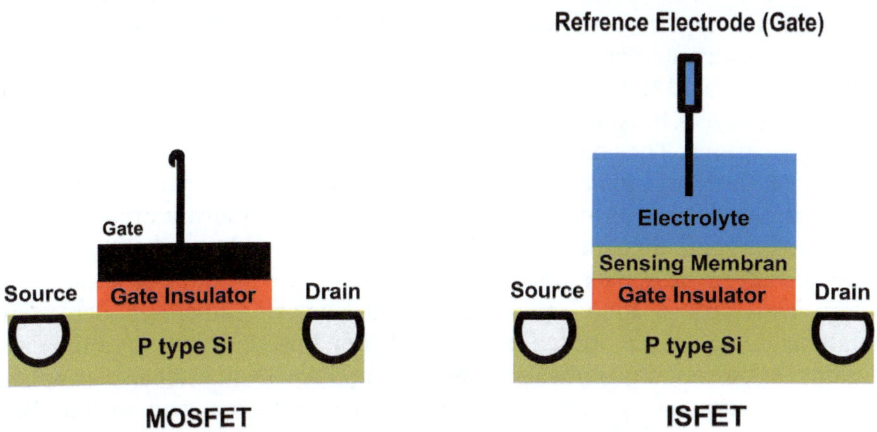

Figure 1.3 The schematic of a metal oxide semiconductor field-effect transistor (MOSFET) and an ion-sensitive field-effect transistor (ISFET).

nanostructures formed through C–C sp^2 hybridization,[47] composed of one or more layers of graphite arranged in the shape of hollow cylindrical tubes. Furthermore, owing to their remarkable characteristics, such as high conductivity, relatively high carrier mobility,[48] and robust chemical stability, with an impressive surface-to-volume ratio that enhances detection sensitivity, CNTs and related structures have found widespread utility in various biosensor applications. They serve as channels in field-effect transistor FET biosensors, proficiently detecting a spectrum of bioanalytes such as nucleic acids, viruses, and structural proteins of various pathogens.[26] Furthermore, graphene and its derivatives, graphene oxide (GO) and reduced graphene oxide (RGO), are also part of the carbon family. Graphene, a single-atom-thick layer of carbon, boasts exceptional properties that position it as a prime candidate for cutting-edge technologies and electronic devices. Its characteristics parallel those of CNTs, making it a prominent player in FET platforms. Researchers have extensively explored its applications, particularly in pathogen detection encompassing bacteria like *Escherichia coli* and viruses including HIV, avian influenza, Ebola, and coronaviruses, among others. This functionality is achieved through the binding of analyte targets to their corresponding receptors.[35,49–52] The resultant alteration in conductivity is attributed to changes in charge carriers on the graphene, GO, and RGO surfaces, induced by the presence of the analyte. For instance, Kakatkar *et al.*[53] demonstrated a graphene-based FET biosensor capable of detecting double-stranded DNA and poly-L-lysine. The sensor registered shifts in the Dirac voltage in response to the binding or non-binding of charged biomolecules to the graphene surface. Reports on RGO biosensors for the label-free detection of peptide-nucleic acid (PNA)-DNA hybridization[54] achieved a detection limit of less than 100 femtomolar (fM). Furthermore, RGO-FET was reported to detect Ebola virus[51] in the concentration range from 2.4×10^{-12} g mL^{-1} to 1.2×10^{-7} g mL^{-1}, with a limit of detection as low as 2.4 pg mL^{-1}.

In spite of the fact that carbon-based materials have great electron mobility, off-state current leakage in graphene-based biosensors may increase due to its near-zero band gap, leading to misleading indications.[10,55] Two-dimensional (2D) transition metal dichalcogenides such as WSe_2 and MoS_2 monolayers have recently emerged as potential materials for biosensing applications. They have been under extensive investigation as channels in BioFET devices, indicating a promising biosensing performance under different bioanalyte contents.[56,57] In addition to the above inorganic-based materials, organic field-effect transistors (OFETs) have emerged as one of the most advanced biosensor platforms owing to their unique signal amplification capabilities, high sensitivity, rapid analysis, label-free detection, small size and simple operation,[58,59] and they have become potential candidates for biosensing applications where they have been extensively applied in the detection of DNA, proteins, antibodies, antigens, and various biological substances.[60–62]

The preparation methods of materials used as channels in BioFETs play a crucial role in their performance. Different methods were employed for this purpose, such as sol–gel, hydrothermal, RF/DC sputtering, chemical vapor deposition, metal organic chemical vapor deposition, and spin coating, among others.[63–65] The crucial step in preparing BioFETs is surface immobilization with bioreceptors (biomolecules), such as antibodies, nucleic acids, or enzymes that can recognize and bind with the target analyte. The selection of bioreceptors is primarily based on the specific analyte that needs to be detected. This layer will enhance the sensitivity, selectivity, reliability, and stability of the biosensors in general. The bioreceptors are normally immobilized on the channel surface of the FET.[57,64,66] For a stable sensing performance of biosensors, bioreceptors must be capable of preserving their structure, function, and biological activity.[32] There are two main immobilization methods; one is the physical immobilization method, in which the bioreceptors are attached to the surface of the FET gate or the channel without the creation of chemical bonds. The most common methods for the physical immobilization process are adsorption, bioaffinity binding, and chelation or metal binding, where bonding occurs through non-covalent interactions. In this process, the bioreceptors can be detached from the transducer material without destroying either the biological activity or the transducer. The process is also known as the reversible immobilization method. On the other hand, the chemical immobilization method can be covalent binding, covalent cross-linking, or entrapment/micro-encapsulation. Here, the process is more complicated and it can be referred to as a functionalization process, which requires a more complex process than immobilizing the probes only.[21] These processes are based on chemical bonding. Here, the bioreceptors cannot be detached without destroying either their biological activity or the transducer. The process is also known as the irreversible immobilization method.[32,67] Figure 1.4 depicts various approaches that have been used widely for bioreceptor immobilization.

1.2.3 Operating Principle of FET Biosensors

A biosensor converts the biological information into a measurable signal that can be displayed, processed, or sent for further analysis.[1,10,22,57] The operation of a biosensor involves a change in analyte concentration, which alters the charge near the sensor interface, inducing a change in effective gate voltage and leading to changes in drain current that can be evaluated from the I–V characteristics. The main parameters that are usually monitored in BioFETs is the drain–source current (I_{DS}) (that represents the current following through the channel) or the shift in the threshold voltage (V_{TH}).[10,22,35,40] Both parameters are related to the analyte contents in a solution. In fact, the FET operates depending on both the gate–source and drain–source voltages. By controlling these two parameters, the FET is said to be operating in one of the three main regions: sub-threshold, linear, and saturation.[10,68] Figure 1.5 depicts the three regions of the FET operation depending on the V_{DS} and V_{GS}.

A FET usually operates in two different modes (regions of operation). We can distinguish between the two operating regions for the FET depending on

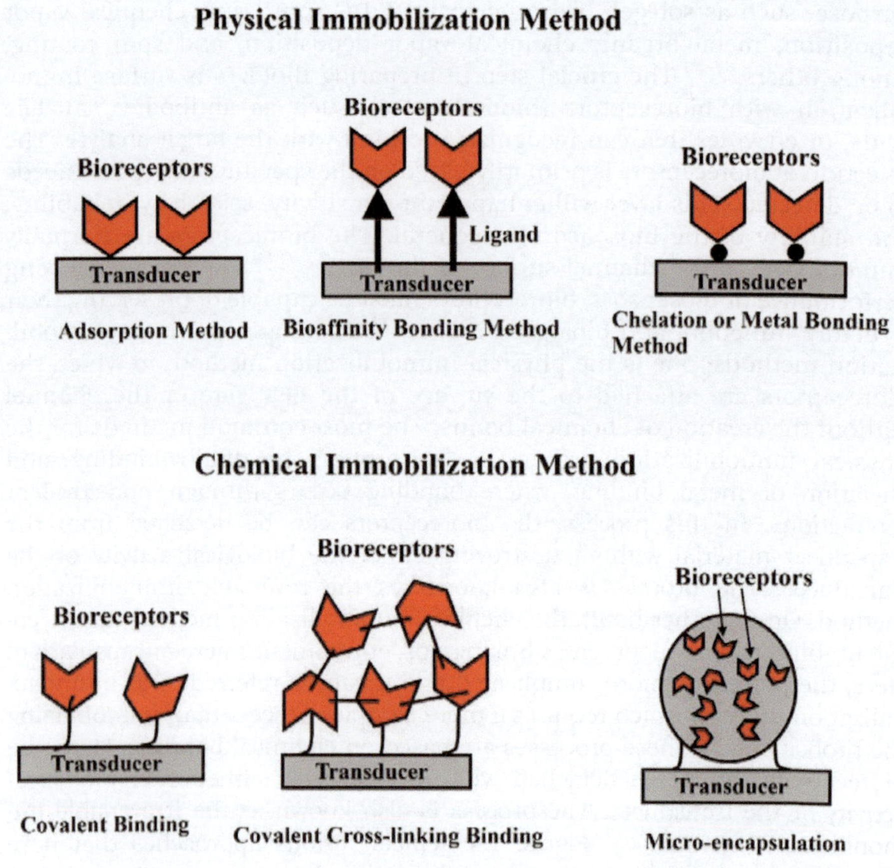

Figure 1.4 Schematic representation of both physical and chemical methods used for bioreceptor immobilization.

the V_{GS} (gate–source voltage) and V_{DS} values used (providing $|V_{GS}|>|V_{TH}|$). For $|V_{DS}|<|V_{GS}|-|V_{TH}|$, the FET is said to be operating in a linear regime. To describe the I_{DS} flowing from the drain to the source,

$$I_{DS}=\alpha\left[2(V_{GS}-V_{TH})V_{DS}-V_{DS}{}^{2}\right] \tag{1.1}$$

where $\alpha=\dfrac{\mu CW}{2L}$ is the conduction parameter, μ is the electron mobility, W is the channel width, L is the channel length, and C is the total gate-semiconductor capacitance per unit area. For the case where $|V_{DS}|>|V_{GS}|-|V_{TH}|$, the FET is said to be operating in the saturation regime. Here, the I_{DS} is referred to as

$$I_{DS}=\alpha(V_{GS}-V_{TH})^{2} \tag{1.2}$$

In the linear region of a BioFET, the output of the BioFET will reflect how the analyte content modifies the V_{TH} (at constant I_{DS} and V_{DS}). This can be

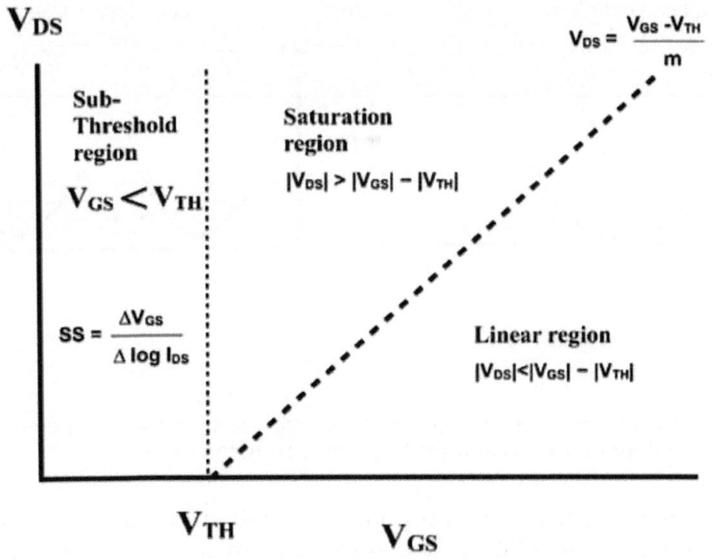

Figure 1.5 The FET operational regions in the V_{DS}–V_{GS} plane, where m represents the body-effect coefficient ($m \geq 1$).[10,68]

achieved by sweeping the gate voltage at constant V_{DS} and recording the I_{DS} output. The relation is known as transfer characteristics, where the measurements are performed to find the relationship between I_{DS} and V_{GS} as a function of the analyte contents. The variation of the analyte contents will result in changes of the threshold voltage V_{TH}, where a horizontal shift in the transfer curve is obtained, which might be attributed to either the transfer of charge between the analyte and the FET channel or a result of electrostatic gating.[10,35,69] In the context of two-dimensional materials, such as graphene-based field-effect transistors (FETs), a significant transition voltage exists that separates the regions of hole-dominant and electron-dominant conduction. This transition voltage corresponds to the point at which I_{DS} achieves its minimal value. This point is known as the Dirac point, and the voltage at this point is known as the Dirac voltage (V_{Dirac}). At the Dirac voltage, there are very few charge carriers in the device. Consequently, it is appropriate for monitoring changes in the device's environment, making it an attractive feature for biosensing applications.[9,57] In the saturation region of the FET operation, I_{DS} can be related to the analytes. Further, the dynamic behavior of I_{DS} *vs.* time can be recorded with the variation of analyte content at fixed values of both V_{GS} and V_{DS}. The I_{DS} *vs.* time measurement is useful for studying the kinetics of analyte–BioFET interactions. However, it does not reveal the physical mechanism of the reactions due to the fixed bias voltages.[33] Typical transfer, the Dirac voltage point, and the I_{DS} curve obtained from a BioFET in both operating regions are depicted in Figures 1.6 and 1.7, respectively. The voltage sensitivity towards the analyte contents can be found in the transfer curve.

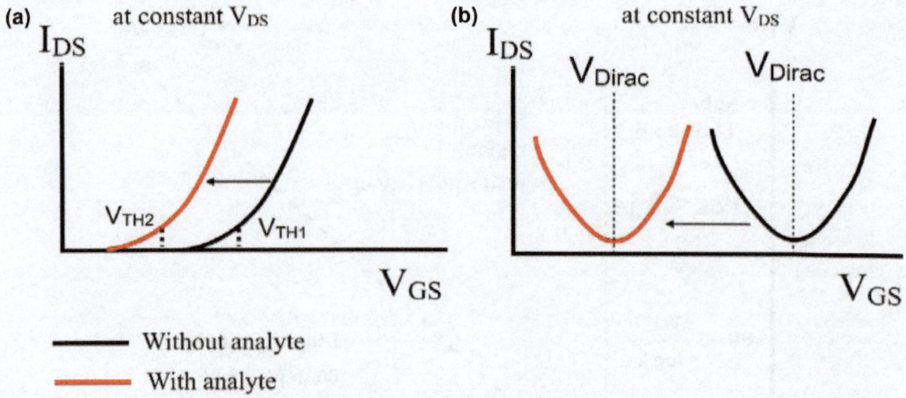

Figure 1.6 Typical transfer characteristics of a FET as a function of the analytes in a solution: (a) V_{TH} and (b) V_{Dirac} for 2D materials.

For practical reasons, it is common practice to utilize an electrolyte as the gate dielectric in the BioFET structure since most of the bioanalytes are found in physiological media (such as blood, serum, plasma, saliva, and urine). The gate is usually replaced by reference electrodes such as Ag/AgCl (silver/silver chloride).[70] However, a conventional reference electrode is unsuitable for miniaturized biosensors, so a pseudo-reference electrode is utilized. In this configuration, the electrolyte will be in direct contact with both the gate and the channel (insulator/semiconductor) layer. The operating principle relies on tuning the electrical current traveling through the channel by applying a gate voltage, which results in the formation of two electrical double layers (EDLs) at the insulator/electrolyte and electrolyte/gate interfaces.[35,71] Since the reaction between the analytes with the bioreceptors (mostly the reactions are in the channel part) will result in charge release, the potential at the gate/electrolyte interface will be modified, thus affecting the current flowing between the drain and source (I_{DS}) due to the V_{TH} shift.[72]

It is important to determine the relationship between the concentration of bioanalytes and the response of the BioFET. This is done by utilizing the electrostatic gating effect. The shift in V_{TH} in electrostatic gating is proportional to the shift in surface potential at the electrolyte–insulator interface (ψ_O). The precise relationship between the biomolecular charge and the change in the surface potential is a complicated process owing to the presence of hydrogen/hydroxide ions, biomolecules, and other ion contents.[34,57] The electrolyte–insulator interface can be represented as a parallel plate capacitor,[9,10] and V_{TH} can be estimated by the change in the charge at the electrolyte–insulator interface/gate surface due to the modulation of analytes (ΔQ), with capacitance C_o, which represents the total capacitance of the BioFET as

$$\Delta V_{TH} = \Delta\psi_o = \frac{\Delta Q}{C_o} \tag{1.3}$$

Alternatively, the change in V_{TH} due to analyte binding can be measured directly from the shift in the $I_{DS}-V_{GS}$ curve.[10] The change in charges at the gate surface caused by the presence of bioanalytes is analogous to applying a voltage to the gate of a MOSFET device. It is known that almost all biomolecule reactions are leading to a local change of the pH value, which can bind or release protons (protonation/deprotonation events).[73–75] Accordingly, the sensing mechanism of biomolecules can be dealt with as a pH sensing mechanism.[16,34,75]

According to the site-binding model,[76] the binding between the hydrogen/hydroxide ions in the electrolyte will bind to hydroxyl groups at the surface of the insulator. These sites are known as binding sites. The binding with hydrogen/hydroxide ions will result in alteration of the electrolyte–insulator interface potential (ψ_o) which can be found through

$$\psi_o = 2.3\left(\frac{kT}{q}\right)\left(\frac{\beta}{\beta+1}\right)\left(\mathrm{pH}_{pzc} - \mathrm{pH}\right) \tag{1.4}$$

where pH_{pzc} is the pH (hydrogen/hydroxide ions) value at which the channel surface is electrically neutral, known as the point of zero charge; β determines the final sensitivity of the gate insulator and depends on surface hydroxyl groups and surface reactivity; k, T, and q are the Boltzmann constant, the absolute temperature and the elementary charge, respectively.

The sensing mechanism of biosensors based on FETs can be divided into three main stages.[77] According to the derivation from Shoorideh and Chui,[77,78] the change of analyte contents will modulate the charge density near the electrolyte–insulator interface of the biosensor (dσ). This modulation in the charges will induce measurable effective gate voltage changes (dV_{gate}). Lastly, the gate voltage modulation leads to drain current variations (dI_{DS}), which can also be evaluated from the transfer characteristics. These steps can be represented mathematically, as depicted in eqn (1.5). The total sensitivity of a sensor can be represented through the total of the three mechanism steps of the detection as[21,77,79]

$$\frac{\mathrm{d}I_{DS}}{I_{DS}} = \left(\mathrm{d}c\frac{\mathrm{d}\sigma}{\mathrm{d}c}\right)\left(\frac{\mathrm{d}V_{gate}}{\mathrm{d}\sigma}\right)\left(\frac{\frac{\mathrm{d}I_{DS}}{\mathrm{d}V_{gate}}}{I_o}\right) \tag{1.5}$$

Here, the analyte concentration is represented by dc, and the steady-state drain current is represented by I_o (when the sensor surface is exposed to a reference (background solution)).

The modification of I_{DS} depends on the charge–exchange properties of the analyte and the sensing membrane. The charges from analytes affect the conductivity of the FET channel due to attractive or repulse electrostatic force.[63,80] Figure 1.7 depicts the working principle of an n-type channel FET. As the BioFET faces the neutral biofluid, the I_{DS} shows a stable output (Figure 1.7(a)). For bioanalytes with negative charges (such as DNA molecules),

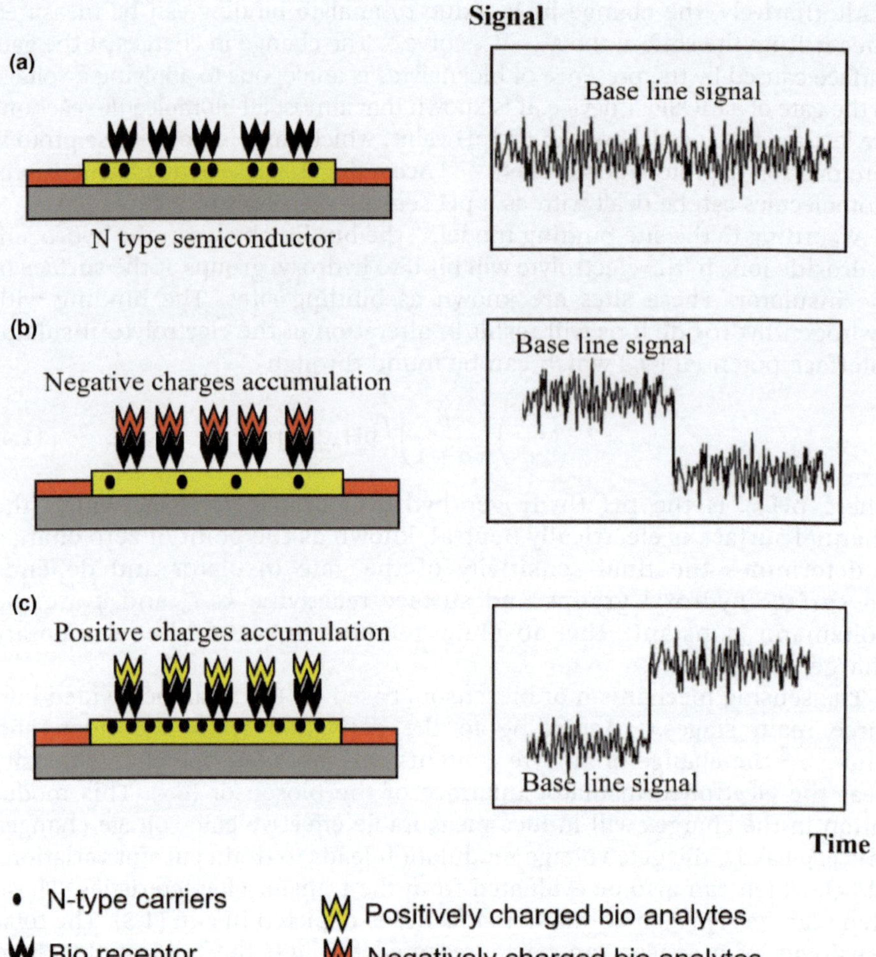

Figure 1.7 Principle of operation of n-type semiconductor biosensors. (a) Initial condition of n-type semiconductor biosensors. (b) As negative charges accumulate on the surface, electrostatic repulsion will decrease conductance. (c) Positive charges accumulate on the surface of the n-type semiconductor, resulting in electrostatic attraction resulting in higher conductance.

this will lead to repulsive electrostatic force between the negatively charged analytes and the n-type charge carriers of the channel (Figure 1.7(b)); consequently, a depletion of charge carriers inside the channel leads to a decrease in the electrical conductance and a drop will be noticed on the output signal (I_{DS}). In the other case, where the bioanalytes release a positive charge (such as small-size protein) (Figure 1.7(c), positive charges accumulate on the surface of the n-type semiconductor, resulting in electrostatic attraction resulting in a higher output signal (I_{DS}).[10,33,35]

1.3 Characteristics of BioFET Sensors – "Performance Criteria"

BioFET performance must be characterized using a variety of criteria, both static and dynamic. For a biosensor system to be effective and powerful, the standards below will be used to optimize the functionality of biosensors in practical settings.[81]

- Sensitivity is defined as the smallest amount of analyte that can be accurately detected or identified in the fewest steps at low concentrations (ng mL^{-1} or fg mL^{-1}) to confirm the presence of analyte traces in the sample.
- Linearity contributes to the precision of the measured outcomes. The higher the detected analyte concentration, the greater the linearity. Furthermore, it is closely linked to the biosensor's resolution, which is defined as the smallest change in the target analyte concentration required for the biosensor to respond.
- Selectivity is an essential characteristic to consider when choosing a bioreceptor for a biosensor. A bioreceptor is capable of detecting a specific target analyte molecule in a mixture of spices and contaminants.
- The biosensor response time is when the device produces a measurable signal or response after introducing the target analyte to the sensor surface. It is the time necessary to reach 90% of the steady-state response.
- Reproducibility is the ability of a sensor to generate a mean value closer to the actual value when the sample is measured repeatedly. It also reflects the accuracy of the sensor.
- Stability is one of the most important characteristics in biosensor applications that require continuous monitoring. Stability is the susceptibility of a biosensor to internal and external environmental disturbances. Furthermore, it is affected by the affinity of the bioreceptor (how well the analyte binds to the bioreceptor) and the degradation of the bioreceptor over time.

1.4 Configuration of FET-based Biosensors

This section provides an overview of different field-effect transistor sensor architectures that have been explored in the field of biosensors. The electrolyte gated FET is the most potential BioFET platform that detects a variety of pathogens.[9] For its importance, Chapter 2 will cover its mechanism and applications in detecting pathogens. For this reason, this platform will be avoided in the current chapter. Here, we will summarize the other types of FET platforms that have been used for pathogens (BioFETs).

1.4.1 Ion-sensitive Field Effect Transistors (ISFETs)

The first work on FET biosensors is initiated by Bergveld in the early 1970s through the introduction of ion-sensitive field effect transistors (ISFETs) in order to measure the value of pH in solution. In an ISFET, the current is produced through the channel surface of the ISFET depending on the ion activities, which could be used to evaluate the acidity of solutions. As shown in Figure 1.3, its configuration is slightly different compared to conventional MOSFETs where the metallic gate is separated and the underlying gate oxide is inserted in an aqueous solution along with a reference electrode (*e.g.*, Ag/AgCl electrode) through which the gate voltage is applied, and the gate dielectric layer is replaced with an electrolyte. Several gate modifications have been proposed to enhance biomolecular interaction detection, such as enzyme-modified field-effect transistors,[82] electrolyte gated FETs (EG-FETs)[35] and others.[22] The gate dielectric could be made from oxide, nitride, graphene, or organic-based membranes, usually deposited between the two electrodes (drain and source). The interaction with biomolecules would occur through the capture of the bioanalytes by the receptors in the electrolyte, thus releasing charges and consequently directly changing the potential of the FET channel surface; carriers will accumulate or deplete, increasing or decreasing the source–drain current.[83]

1.4.2 Separative Extended Gate Field-effect Transistors (SEG-FETs)

Under ISFET architectures, the detection of analyte contents is achieved by direct contact with the gate dielectric membrane. This will potentially deteriorate and reduce the operational lifetime of the device due to direct contact of the channel with the analytes. An alternative approach was introduced by Spiegel *et al.*[84] In their configuration, an external electrode was attached to the gate of a separated MOSFET. The configuration is known as separative extended-gate FET (SEG-FET), also known as an extended-gate field-effect transistor (EG-FET).[85] It is the most straightforward FET structure that could be used for general (bio)chemical sensing.[86] The SEG-FET consists of two separate parts: a sensing (the electrode) unit and a transducer unit (the MOSFET). The sensing part is immersed into the test solution, whereby the MOSFET is kept in a dry environment. Several MOSFET models have been used for this task, and one of the most commercial ICs used is CD4007UB.[87] This configuration in which the sensing part is separated from the transducer becomes suitable for POC applications.[88] It has been used for label-free bacterial detection such as *E. coli*.[89] Here, the sensing area of the external electrode was functionalized with porphyrins and immersed in liquid containing bacteria. The negative charge on the surface induced by *E. coli* cells will change the surface potential of the electrode reflected directly in the electrical characteristics of the SEG-FET. The SARS-CoV-2 genome[88] was detected in wastewater samples with the same platform. The limit of detection was 0.31×10^{-3} ng μL^{-1}, and the device provides its results within a short measurement time. Furthermore, SEG-FETs have also been employed to detect DNA,[90–92] proteins,[93] and viruses.[94,95] The SEG-FET structure possesses the advantages of a straightforward operation and cost-effectiveness in the context of personal healthcare devices. Figure 1.8 depicts

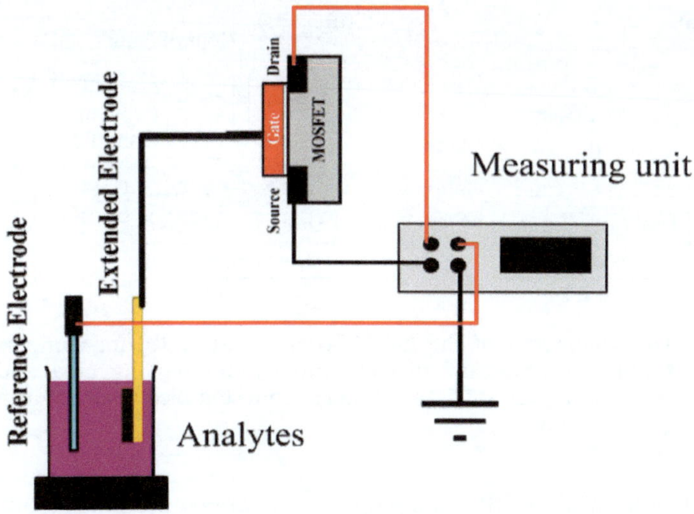

Figure 1.8 Schematic diagrams of the SEG-FET platform. The sensing electrode is attached to the gate terminal of a commercial MOSFET which functions as a transducer. The sensing behavior is recorded with respect to the reference electrode. The measurements are obtained using a parameter analyzer.

a typical measurement system employing an extended electrode for the SEG-FET platform.

1.4.3 Floating-gate FET Sensors

Floating gate field effect transistors (FG-FETs)[69] are another interesting FET platform for bio- and chemical sensing. The platform has two gates and one gate serves as a sensing gate, which will yield a response from the desired target, and the other as a control gate (used for biasing). Both gates operate similarly and are capacitively coupled to a shared floating gate.[22] With this configuration, the floating electrode protects the oxide sensing region on top of the channel.[34] A floating-gate FET sensor can detect changes in electric charge caused by various stimuli, such as the presence of certain biomolecules.[96,97] The magnitude of the current flowing through the transistor is dependent on the control voltage applied and the charges present on the sensing surface.[98] Any variations directly influence the electrical characteristics of the field-effect transistor (FET) in charge. In contrast to the ISFET, the reference electrode is not required. A floating-gate FET sensor can be used to detect and quantify biological molecules, such as DNA,[97] proteins, and antibodies.[96] The use of a floating gate allows for highly sensitive detection of these molecules with high specificity and selectivity.[99]

In a typical biosensor application, the surface of the floating-gate FET (the sensing gate) is coated with a bioreceptor, such as an antibody[96] or a DNA probe,[100,101] that can selectively bind to the target molecule. When the

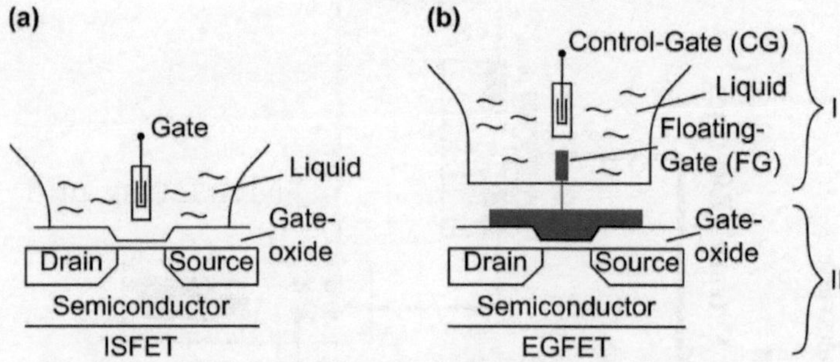

Figure 1.9 The schematics of the ISFET (a) and FG-FET (b) for comparison. The FG-FET is physically divided into (I) the sensing area and (II) the transducer (field-effect transistor) connected electrically by the floating-gate electrode.[89]

target molecule binds to the bioreceptor, it changes the charge on the gate and consequently modulates the threshold voltage of the transistor that can be detected as a change in current flow through the transistor. The final threshold voltage of the transistor can be estimated according to the equation:

$$V_{FTH} = V_{TH} - \frac{Q_{Total}}{C_{GC} + C_{FB}} \tag{1.6}$$

where V_{TH} is the MOSFET threshold voltage, C_{GC} is the control capacitor, C_{FB} represents the parasitic capacitance, and Q_{Total} is the total amount of the charge immobilized on the sensing pad.

The sensitivity of a floating-gate FET biosensor depends on the amount of target molecule that binds to the bioreceptor, as well as the sensitivity of the floating gate to change with the accumulating charges. One advantage of floating-gate FET biosensors is that they can be easily integrated into electronic devices for signal processing and data analysis.[102,103] This allows for real-time monitoring of biological processes and rapid detection of target molecules in complex biological samples, such as blood or saliva.[96] In terms of applications, they can be used to detect pathogens in food samples and in patients, screen for genetic mutations in DNA samples[102,103] and viruses such as the coronavirus.[96] Figure 1.9 depicts the schematics of both the ISFET and FG-FET for comparison.

1.4.4 Dielectric-modulated Field Effect Transistors (DM-FETs)

While traditional biosensors have limitations, such as low sensitivity and difficulty in detecting neutral biomolecules, DM-FET devices offer a more efficient solution. These devices are able to detect both charged and neutral biomolecules in small sample volumes with high sensitivity and

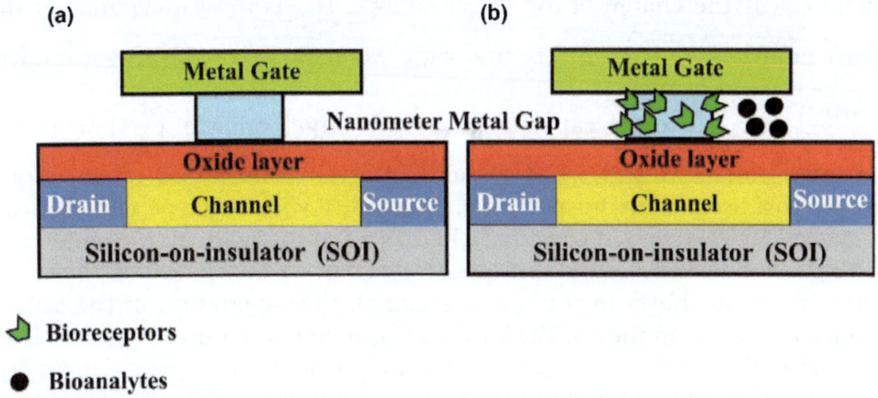

Figure 1.10 Schematic illustration of the structure of a DM-FET biosensor. (a) The nanogap etching process and (b) the nanogap-filling phenomenon.

fast screening.[104,105] A DM-FET biosensor (Figure 1.10(a) and (b)) consists of a nanogap positioned at the gate dielectric edge where it formed between the gate and gate oxide on both sides (source and drain junction) to accommodate the target analytes.

DM-FET biosensors can be categorized as planar or vertical nanogap devices, with the latter being preferred for their higher sensitivity.[106] DM-FET devices work by responding to the dielectric constant and charge effect of the biomolecule introduced into the cavity (gap), with the charge effect dominating for negatively charged biomolecules like DNA[105,107] and the dielectric constant effect affecting sensitivity for neutral biomolecules such as biotin–streptavidin binding.[108,109] The primary distinction between a DM-FET and a standard MOSFET is the cutting of the dielectric layer at which the biomolecules are connected.[110] The length of the cavity is crucial in affecting device parameters such as sensitivity, responsivity, electric field, and transconductance.[111] The confined biomolecules occupy a significant percentage of the cavity volume, allowing for efficient immobilization and high sensitivity. The absorption of biomolecules changes the electrostatic coupling between the gate and channel, resulting in an increase in threshold voltage compared to an air-filled gap.[110,111] Changes in threshold voltage can detect the presence and orientation of biomolecules in the gap, with a shift towards positive voltage indicating the presence of negatively charged molecules. The sensitivity of the DM-FET device is a function of the target analyte charge and the modulation effect of the gate dielectric, allowing for the detection of negatively, positively, and neutrally charged biomolecules. The ΔV_{TH} can be expressed as[79]

$$\Delta V_{TH} = \frac{Q_{TA}}{\dfrac{1}{C_{TA}} + \dfrac{1}{C_{OX}} + \dfrac{1}{C_A}} \tag{1.7}$$

where Q_{TA} is the charge of the target analyte, C_{TA} is the capacitance of the target analyte $\left(\frac{\varepsilon_A \varepsilon_{TA}}{l_{TA}}\right)$, C_{OX} is the oxide capacitance of the target analyte $\left(\frac{\varepsilon_A \varepsilon_{OX}}{l_{OX}}\right)$, C_A is the air capacitance of the target analyte $\left(\frac{\varepsilon_A}{l_A}\right)$, ε is the dielectric constant, and l_{TA}, l_{OX} and l_A are the thicknesses of the target analyte, oxide, and air, respectively. A DM-FET with a p-type channel can detect negatively charged proteins, whereas a DM-FET with an n-channel can distinguish between neutral and positively charged biomolecules.

Due to these effects in the n-type channel, the integration of the sensor surface and an amplifier in the shift in V_{TH} is not collaborative, but it is in the same direction for a p-type channel. Due to its great sensitivity, a p-channel DM-FET is preferred over an n-channel DM-FET. DM-FETs have been successfully applied for various pathogen sensing applications, such as SARS-CoV-2,[107] DNA[110] and avian influenza.[112]

1.4.5 Dual-gate FET Sensors

Recently, dual-gate field-effect transistors (DG-FETs) emerged as advanced biosensors. The dual-gate system utilizes two independent gate dielectrics: the top gate and the bottom gate. This will result in the amplification of the sensor signals and greatly improve the performance of sensor systems by enhancing the sensitivity and detection capabilities.[22] One of the gates is the sweeping gate that measures the voltage signal, and the second gate is the biasing gate (supporting gate). The biasing gate will ensure the sensor operates in a sensitive region.[113] Figure 1.11 depicts the schematic of a DG-FET. With this platform, it was reported that the sensitivity of the DG-FET for pH has

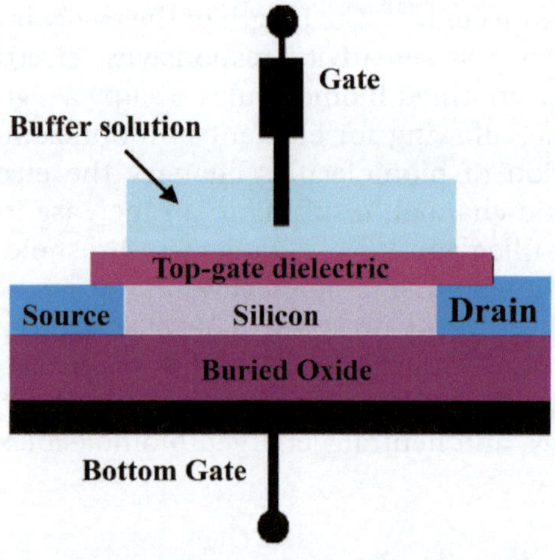

Figure 1.11 Schematic of a typical dual-gate FET with a liquid gate (top gate) and a bottom gate which control the potential of the silicon channel *via* a buffer solution/top dielectric and buried oxide, respectively.

exceeded the Nernstian limit from 59.16 mV pH^{-1} up to 304.12 mV pH^{-1}.[114] This reflects the enhancement of the device by adding the extra gate. The DG-FET has been further enhanced through adding a gap at the gate dielectric edge, which will behave as a dual-gate dielectric modulated FET.[115,116] The platform has been successfully used for the detection of SARS-CoV-2,[117,118] nucleic acid such as miRNA,[119] DNA[116] and proteins.[120]

1.5 Current State of Affairs in FET-based Biosensors

Field effect transistor-based biosensors have garnered considerable attention and have been the subject of intensive research in recent years. The field of FET-based biosensing continues to advance rapidly. As researchers strive to improve the performance, sensitivity, and specificity of these devices, new opportunities for their use in different applications are emerging. The utilization of BioFETs presents numerous benefits, including label-free biomolecule sensing, enhanced sensitivity, rapid detection, ultra-low limit of detection, compatibility with the conventional CMOS manufacturing process, adaptability to read-out circuits for on-chip integration, design flexibility, and compatibility with microfluidic systems. The latter has been a breakthrough, which allows for precise sample handling and control, enabling real-time and continuous monitoring of analytes. This integration not only enhances the performance of FET-based biosensors but also reduces the sample volume and minimizes the risk of cross-contamination. The advancements in the field of BioFETs lead to the development of a variety of biosensors that can overcome the limitations of other types of biosensors, such as optical and electrochemical-based biosensors. Research articles highlight the progress made in the field of BioFETs, specifically focusing on their applications in various areas such as infectious diseases, biomarker detection, and biomedical applications. Further studies depict the improvements in the detection sensitivity of BioFETs by utilizing different bioreceptors such as aptamers, making them well-suited for varieties of diagnostics. An important development in the field is the substitution of conventional insulating layer materials with electrolytes, specifically electrolyte-gated FETs, which has facilitated the broadening of applications for BioFETs. With ongoing research and development, they are expected to play a crucial role in the future of biosensing technologies.

Several BioFET biosensors have a wide range of structural designs and sensing materials, resulting in numerous combinations of biosensors. These design variations have improved sensor performance and device parameters, expanding the field of application. Sensing materials with exceptional properties, such as high charge mobility or mechanical strength, such as graphene FETs, have added diversity to the field of FET biosensors. Nanowire FETs offer a broad limit of detection and high sensitivity. Research is focused on incorporating nanotechnology by exploiting different nanomaterials and nanostructures for miniature and highly efficient BioFET devices. Multidisciplinary approaches, combining knowledge from different fields, have opened up opportunities for novel sensor architectures. Due to their advantages, such as electronic output signals, high sensitivity, integration flexibility, low-cost fabrication, and disposable strip-type sensing systems, these devices became suitable for point-of-care (POC) testing applications. Consequently, BioFETs have become competitive candidates for POC applications compared to bulky optical-based biosensor instruments.

Despite the improvements in the technology, there are challenges in the practical implementation of BioFETs due to variations in the FET structures, target analyte properties, sensing components, and testing environments. The lack of a common design rule for BioFETs hinders their technical maturation and commercial acceptance. Once the device performance parameters are optimized, highly efficient and reliable biosensors can be designed.

1.6 Conclusion

In conclusion, the field of FET biosensors has undergone tremendous development in recent years, thanks to their scalability, high sensitivity, real-time detection, inherent amplification, and low power requirements. Researchers have successfully detected a wide range of biomolecular targets of varying sizes, such as nucleotides (RNA and DNA), amino acids, antibodies and antigens, to diagnose various diseases. Different types of FET-based biosensors, including ISFETs, SEG-FETs, and FG-FETs, have been used to achieve these results. Despite facing some challenges, FET-based biosensors hold great promise for healthcare applications, and they could become the next generation of biomedical detection devices due to their robust chemical properties and customizable biosensing functions. Continued research and development could improve the versatility and effectiveness of these devices.

Acknowledgements

This work was supported by the Universiti Kebangsaan Malaysia (UKM) through matching short-term grant number GPS-DPK-2023-002 (UKM).

References

1. N. Bhalla, P. Jolly, N. Formisano and P. Estrela, *Essays Biochem.*, 2016, **60**, 1–8.
2. L. C. Clark Jr and C. Lyons, *Ann. N. Y. Acad. Sci.*, 1962, **102**, 29–45.
3. R. S. Selvarajan, R. A. Rahim, B. Y. Majlis, S. C. B. Gopinath and A. A. Hamzah, *Sensors*, 2020, **20**, 2642.
4. U. Chadha, P. Bhardwaj, R. Agarwal, P. Rawat, R. Agarwal, I. Gupta, M. Panjwani, S. Singh, C. Ahuja, S. K. Selvaraj, M. Banavoth, P. Sonar, B. Badoni and A. Chakravorty, *J. Ind. Eng. Chem.*, 2022, **109**, 21–51.
5. B. T. Murti, A. D. Putri, Y.-J. Huang, S.-M. Wei, C.-W. Peng and P.-K. Yang, *RSC Adv.*, 2021, **11**, 20403–20422.

6. A. Haleem, M. Javaid, R. P. Singh, R. Suman and S. Rab, *Sens. Int.*, 2021, **2**, 100100.
7. A. A. Hamzah and S. Nadzirah, in *Encyclopedia of Sensors and Biosensors*, ed. R. Narayan, Elsevier, Oxford, 1st edn, 2023, pp. 209–217.
8. D. Sadighbayan, M. Hasanzadeh and E. Ghafar-Zadeh, *TrAC, Trends Anal. Chem.*, 2020, **133**, 116067.
9. N. H. Al-Hardan, M. Firdaus-Raih, M. A. A. Hamid and A. Jalar, *Crit. Rev. Solid State Mater. Sci.*, 2024, **49**, 1–27.
10. B. M. Lowe, K. Sun, I. Zeimpekis, C.-K. Skylaris and N. G. Green, *Analyst*, 2017, **142**, 4173–4200.
11. E. Stern, J. F. Klemic, D. A. Routenberg, P. N. Wyrembak, D. B. Turner-Evans, A. D. Hamilton, D. A. LaVan, T. M. Fahmy and M. A. Reed, *Nature*, 2007, **445**, 519–522.
12. E. Macchia, K. Manoli, C. Di Franco, R. A. Picca, R. Österbacka, G. Palazzo, F. Torricelli, G. Scamarcio and L. Torsi, *ACS Sens.*, 2020, **5**, 1822–1830.
13. J. Gao, Y. Gao, Y. Han, J. Pang, C. Wang, Y. Wang, H. Liu, Y. Zhang and L. Han, *ACS Appl. Electron. Mater.*, 2020, **2**, 1090–1098.
14. V. R. Samuel and K. J. Rao, *Biosens. Bioelectron.: X*, 2022, **11**, 100216.
15. M. Eswaran, B. Chokkiah, S. Pandit, S. Rahimi, R. Dhanusuraman, M. Aleem and I. Mijakovic, *Small Methods*, 2022, **6**, 2200809.
16. M. J. Schöning and A. Poghossian, *Analyst*, 2002, **127**, 1137–1151.
17. S. S. A. Karim, S. Nadzirah, J. Kazmi, R. A. Rahim, C. F. Dee, A. A. Hamzah and M. A. Mohamed, *J. Mater. Sci.*, 2021, **56**, 15344–15353.
18. V. Pachauri and S. Ingebrandt, *Essays Biochem.*, 2016, **60**, 81–90.
19. C.-A. Vu and W.-Y. Chen, *Sensors*, 2019, **19**, 4214.
20. J.-J. Xu, X.-L. Luo and H.-Y. Chen, *Front. Biosci.-Landmark*, 2005, **10**, 420–430.
21. Y.-C. Syu, W.-E. Hsu and C.-T. Lin, *ECS J. Solid State Sci. Technol.*, 2018, **7**, Q3196–Q3207.
22. M. Kaisti, *Biosens. Bioelectron.*, 2017, **98**, 437–448.
23. P. Malik, R. Gupta, V. Malik and R. K. Ameta, *Meas.: Sens.*, 2021, **16**, 100050.
24. K. W. Cho, S.-H. Sunwoo, Y. J. Hong, J. H. Koo, J. H. Kim, S. Baik, T. Hyeon and D.-H. Kim, *Chem. Rev.*, 2022, **122**, 5068–5143.
25. Y. Lee, J. Kim, J. H. Koo, T.-H. Kim and D.-H. Kim, *Korean J. Chem. Eng.*, 2018, **35**, 1–11.
26. C.-S. Lee, R. E. A. Gwyther, M. Freeley, D. Jones and M. Palma, *ChemBioChem*, 2022, **23**, e202200282.
27. A. A. Hamzah, R. S. Selvarajan and B. Y. Majlis, *Sains Malays.*, 2017, **46**, 1125–1139.
28. A. Nehra and K. Pal Singh, *Biosens. Bioelectron.*, 2015, **74**, 731–743.
29. G. Fisher, M. R. Seacrist and R. W. Standley, *Proc. IEEE*, 2012, **100**, 1454–1474.
30. S. M. Sze, Y. Li and K. K. Ng, *Physics of semiconductor devices*, John Wiley & Sons, 2021.

31. B.-R. Li, C.-C. Chen, U. R. Kumar and Y.-T. Chen, *Analyst*, 2014, **139**, 1589–1608.
32. S. Liébana and G. A. Drago, *Essays Biochem.*, 2016, **60**, 59–68.
33. A. Béraud, M. Sauvage, C. M. Bazán, M. Tie, A. Bencherif and D. Bouilly, *Analyst*, 2021, **146**, 403–428.
34. A. Panahi, D. Sadighbayan, S. Forouhi and E. Ghafar-Zadeh, *Biosensors*, 2021, **11**, 103.
35. B. Shkodra, M. Petrelli, M. A. Costa Angeli, D. Garoli, N. Nakatsuka, P. Lugli and L. Petti, *Appl. Phys. Rev.*, 2021, **8**, 041325.
36. M. Kaisti, Q. Zhang, A. Prabhu, A. Lehmusvuori, A. Rahman and K. Levon, *IEEE Trans. Electron Devices*, 2015, **62**, 2628–2635.
37. P. Bergveld, *IEEE Trans. Biomed. Eng.*, 1972, **19**, 342–351.
38. V. K. Khanna, *Sens. Rev.*, 2013, **33**, 228–237.
39. S. Cao, P. Sun, G. Xiao, Q. Tang, X. Sun, H. Zhao, S. Zhao, H. Lu and Z. Yue, *Electrochem. Sci. Adv.*, 2022, **3**, e2100207.
40. X. Ma, R. Peng, W. Mao, Y. Lin and H. Yu, *Electrochem. Sci. Adv.*, 2023, **3**, e2100163.
41. S. K. Arya, C. C. Wong, Y. J. Jeon, T. Bansal and M. K. Park, *Chem. Rev.*, 2015, **115**, 5116–5158.
42. Z. Zhu, T. Yasui, Q. Liu, K. Nagashima, T. Takahashi, T. Shimada, T. Yanagida and Y. Baba, *Micromachines*, 2021, **12**, 642.
43. M. Shariati, *Biosens. Bioelectron.*, 2018, **105**, 58–64.
44. S. Jamasb, *IEEE Sens. J.*, 2021, **21**, 4705–4712.
45. D. Kumar, S. Jit, S. Sinha, R. Sharma and R. Mukhiya, *J. Electrochem. Soc.*, 2021, **168**, 107510.
46. J. Gustavsson, G. Altankov, A. Errachid, J. Samitier, J. A. Planell and E. Engel, *J. Mater. Sci.: Mater. Med.*, 2008, **19**, 1839–1850.
47. H. Meskher, H. C. Mustansar, A. K. Thakur, R. Sathyamurthy, I. Lynch, P. Singh, T. K. Han and R. Saidur, *Nanoscale Adv.*, 2023, **5**, 992–1010.
48. R. Maheswaran and B. P. Shanmugavel, *J. Electron. Mater.*, 2022, **51**, 2786–2800.
49. M. A. Zamzami, G. Rabbani, A. Ahmad, A. A. Basalah, W. H. Al-Sabban, S. N. Ahn and H. Choudhry, *J. King Saud Univ., Sci.*, 2022, **34**, 102137.
50. A. Roberts, N. Chauhan, S. Islam, S. Mahari, B. Ghawri, R. K. Gandham, S. S. Majumdar, A. Ghosh and S. Gandhi, *Sci. Rep.*, 2020, **10**, 14546.
51. X. Jin, H. Zhang, Y.-T. Li, M.-M. Xiao, Z.-L. Zhang, D.-W. Pang, G. Wong, Z.-Y. Zhang and G.-J. Zhang, *Microchim. Acta*, 2019, **186**, 223.
52. H.-M. So, D.-W. Park, E.-K. Jeon, Y.-H. Kim, B. S. Kim, C.-K. Lee, S. Y. Choi, S. C. Kim, H. Chang and J.-O. Lee, *Small*, 2008, **4**, 197–201.
53. A. Kakatkar, T. Abhilash, R. De Alba, J. M. Parpia and H. G. Craighead, *Nanotechnology*, 2015, **26**, 125502.
54. B. Cai, S. Wang, L. Huang, Y. Ning, Z. Zhang and G.-J. Zhang, *ACS Nano*, 2014, **8**, 2632–2638.
55. N. H. Al-Hardan, M. A. A. Hamid, L. K. Keng, E. M. Al-Khalqi, A. Jalar and W. S. Chiu, *J. Mater. Sci.: Mater. Electron.*, 2021, **32**, 9431–9439.

56. Y. Li, Z. Peng, N. J. Holl, M. R. Hassan, J. M. Pappas, C. Wei, O. H. Izadi, Y. Wang, X. Dong, C. Wang, Y.-W. Huang, D. Kim and C. Wu, *ACS Omega*, 2021, **6**, 6643–6653.
57. S. Chen, Y. Sun, X. Fan, Y. Xu, S. Chen, X. Zhang, B. Man, C. Yang and J. Du, *J. Nanobiotechnol.*, 2023, **21**, 144.
58. W. Tang, Y. Fu, Y. Huang, Y. Li, Y. Song, X. Xi, Y. Yu, Y. Su, F. Yan and X. Guo, *npj Flexible Electron.*, 2022, **6**, 18.
59. J. Tao, W. Sun and L. Lu, *Biosens. Bioelectron.*, 2022, **216**, 114667.
60. S. Lai, M. Barbaro and A. Bonfiglio, *Sens. Actuators, B*, 2016, **233**, 314–319.
61. K. Ditte, T. A. Nguyen Le, O. Ditzer, D. I. Sandoval Bojorquez, S. Chae, M. Bachmann, L. Baraban and F. Lissel, *ACS Biomater. Sci. Eng.*, 2023, **9**, 2140–2147.
62. J. Song, J. Dailey, H. Li, H.-J. Jang, L. Russell, P. Zhang, P. C. Searson, J. T.-H. Wang, A. D. Everett and H. E. Katz, *Adv. Funct. Mater.*, 2018, **28**, 1802605.
63. T. Manimekala, R. Sivasubramanian and G. Dharmalingam, *J. Electron. Mater.*, 2022, **51**, 1950–1973.
64. R. Hao, L. Liu, J. Yuan, L. Wu and S. Lei, *Biosensors*, 2023, **13**, 426.
65. L. Petti, N. Münzenrieder, C. Vogt, H. Faber, L. Büthe, G. Cantarella, F. Bottacchi, T. D. Anthopoulos and G. Tröster, *Appl. Phys. Rev.*, 2016, **3**, 021303.
66. M. A. Morales and J. M. Halpern, *Bioconjugate Chem.*, 2018, **29**, 3231–3239.
67. M. Asal, Ö. Özen, M. Şahinler, H. T. Baysal and İ. Polatoğlu, *Sens. Rev.*, 2019, **39**, 377–386.
68. Y. Taur and T. H. Ning, *Fundamentals of modern VLSI devices*, Cambridge University Press, 2021.
69. K. D. Dorfman, D. Z. Adrahtas, M. S. Thomas and C. D. Frisbie, *Biomicrofluidics*, 2020, **14**, 011301.
70. S. Sinha and T. Pal, *Electrochem. Sci. Adv.*, 2022, **2**, 2100147.
71. S. H. Kim, K. Hong, W. Xie, K. H. Lee, S. Zhang, T. P. Lodge and C. D. Frisbie, *Adv. Mater.*, 2013, **25**, 1822–1846.
72. J. Muñoz, F. Leonardi, T. Özmen, M. Riba-Moliner, A. González-Campo, M. Baeza and M. Mas-Torrent, *J. Mater. Chem. C*, 2019, **7**, 14993–14998.
73. J. Mongan and D. A. Case, *Curr. Opin. Struct. Biol.*, 2005, **15**, 157–163.
74. C. Andronescu and W. Schuhmann, *Curr. Opin. Electrochem.*, 2017, **3**, 11–17.
75. J. C. Dutta, in *Advanced Materials and Techniques for Biosensors and Bioanalytical Applications*, ed. P. Goswami, Taylor & Francis Group, LLC, 1st edn., 2021, pp. 283–296.
76. D. E. Yates, S. Levine and T. W. Healy, *J. Chem. Soc., Faraday Trans. 1*, 1974, **70**, 1807–1818.
77. K. Shoorideh and C. O. Chui, *IEEE Trans. Electron Devices*, 2012, **59**, 3104–3110.

78. K. Shoorideh and C. O. Chui, *Proc. Natl. Acad. Sci. U. S. A.*, 2014, **111**, 5111–5116.
79. T. Wadhera, D. Kakkar, G. Wadhwa and B. Raj, *J. Electron. Mater.*, 2019, **48**, 7635–7646.
80. Q. Liu, C. Zhao, M. Chen, Y. Liu, Z. Zhao, F. Wu, Z. Li, P. S. Weiss, A. M. Andrews and C. Zhou, *iScience*, 2020, **23**, 101469.
81. D. R. Thévenot, K. Toth, R. A. Durst and G. S. Wilson, *Biosens. Bioelectron.*, 2001, **16**, 121–131.
82. M. J. Schöning and A. Poghossian, *Electroanalysis*, 2006, **18**, 1893–1900.
83. M. Yuqing, G. Jianguo and C. Jianrong, *Biotechnol. Adv.*, 2003, **21**, 527–534.
84. J. van der spiegel, I. Lauks, P. Chan and D. Babic, *Sens. Actuators*, 1983, **4**, 291–298.
85. A. S. Pullano, D. C. Critello, I. Mahbub, T. N. Tasneem, S. Shamsir, K. S. Islam, M. Greco and S. A. Fiorillo, *Sensors*, 2018, **18**, 4042.
86. J. Kwon, B.-H. Lee, S.-Y. Kim, J.-Y. Park, H. Bae, Y.-K. Choi and J.-H. Ahn, *ACS Sens.*, 2019, **4**, 1724–1729.
87. A. Das, D. H. Ko, C.-H. Chen, L.-B. Chang, C.-S. Lai, F.-C. Chu, L. Chow and R.-M. Lin, *Sens. Actuators, B*, 2014, **205**, 199–205.
88. B. E. Alvarez-Serna, R. G. Ramírez-Chavarría, E. Castillo-Villanueva, J. Carrillo-Reyes, R. M. Ramírez-Zamora, G. Buitrón and L. Alvarez-Icaza, *Talanta*, 2023, **253**, 124060.
89. L. Könemund, L. Neumann, F. Hirschberg, R. Biedendieck, D. Jahn, H.-H. Johannes and W. Kowalsky, *Sci. Rep.*, 2022, **12**, 4397.
90. Y. Ishige, M. Shimoda and M. Kamahori, *Jpn. J. Appl. Phys.*, 2006, **45**, 3776.
91. C.-P. Chen, A. Ganguly, C.-Y. Lu, T.-Y. Chen, C.-C. Kuo, R.-S. Chen, W.-H. Tu, W. B. Fischer, K.-H. Chen and L.-C. Chen, *Anal. Chem.*, 2011, **83**, 1938–1943.
92. A. Purwidyantri, L. Kamajaya, C.-H. Chen, J.-D. Luo, C.-C. Chiou, Y.-C. Tian, C.-Y. Lin, C.-M. Yang and C.-S. Lai, *J. Electrochem. Soc.*, 2018, **165**, H3170.
93. D.-S. Kim, J.-E. Park, J.-K. Shin, P. K. Kim, G. Lim and S. Shoji, *Sens. Actuators, B*, 2006, **117**, 488–494.
94. N. C. S. Vieira, A. Figueiredo, J. F. dos Santos, S. M. Aoki, F. E. G. Guimarães and V. Zucolotto, *Anal. Methods*, 2014, **6**, 8882–8885.
95. J. Kwon, Y. Lee, T. Lee and J.-H. Ahn, *Anal. Chem.*, 2020, **92**, 5524–5531.
96. H.-J. Jang, X. Sui, W. Zhuang, X. Huang, M. Chen, X. Cai, Y. Wang, B. Ryu, H. Pu, N. Ankenbruck, K. Beavis, J. Huang and J. Chen, *ACS Appl. Mater. Interfaces*, 2022, **14**, 24187–24196.
97. S. P. White, K. D. Dorfman and C. D. Frisbie, *Anal. Chem.*, 2015, **87**, 1861–1866.
98. C. Baozhen, T. Chengwu, S. William and S. Pandey, *2008 IEEE International Conference on Electro/Information Technology*, 2008.
99. M. S. Thomas, D. Z. Adrahtas, C. D. Frisbie and K. D. Dorfman, *ACS Sens.*, 2021, **6**, 1910–1917.

100. B. Veigas, E. Fortunato and P. V. Baptista, *Sensors*, 2015, **15**, 10380–10398.
101. B. Kim, J. Lee, S. Namgung, J. Kim, J. Y. Park, M.-S. Lee and S. Hong, *Sens. Actuators, B*, 2012, **169**, 182–187.
102. M. Barbaro, A. Caboni, D. Loi, S. Lai, A. Homsy, P. D. van der Wal and N. F. de Rooij, *Sens. Actuators, B*, 2012, **171–172**, 148–154.
103. M. Barbaro, A. Bonfiglio, L. Raffo, A. Alessandrini, P. Facci and I. Barák, *Sens. Actuators, B*, 2006, **118**, 41–46.
104. R. Das, A. Chattopadhyay, M. Chanda, C. K. Sarkar and C. Bose, *Silicon*, 2022, **14**, 9417–9430.
105. A. Sariki, K. Venkata Rao, L. Chandrasekar, R. R. Shaik and K. P. Pradhan, *AEU – Int. J. Electron. Commun.*, 2021, **137**, 153791.
106. M. Kumari, N. K. Singh and M. Sahoo, *Sci. Rep.*, 2022, **12**, 21115.
107. S. Yadav, A. Gedam and S. Tirkey, *IEEE Sens. J.*, 2021, **21**, 14483–14490.
108. H. Im, X.-J. Huang, B. Gu and Y.-K. Choi, *Nat. Nanotechnol.*, 2007, **2**, 430–434.
109. X. Zhang, T. Liu, A. Boyle, A. Bahreman, L. Bao, Q. Jing, H. Xue, R. Kieltyka, A. Kros, G. F. Schneider and W. Fu, *Adv. Mater.*, 2022, **34**, 2106666.
110. C.-H. Kim, C. Jung, H. G. Park and Y.-K. Choi, *BioChip J.*, 2008, **2**, 127–134.
111. C.-H. Kim, C. Jung, K.-B. Lee, H. G. Park and Y.-K. Choi, *Nanotechnology*, 2011, **22**, 135502.
112. K. N. Singh and P. K. Dutta, *Silicon*, 2022, **14**, 11303–11313.
113. J.-H. Ahn, B. Choi and S.-J. Choi, *J. Appl. Phys.*, 2020, **128**, 080903.
114. J.-H. Jeon and W.-J. Cho, *ACS Omega*, 2020, **5**, 12809–12815.
115. S. S. Mohanty, S. Mishra, M. Mohapatra and G. P. Mishra, *Trans. Electr. Electron. Mater.*, 2022, **23**, 156–163.
116. M. K. Anvarifard, Z. Ramezani and I. S. Amiri, *Mater. Sci. Eng., C*, 2020, **117**, 111293.
117. S. Kumar, R. K. Chauhan and M. Kumar, *Silicon*, 2022, **14**, 11453–11462.
118. J. Kim, S. Jeong, S. Sarawut, H. Kim, S. U. Son, S. Lee, G. Rabbani, H. Kwon, E.-K. Lim, S. N. Ahn and S.-H. K. Park, *Lab Chip*, 2022, **22**, 899–907.
119. C.-H. Huang, W.-T. Huang, T.-T. Huang, S.-H. Ciou, C.-F. Kuo, A.-H. Hsieh, Y.-S. Hsiao and Y.-J. Lee, *ACS Appl. Electron. Mater.*, 2021, **3**, 4300–4307.
120. S. Park, M. Kim, D. Kim, S. H. Kang, K. H. Lee and Y. Jeong, *Biosens. Bioelectron.*, 2020, **147**, 111737.

CHAPTER 2

Electrolyte-gated FET Biosensors

B. PIRO

ITODYS, CNRS, Université Paris Cité, Paris, France
Email: benoitpiro.diderot@gmail.com

2.1 Introduction

Electrolyte-gated field-effect transistors (EGFETs) have been successful in the field of biosensing due to their easy integration into the electronics of any sensor and versatile transduction pathways. They can convert a small change in potential, resulting from a molecular recognition at an interface, into a significant change in current. This current can be converted into an amplified potential difference to be processed by analog-to-digital converters (ADCs) and recorded or transmitted wirelessly to a remote data centre for further processing. EGFETs can significantly amplify the input signal by 10, 100, or 1000 times, a feature that conventional electrochemical transducers can only provide through specific and complex molecular amplification strategies. Compared to conventional field-effect transistors (FETs) which use humidity-sensitive semiconducting materials and operate under several volts or tens of volts, EGFETs are stable in aqueous media, operate at low voltages below 1 V, and are thus suitable for transducing biological events such as changes in pH, ionic strength, and local ionic density, resulting in changes in local electrical potential at an interface, into a voltage or a current. EGFETs can even detect biological events that do not produce any electron transfer and therefore any current by themselves. EGFETs are true field-effect transistors, distinct from ECTs (electrochemical transistors), that

Detection Science Series No. 24
Field-effect Transistor Biosensors for Rapid Pathogen Detection
Edited by Naif H. M. Al-Hardan, Azman Jalar, Muhammad Azmi Abdul Hamid and Mohd Firdaus Raih

Published by the Royal Society of Chemistry, www.rsc.org

are based on the change of the oxidation state of a conducting polymer as the active material, which typically isolates in its reduced state and conducts in its oxidized state.

2.2 General Concepts of Electrolyte-gated Field-effect Transistors

A field-effect transistor (FET) is a four-terminal device made of an electrically active material, such as a semiconductor. However, degenerated semiconductors or even conductors can also be used (this will be explained later). The active material is between two conducting terminals called the source and drain, and when a voltage difference (usually named V_{DS} for source–drain voltage) is applied, a current, I_{DS}, can flow through it. In the case of thin-film transistors (TFTs), the active material is placed between two other terminals: a body (also known as the substrate) and a conducting gate, thus forming a sandwich structure. The conducting gate is isolated from all other contacts by a thin layer of dielectric material, preventing any stationary current from flowing through the gate (if a stationary current takes place, it is called a leakage current, generally having negative effects on the transistor functioning). In general, the body and source of FETs are short-circuited, so they are regarded as three-terminal devices. Nevertheless, if a 'body effect' is required, the body contact acts as a second gate; it can be polarized to either amplify or counteract the voltage applied to the primary gate contact. Figure 2.1 depicts the basic structure of a field effect transistor (FET).

Various geometries are possible: (i) the gate could be located on the top (top-gate) or (ii) at the bottom of the semiconductor (bottom-gate). Source and drain contacts can be situated on the top of the semiconductor (top-contact) or below it (bottom-contact). Upon applying a voltage difference, V_{GS}, between the gate and the body (body and source being short-circuited), an electric field is generated, oriented from the high voltage contact to the low voltage contact, through the active semiconducting layer. Normally, for an n-type FET as exemplified in Figure 2.1, the gate needs to be positively polarized in relation to the body, in order for the electric field ($\vec{E}$) to be directed from the gate towards the body. The electric field will alter the density of charge carriers between the source and drain, and consequently the conductivity of the semiconductor will also change, resulting in the current flowing between the source and drain. A more comprehensive explanation will follow in Section 2.2.3.

If the active material is a semiconductor, the charge carriers can be electrons or holes for n-type or p-type semiconductors, respectively. In general, semiconductors with a low highest occupied molecular orbital (HOMO) level are used as p-type semiconductors, while those with a high HOMO level are used as n-type semiconductors. As we will see below, most semiconductor-based EGFETs use p-type materials for reactivity and stability reasons.

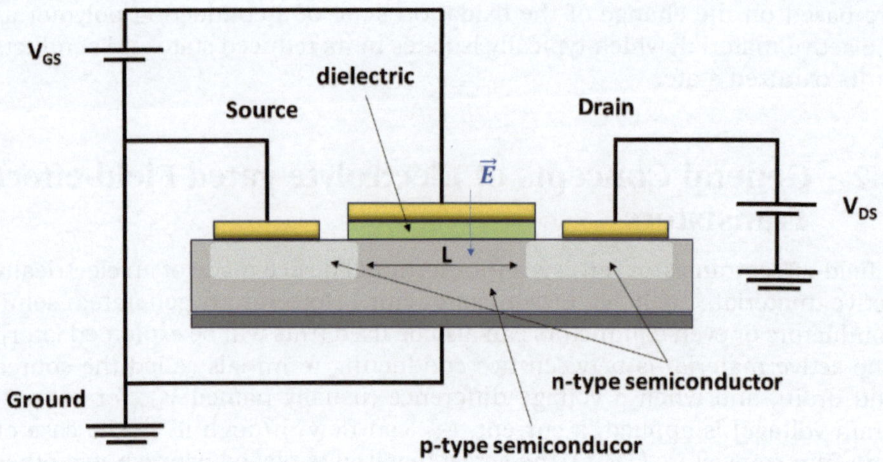

Figure 2.1 Schematic of a conventional field-effect transistor (FET) which includes essential elements such as the active material (in this case, a p-type semiconductor, where the majority charge carriers for the material's conductivity are electron holes, *i.e.* holes), a dielectric material (any form of an inorganic or an organic insulator), and four terminals named the gate, source and drain (connected to the semiconductor *via* boxes of inverse doping compared to the semiconductor) and the body. The gap between the source and drain is called the 'channel' of length *L*. The illustrated FET is a top gate, top contact one.

However, conductive materials can also be used, most commonly in the form of 2D materials; this will be discussed in detail in Sections 2.2.3 and 2.2.4.

As explained above, in conventional FETs, the gate is separated from the active material by a solid dielectric, including polymers or insulating oxides. In EGFETs, the dielectric is replaced by a solid or liquid electrolyte, thus containing mobile ions; we will see in this chapter that for biosensing applications it is advantageous to work with water as the electrolyte, since it is the natural environment for biological targets and receptors, so it can act both as an electrolyte and as a carrier medium for the analyte. We will also see that hydrogels are well adapted materials, in combination with or without a liquid electrolyte. Figure 2.2 shows the simplest structure of an EGFET.

In all cases, the application of a voltage to the gate electrode leads to the polarization of all the interfaces between the gate and the substrate (body), in particular at the interface between the electrolyte and the gate and between the electrolyte and the active material: two so-called electrical double layers are formed.

2.2.1 Basics in Electrical Double Layers

In a conventional FET, the field effect is due to the presence of a thin film (of thickness d) of dielectric between the gate electrode and the active material.

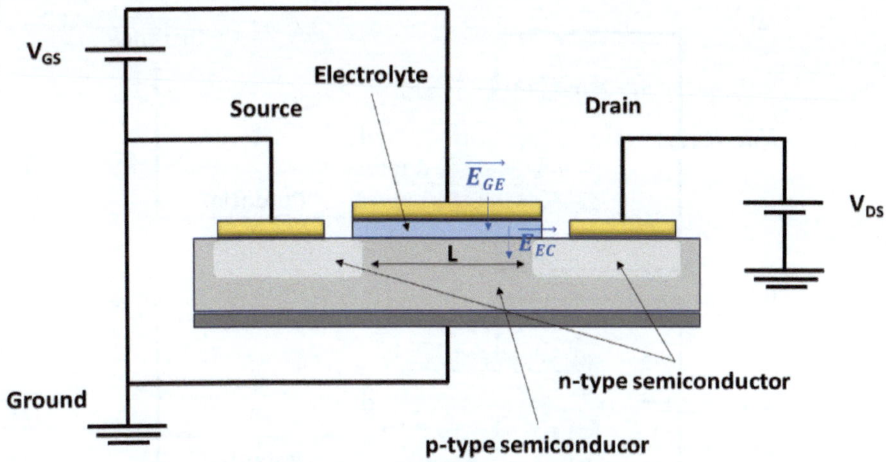

Figure 2.2 Schematic of the simplest EGFET. Compared to a conventional FET, the dielectric material is replaced by a solid or liquid electrolyte. When a voltage is applied to the gate, the gate/electrolyte and electrolyte/active material are polarized, creating two strong local electric fields, E_{GE} at the gate/electrolyte interface and E_{EC} at the electrolyte/channel interface, the latter causing an increase in charge carrier concentration in the channel.

When a potential difference is applied across the dielectric film between the gate and the substrate, a transient current briefly flows to charge the interfaces, but no steady current flows through the material because it has no charge carriers. Dipoles are formed (*e.g.*, from polarized covalent bonds or any entity within the dielectric material that has a dipole moment) and adopt an anisotropic orientation rather than a random one. On average, positive poles are oriented towards the negative electrode, while negative poles are oriented towards the positive electrode. The electric field is therefore progressively weakened within the dielectric, causing a voltage drop across the dielectric (Figure 2.3(a)). In other words, the electric field is distributed throughout the thickness of the dielectric. At steady state, at zero gate current, the norm of the electric field $\vec{E}$ is given by eqn (2.1). This means that, for a given gate voltage V_{GS}, the thinner the dielectric, the higher the electric field.

$$\overrightarrow{\|E\|} = \frac{V_{GS}}{d} \tag{2.1}$$

EGFETs are different. When a potential difference is applied across the electrolyte between the gate and the substrate, an electrical double layer (EDL) quickly forms at any interface where the electrolyte is in contact with a solid. This is due to a redistribution of mobile ions within the electrolyte. If the ionic conductivity is sufficient, there is no voltage drop in the bulk of the electrolyte because the anions and cations remain evenly distributed, but most, if not all, of the voltage drop occurs at the two interfaces (Figure 2.3(b)).

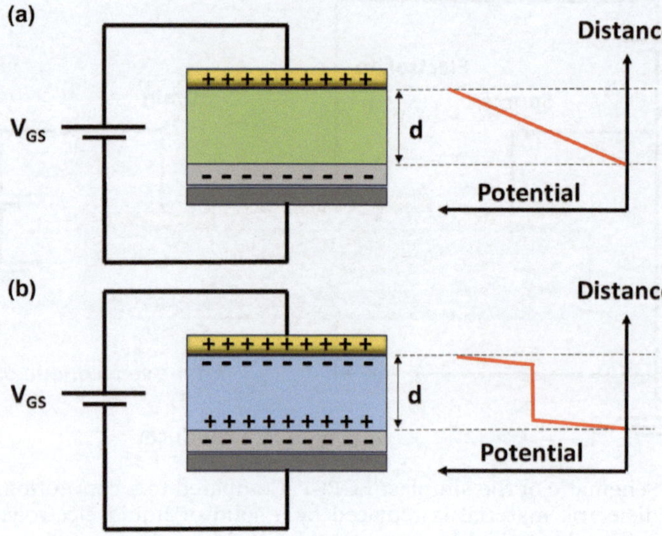

Figure 2.3 Schematics of a voltage drop across a dielectric (a) and across an electrolyte (b). In the latter case, the voltage drop occurs only at the two interfaces.

In contrast to the case of a dielectric, two different electric fields are therefore generated at the two interfaces (Figure 2.3(b)). The total electric field $\vec{E}$ can be decomposed into two components, $\vec{E}_{GE}$ and $\vec{E}_{EC}$ for the gate/electrolyte (GE) and electrolyte/channel (EC) interfaces, respectively, as given by eqn (2.2a):

$$\vec{E} = \vec{E}_{GE} + \vec{E}_{EC} \tag{2.2a}$$

$$\overrightarrow{\|E_{GE}\|} = \frac{V_{GE}}{d_{GE}} \quad \text{and} \quad \overrightarrow{\|E_{EC}\|} = \frac{V_{EC}}{d_{EC}} \tag{2.2b}$$

where V_{GE} and V_{EC} are the voltage drops across the gate/electrolyte and electrolyte/channel interfaces, respectively, and d_{GE} and d_{EC} are the thicknesses of the gate/electrolyte and electrolyte/channel double layers, respectively. The voltage drops are derived from the amount of trapped charges in the bilayer Q and the interfacial capacitance C, so that $V_{GE} = Q_{GE}/C_{GE}$ and $V_{EC} = Q_{EC}/C_{EC}$. The EDL capacitance C depends on its thickness d, the interface area A, the vacuum permittivity ε_0 and the relative permittivity of the medium ε_r according to eqn (2.3):

$$C = \frac{\varepsilon_0 \varepsilon_r A}{d} \tag{2.3}$$

In other words, EGFETs benefit from the high EDL capacitances generated by the use of an electrolyte, on the order of several $\mu F\,cm^{-2}$ or even tens of $\mu F\,cm^{-2}$ (for water, it is around 80) compared to tens of $nF\,cm^{-2}$ for conventional polymer dielectrics. This allows high electric fields to be obtained at low gate voltages, meaning that EGFETs can generally be operated below 1 V with water as the electrolyte, compared to tens of volts for conventional FETs. This is discussed in more detail in Section 2.2.4.

The Helmholtz description of the EDL, improved by Bockris in the 1960s, allows a better understanding of the role of EDLs in EGFETs. In this model, the solid interface is first covered by a homogeneous layer of solvent molecules, typically water molecules, which are anisotropically aligned along the electric field, defining the so-called internal Helmholtz plane (IHP). Solvated ions (*e.g.*, cations if the electrode is negatively polarized) from the electrolyte sit directly on this IHP, forming the outer Helmholtz plane (OHP); above these two layers is the diffuse layer where cations are free in solution, their concentration being greater than that of the bulk electrolyte and decreasing exponentially with distance. The equation giving the potential can be written as in eqn (2.4):

$$\Psi(z) = \Psi^{d} \exp(-z/\lambda_{\mathrm{d}}) \tag{2.4}$$

where $\Psi(z)$ is the potential at a distance z from the interface, Ψ_{d} is the Stern potential, *i.e.*, the potential at the OHP, and λ_{d} is the so-called Debye length, the value of which roughly defines the thickness of the EDL. The Debye length, which is only a few nanometres in aqueous electrolytes, is related to the ionic strength I according to eqn (2.5a), simplified to eqn (2.5b) in water:

$$\lambda = \sqrt{\frac{\varepsilon_0 \varepsilon_{\mathrm{r}} k_{\mathrm{B}} T}{2e^2 I}} \tag{2.5a}$$

$$\lambda = \sqrt{\frac{0.1}{I}} \tag{2.5b}$$

where λ and I are the Debye length and ionic strength, expressed in m and $\mathrm{m^{-3}}$, respectively, in eqn (2.5a) and in nm and $\mathrm{mol\,L^{-1}}$, respectively, in eqn (2.5b). k_{B} is the Boltzmann constant and e is the electron charge. For a deeper understanding of the EDL, the reader is referred to the review "Understanding the Electric Double-Layer Structure, Capacitance, and Charging Dynamics" by Wu.[1]

With these basics in mind, we can now move on to the most important part of EGFETs: the electrolyte.

2.2.2 Electrolytes

An electrolyte contains mobile ions. It is electrically conductive, but this conductivity is not due to the movement of electrons or holes, but to the movement of cations and anions. With no voltage applied, these ions are uniformly distributed throughout the bulk of the electrolyte, with negatively solvated anions compensating for positively solvated cations to make the electrolyte neutral throughout its volume. For example, if two noble metal electrodes are immersed in this electrolyte and a voltage is applied between them, the one with the highest voltage will have a deficit of electrons (and be positively charged), while the other will have an excess of electrons (and be negatively charged). Ions will migrate (*i.e.*, move along the electric field), anions going to the positively charged electrode and cations going to the negatively charged electrode, as explained in Section 2.2.1. Electrolytes can be liquid or solid, as explained below.

Historically, EGFETs, which mimic conventional FETs based on solid dielectrics, were first reported using solid electrolytes such as polyelectrolytes, *i.e.*, polymers carrying ionizable groups, whether negative or positive, for example, blends of polyacrylic acid (PAA) with polyethylene oxide (PEO) or polyvinyl alcohol (PVA), PAA as the polyanionic proton conductor and PEO or PVA as the hydrophilic phase to provide the highest proton mobility. Polystyrene sulphonate (PSS) is also widely used. In addition to blends, which are easier to obtain, copolymers can also be used, such as poly(vinyl phosphonic acid-*co*-acrylic acid).[2] Polyethyleneimine (PEI), a polycationic polymer, can also be used.[3] However, as most active materials used in EGFETs are p-type materials, negatively charged polymers are most commonly used to gate holes (positive charge carriers) into the material. The use of negatively charged polymers with p-type active materials also ensures that anions, which are immobile on large polymer chains, cannot penetrate into the p-type material and "trap" holes.

In addition to this advantage, such polyelectrolytes offer rapid formation of the EDL. Indeed, as shown in Figure 2.4 for a p-type EGFET, if the negative

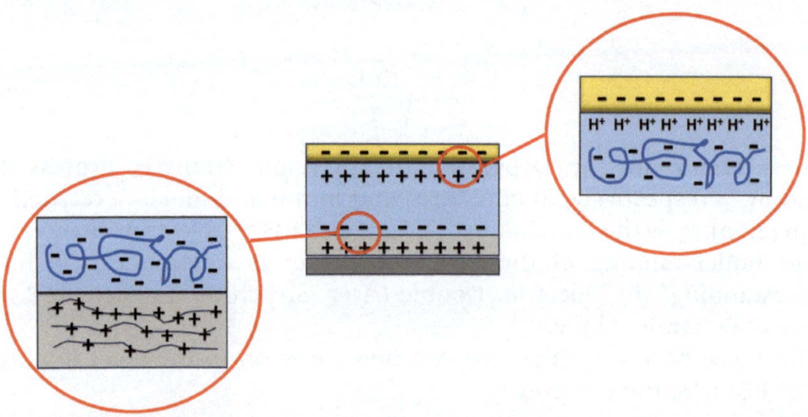

Figure 2.4 Schematic of an EGFET with a solid polyelectrolyte (in blue) between the gate (yellow) and the active material (here p-type, in grey). The polyelectrolyte carries fixed negative charges on the low pK_a side groups of the polymer chains (*e.g.* sulphonate, carboxylate, phosphonate…) together with mobile protons. When a negative voltage is applied to the gate, there is a rapid redistribution of ions: protons go to the gate, leaving immobile negative charges on the immobile polymer chains, therefore close to the active material. This drains holes in the semiconducting polymer, forming a conductive channel. Reproduced from ref. 2 with permission from John Wiley & Sons, Copyright 2007 Wiley-VCH Verlag GmbH & Co. KGaA, Weinheim.

charges carried by the side groups of the polymer chains are immobile, the protons coming from the Brönsted acidity of these side groups are highly mobile. When a negative V_{GS} is applied, the protons are redistributed and go to the gate to form an EDL; as a result, the immobile negative charges left near the active material drain holes, forming a conductive channel.[2]

Therefore, the use of such solid electrolytes allows a fast switching rate from the on (conductive) to the off (insulating) state, which is useful for certain applications that require fast operations. However, solid electrolytes may not be ideal media for biosensing applications because (i) the diffusion rate of analytes in such a matrix is infinitesimally low and (ii) (bio)chemical processes are quite slow and kinetically limiting, so fast switching transistors are not required.

However, some other types of condensed materials can be used as electrolytes, such as hydrogels, ion gels and biomembranes. Regarding hydrogels, *i.e.*, insoluble polymer matrices swollen with water, "smart" hydrogels can be incorporated into EGFETs, *i.e.*, hydrogels with the property of changing their physicochemical properties upon application of a stimulus. Examples are pH-responsive hydrogels[4] or DNA-based hydrogels.[5] pH-responsive hydrogels have the property of shrinking or swelling depending on the pH. This is the case of polyacrylic acid (PAA), for example, which swells at pH above the pK_a of the acrylic acid groups ($pK_a = 4.25$) due to electrostatic repulsion of the $-COO^-$ groups from neighboring chains and shrinks at lower pH when the $-COOH$ groups no longer repel and that their hydrophobic character leads to solvent expulsion from the polymer matrix. The shrinkage of the $-COOH$ groups results in a locally high concentration of protons; upon application of a negative gate voltage, these protons are able to migrate to the gate, forming a dense EDL, while negatively deprotonated PAA chains attract holes to the hydrogel/active material interface, which becomes conductive. Conversely, swelling of the deprotonated hydrogel leads to a decrease in the concentration of holes in the active material, *i.e.*, a decrease in current. Such pH-responsive hydrogels are then capable of carrying out a wide variety of enzymatically catalysed bioreactions where the enzyme activity leads to the production of protons, for example. This will be discussed in detail in Section 2.5.

DNA-based hydrogels are also particularly suitable as electrolytes for EGFETs. Indeed, DNA is itself a polyelectrolyte, carrying negative phosphate groups along the entire strand. Any hybridization or dehybridization of these DNA (or RNA) strands, which are known to change the conformation of the strands (at pH 7 and an ionic strength of about 0.15 M, dsDNA is statistically rigid over a long distance of several tens of nanometres, whereas ssDNA is somewhat flexible with much shorter rigid segments of less than a nanometre), will affect the charge density near the transistor interface, *i.e.* the density of carriers in the active material.

Ion gels can also be used in EGFETs. They consist of a mixture of an ionic liquid and a polymer, *e.g.*, the ionic liquid 1-ethyl-3-methylimidazolium bis(trifluoromethylsulfonyl)imide ([EMIM][OTf]) encapsulated in polyvinyl alcohol (PVA) as a hydrophilic material and poly(ethyl methacrylate) (PEMA) as a crosslinker.[6] Again, however, the diffusion coefficients of biomolecules within such electrolytes are infinitesimal, making them unsuitable for most biosensing applications.

In any case, most biosensing applications use EGFETs with liquid electrolytes, such as biological buffers (PBS, citrate, *etc.*), as will be shown in Section 2.6. The reader interested in electrolytes can refer to the review by Wang *et al.* (2021) for more details.[7]

2.2.3 Charge Transport in Conducting and Semiconducting Materials

The purpose of this section is not to provide a comprehensive course on charge transport in materials, but to recall the basic rules to understand the operation of FETs, including EGFETs, using simplified models.

Materials can be broadly classified into three categories: insulating (dielectric), conducting and semiconducting. In insulating materials, electrons are confined to low energy levels, are involved in interatomic interactions and are therefore localized (these levels belong to the so-called "valence band", VB); they cannot move over long distances and therefore cannot participate in electrical conduction. In conducting materials, localized electrons are still present and completely fill the VB, but some additional electrons have sufficient energy to sit in the higher energy "conduction band" (CB) and become delocalized over long distances. Such mobile electrons are then called "charge carriers" and their concentration (or density) in the material is called n (expressed in cm^{-3}). In semiconducting materials, some electrons from the VB have sufficient energy to escape their localized state and reach the CB, leaving electron holes in the VB, more simply called "holes". In this case, in addition to an electron density n in the CB, there is a hole density p in the VB. Although a hole is actually an electron deficiency, a simple physical model treats holes as positively charged particles. The Boltzmann statistics suggests that the product $n \times p$ is a constant, so that n is low when p is high and *vice versa.*

The conductivity σ of a material is determined by both the density of the charges and their mobility μ, *i.e.*, the ability of a type of charge to move along an electric field in the material, according to eqn (2.6a), where e is the elementary charge:

$$\sigma_{\mathrm{n}} = en\mu_{\mathrm{n}} \quad \text{and} \quad \sigma_{\mathrm{p}} = e\mu_{\mathrm{p}}$$
$$\sigma = \sigma_{\mathrm{n}} + \sigma_{\mathrm{p}} = e(n\mu_{\mathrm{n}} + p\mu_{\mathrm{p}}) \tag{2.6a}$$

Ohm's law gives the current density $\vec{j}$ (the current divided by the section of the conducting or semiconducting material) according to eqn (2.6b):

$$\vec{j_n} = \sigma_n \vec{E} \qquad \vec{j_p} = \sigma_p \vec{E} \qquad \vec{j} = \sigma \vec{E} \tag{2.6b}$$

where $\vec{E}$ is the electric field and $\vec{j_{\mathrm{n}}}$ and $\vec{j_{\mathrm{p}}}$ are the current densities due to electrons and holes, respectively. Some materials are p-type ($p \gg n$), some are n-type ($n \gg p$), and some are both, *i.e.*, ambipolar ($p \sim n$); these are discussed in Section 2.3.

In an EGFET, when a given gate voltage is applied, either electrons or holes are concentrated at the electrolyte/active material interface, forming an n-channel or a p-channel between the source and drain electrodes. Some active materials are discussed in Section 2.3. The operating principles of EGFETs are recalled in Section 2.2.4.

2.2.4 Geometries and Operating Principles

2.2.4.1 Geometries

In Section 2.2.2, a top-gate EGFET geometry was shown. There are actually a number of different geometries, as shown in Figure 2.5: top-gate, bottom-gate, side-gate and extended-gate.

The distribution of potentials between the gate and the active material in the channel depends on the geometry. Figure 2.6 shows three examples corresponding to the most common top-gate architectures: (a) without a biomembrane, (b) with a biomembrane on the gate electrode and (c) with a biomembrane on the active material.

In all cases in Figure 2.6, there is a potential drop across each interface. As explained in Section 2.2.1, this potential drop depends on the capacitance of the interface: the higher the capacitance, the lower the voltage drop. Typically, in case (a), for equal gate and active material areas, because a bare noble metal gate (*e.g.* gold or platinum) presents a capacitance of 10–30 $\mu F\,cm^{-2}$ in water, whereas that of the active material (*e.g.* an organic semiconductor) is about 1 $\mu F\,cm^{-2}$, the voltage drop at the electrolyte/active material interface (denoted by $\Delta V_{E/M}$) is higher than that at the gate/electrolyte interface (denoted by $\Delta V_{G/E}$). For a biofunctionalised gate (case b), in addition to the voltage drop at the biomembrane/electrolyte interface, there may be a voltage drop across the biolayer (biomembrane). As a result, the voltage drop $\Delta V_{E/M}$ at the electrolyte/channel interface is relatively small. For a bare gate and functionalized active material (case c), for the same reasons, most of the voltage drop occurs at the electrolyte/biomembrane interface, together with a possible voltage drop across the biomembrane.

These potential distributions have a major influence on the functioning of the EGFET and its ability to act as a biosensor for two reasons: (i) the electric field that increases the charge carrier density in the active materials is high when the voltage drop is high; a low voltage drop at the channel interface results in a low charge carrier density, *i.e.* a low current; (ii) at high ionic strength, the Debye length is short, which means that any molecular event that occurs beyond this distance, in a region of uniformly distributed ions and where the potential is constant, will not be sensed.

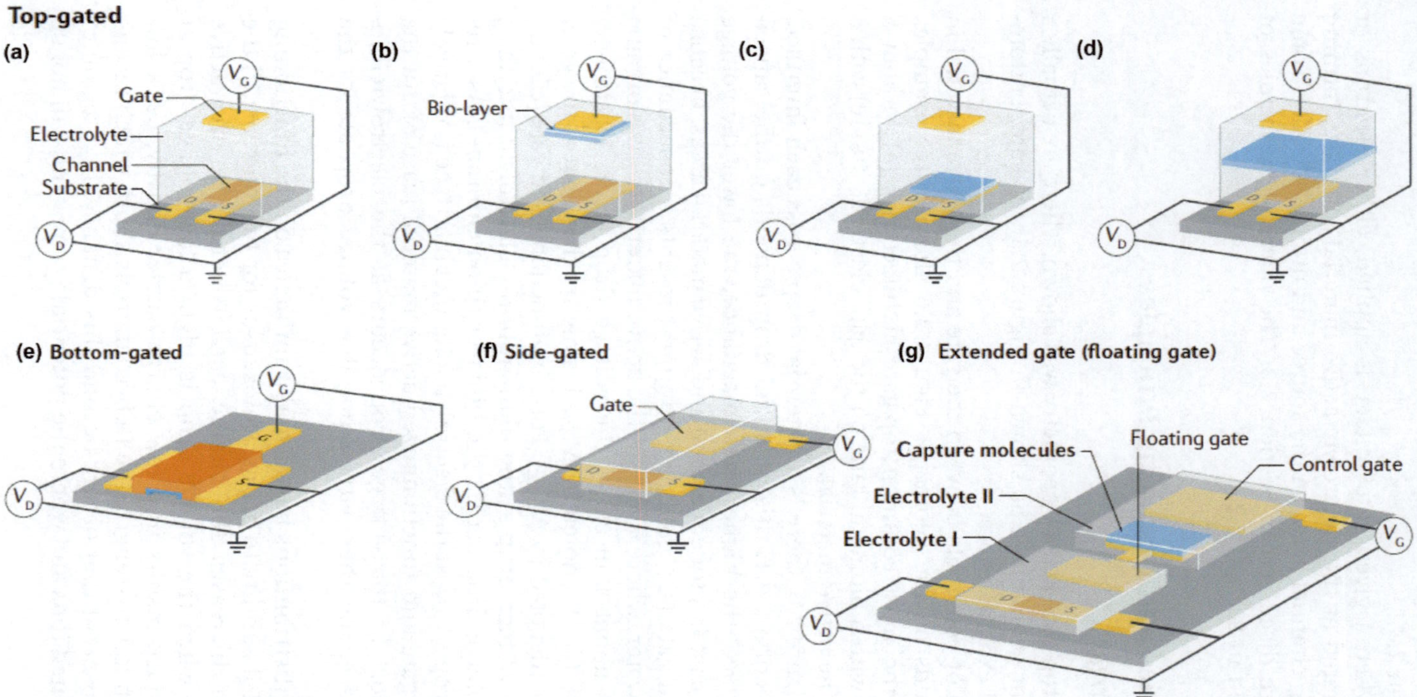

Figure 2.5 Different EGFET geometries. Top-gate structures correspond to gates above the electrolyte (a). For sensing, the functionalized bioactive layer can be on top of the gate (b), on top of the active layer (c) or in between (d). The bottom-gate structure (e) corresponds to the gate below the active material. This means that the electrolyte is trapped between the gate and the active material and therefore cannot be an accessible bio-interface. Side-gate structures include all structures where the source, drain and gate are coplanar, with the electrolyte covering the plane (f). Extended-gate devices correspond to structures with two capacitance-coupled gates and independent electrolytes, where polarization of the first gate induces polarization of the second gate (g). Reproduced from ref. 8 with permission from Springer Nature, Copyright 2021.

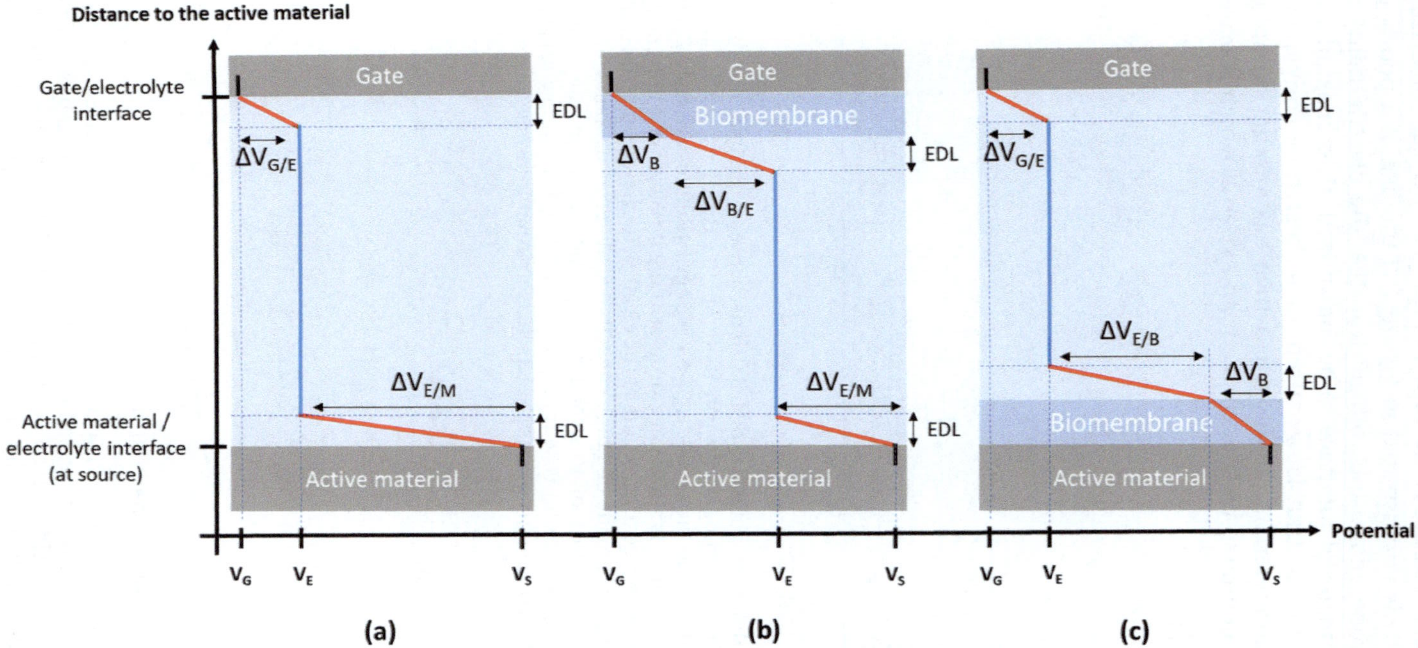

Figure 2.6 Three examples of the potential distribution between the gate electrode and the active material (at the source electrode) for an EGFET (a) without a biomembrane (bare gate), (b) with a biomembrane on the gate electrode and (c) with a biomembrane on the active material. $\Delta V_{G/E}$ is the voltage drop at the gate/electrolyte interface, $\Delta V_{E/M}$ is the voltage drop at the electrolyte/active material interface and $\Delta V_{B/E}$ or $\Delta V_{E/B}$ is the voltage drop at the electrolyte/biomembrane interface. ΔV_B is the voltage drop across the biomembrane. V_G, V_E and V_S are the gate, electrolyte and source potentials, respectively.

Another important feature is the nature of the gate, whether polarizable or non-polarizable. This is illustrated in Figure 2.7: the use of a poorly polarizable gate electrode allows the voltage drop (*i.e.* the electric field) at the electrolyte/active material interface to be maximized, thus maximizing the carrier density and the transistor current. The best case corresponds to the use of a pseudo-reference as the gate, which has another advantage: it allows the gate potential to be kept constant and known, which is extremely

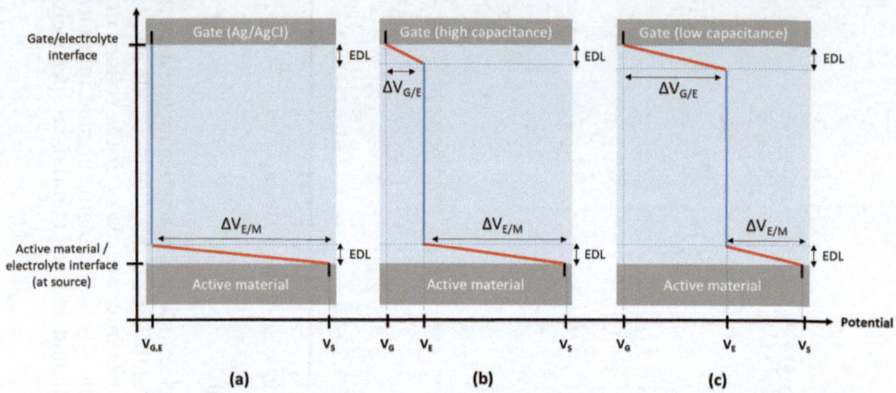

Figure 2.7 Three different potential distributions between the gate and the active material depending on the gate material. In case (a), the gate is an ideally non-polarizable electrode (*e.g.*, an Ag/AgCl pseudo-reference); this implies that there is no voltage drop at the gate/electrolyte interface, and the entire voltage drop occurs at the electrolyte/channel interface. In case (b), the gate is made of a high capacitance material: a small voltage drop occurs, and most of the voltage drop still occurs at the electrolyte/channel interface. In case (c), on the other hand, for a low capacitance gate, the voltage drop at the electrolyte/channel interface is small.

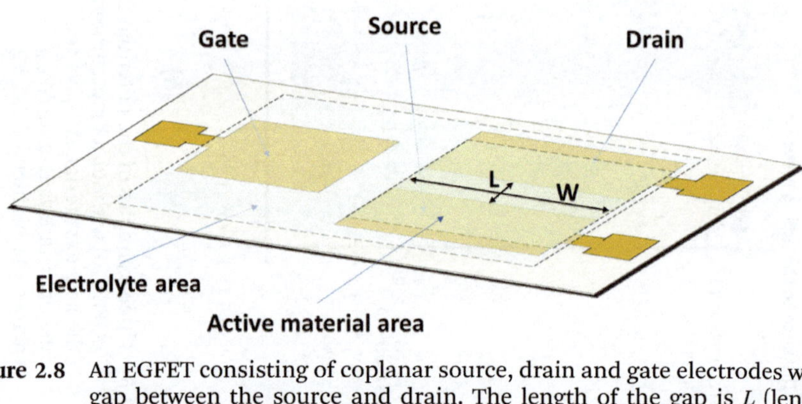

Figure 2.8 An EGFET consisting of coplanar source, drain and gate electrodes with a gap between the source and drain. The length of the gap is L (length), while W is its width. In an EGFET, provided the electrolyte is sufficiently ionically conductive, the distance between the gate and the source (or drain) is of little importance.

important when (bio)electrochemical reactions take place in the electrolyte or in the biomembrane.

2.2.4.2 Operating Principles

Let us consider a simple EGFET with coplanar source, drain and gate electrodes, as shown in Figure 2.8.

EGFETs have typical *I–V* curves, called output and transfer curves, like any FET; examples of such curves are shown in Figure 2.9, showing the different regions of the output curves (ohmic, quadratic, saturation) and the general shape of a transfer curve. Between the quadratic and saturation regimes, a pinch-off point may be placed, corresponding to the state where the charge carriers start to be depleted at the channel/drain interface. The threshold voltage V_{Th} is usually found at the intercept between the tangent at the plot of $\sqrt{I_{DS}} = f(V_{GS})$ and the *x*-axis for $I_{DS} = 0$.

The carrier mobility μ has already been defined (Section 2.2.3), while the total capacitance C of the device is defined by $C^{-1} = C_{G/E}{}^{-1} + C_{E/M}{}^{-1}$, where $C_{G/E}$ and $C_{E/M}$ are the capacitances of the gate/electrolyte interface and the electrolyte/channel interface, respectively. The drain current is described by the following equations:

$$I_{DS} = \frac{W}{L}\mu C\left(V_{GS} - V_{Th} - \frac{V_{DS}}{2}\right)V_{DS} \quad \text{for} \quad V_{GS} - V_{Th} \geq V_{DS} \tag{2.7a}$$

$$I_{DS} = \frac{W}{2L}\mu C(V_{GS} - V_{Th})^2\,(1 + \lambda)V_{DS} \quad \text{for} \quad V_{GS} - V_{Th} < V_{DS} \tag{2.7b}$$

where I_{DS} is the drain current, W is the width of the channel, L is its length, μ is the mobility of the charge carriers involved, C is the total capacitance, V_{GS} is the gate voltage, V_{DS} is the drain voltage, V_{Th} is the threshold voltage and λ is a parameter (unit: inverse of a length) corresponding to the channel length modulation at saturation, when I_{DS} no longer depends much on V_{DS}.

For sensing applications, it is very important to consider the sensitivity of the EGFET, *i.e.*, how sensitive the drain current is to a change in the gate voltage. This sensitivity is defined by the transconductance g_m, according to eqn (2.8a–c):

$$g_m = \frac{\partial I_{DS}}{\partial V_{GS}} \quad \text{at constant } V_{DS} \tag{2.8a}$$

$$g_m = \frac{W}{L}\mu C\ V_{DS} \quad \text{in the linear regime} \tag{2.8b}$$

$$g_m = \frac{W}{L}\mu C(V_{GS} - V_{Th}) \quad \text{in the saturation regime} \tag{2.8c}$$

For constant V_{GS} and V_{DS}, eqn (2.8b and c) show that the sensitivity depends on the capacitance, *i.e.*, any change in capacitance at the gate/electrolyte or

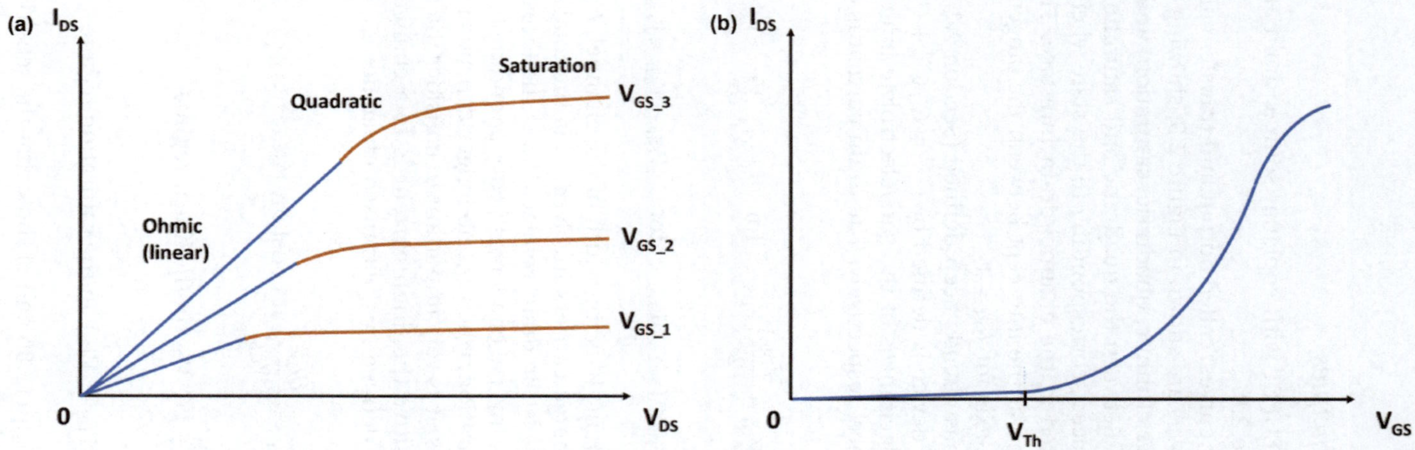

Figure 2.9 (a) Output curves showing the drain current I_{DS} as a function of the drain voltage V_{DS} for different gate voltages V_{GS} ($V_{GS\_1} < V_{GS\_2} < V_{GS\_3}$). It can be divided into three sections: the ohmic (linear) region for low drain voltages, the saturation region for high drain voltages and the quadratic region for intermediate drain voltages. (b) Transfer curve, for a fixed drain voltage, giving the drain current as a function of the gate voltage. Roughly speaking, the current starts to increase from a threshold voltage (V_{Th}).

electrolyte/active material interface, leads to a change in transconductance and hence a change in drain current.

V_{Th}, corresponding to the gate voltage needed to sufficiently increase the charge carrier density to create a conductive channel, needs further explanation. In contact with the electrolyte, it depends on several parameters, as detailed in eqn (2.9a):

$$V_{Th} = 2\varphi_f + \chi^{sol} - \frac{\Phi_{material}}{q} - \frac{Q_{total}}{C} - \psi + E_{ref} \tag{2.9a}$$

where φ_f is the Fermi level (in potential), χ^{sol} is the surface dipole potential of the electrolyte, $\Phi_{material}$ is the work function of the active material, Q_{total} is the total charge at the interface and within the active material, C is the total capacitance of the electrolyte/active material interface, ψ is the potential depending on chemical interactions at the active material surface, and E_{ref} is the reference electrode potential. Therefore, eqn (2.9a) can be simplified to eqn (2.9b):

$$V_{Th} = \text{Constant} - \psi \tag{2.9b}$$

where ψ is the only variable depending on the chemical environment of the active material. V_{Th} is therefore the critical parameter that makes EGFETs good chemical sensors.

2.3 Some Active Materials Used in EGFETs

The active materials used in EGFETs fall into two categories: (i) organic and inorganic semiconductors, or (ii) conductors with 1D or 2D structures such as nanowires or nanosheets. These materials are described in detail below.

2.3.1 Organic and Inorganic Semiconductors

2.3.1.1 Organic Semiconductors

Organic semiconductors are generally large conjugated molecules, up to and including conjugated polymers, *i.e.*, they have long-range electronic conjugation due to an alternation of single and double bonds. In order to provide a certain rigidity to maintain this conjugation over a sufficiently long distance, these molecules are designed to be partially crystalline at room temperature (polymers are only partially crystalline; their degree of crystallinity is usually between 10 and 80%); for this reason, they often contain aromatic rings (benzene, thiophene or pyrrole). Some examples are shown in Figure 2.10(A)–(C). EGFETs made of these organic materials are named EGOFETs for electrolyte-gated organic FETs.

Since these organic materials are semiconductors, this means that they have a significant energy gap between their HOMO (highest occupied molecular orbital) and LUMO (lowest unoccupied molecular orbital).

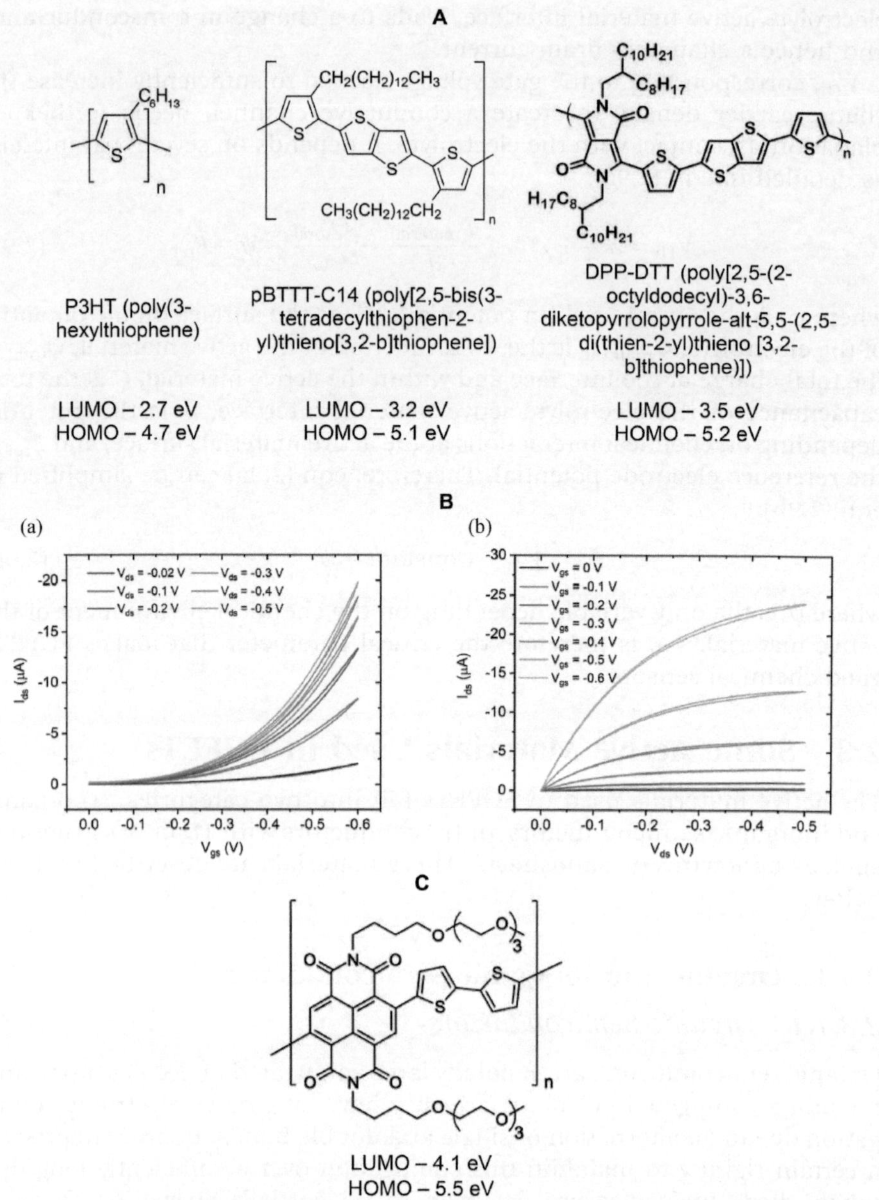

Figure 2.10 (A) Some conjugated polymers or molecules used in EGOFETs. (B) Typical (a) transfer and (b) output curves for a p-type P3HT EGOFET. Electrolyte: water. Ag/AgCl pseudo-reference as the gate. Reproduced from ref. 9, https://doi.org/10.3390/s19204497, under the terms of the CC BY 4.0 license, http://creativecommons.org/licenses/by/4.0/. (C) The n-type conjugated polymer poly(NDI-C4-TEGMe-T2). Reproduced from ref. 10, https://doi.org/10.1002/aelm.202200573, under the terms of the CC BY 4.0 license, http://creativecommons.org/licenses/by/4.0/.

Most polymers used in EGOFETs are p-type, with low HOMO energies corresponding to the low Fermi level of the gold source and drain contacts (5.2 eV); these polymers also have the advantage of being relatively stable to water and oxygen.

However, some n-type semiconductors have recently been reported for EGOFETs, such as poly(NDI-C4-TEGMe-T2), a derivative of poly[*N*,*N*′-bis(2-octyl-dodecyl)-1,4,5,8-napthalenedicarboximide-2,6-diyl]-*alt*-5,5′-(2,2′-bithiophene) (PNDIT2), to which oligoethylene glycol side chains are attached (Figure 2.10(C)).

However, organic semiconductors are generally not stable enough for long-term use (a few hundred hours in aqueous electrolyte for p-type only and a few tens of hours at best for n-type). For this reason, n-type inorganic semiconductors have been used in EGFETs, as described below.

2.3.1.2 *Inorganic Semiconductors*

Various inorganic semiconductors have been used in EGFETs, such as ZnO, In_2O_3, InGaZnO (IGZO), and WO_3 (see Table 2.1). These are n-type semiconductors with excellent mobility. The power and transfer curves obtained with these inorganic materials are very similar to those obtained with organic semiconductors (Figure 2.11).

Table 2.1 Some n-type inorganic oxides used in EGFETs, with their corresponding conduction and valence band energy levels.

Oxide	ZnO	In_2O_3	IGZO	WO_3
Band diagram	$E_C = -4.6$ eV $E_V = -7.7$ eV	$E_C = -3.7$ eV $E_V = -7.3$ eV	$E_C = 4.2$ eV $E_V = 7.2$ eV	$E_C = 4.2$ eV $E_V = 7.2$ eV

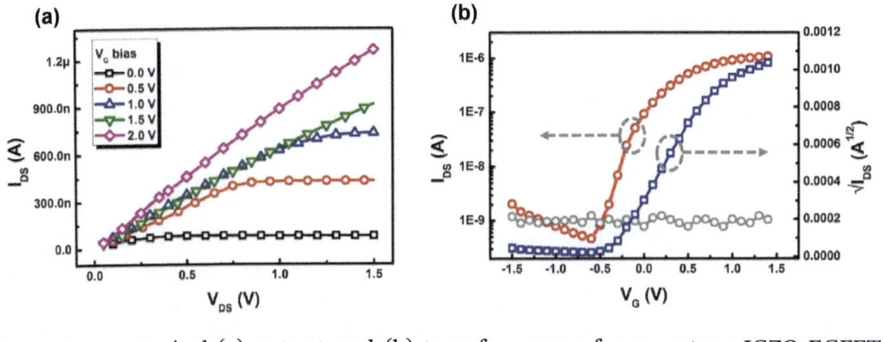

Figure 2.11 Typical (a) output and (b) transfer curves for an n-type IGZO EGFET. Electrolyte: de-ionized water. Ag/AgCl pseudo-reference as the gate. Reproduced from ref. 11 with permission from Elsevier, Copyright 2018.

2.3.2 Nanostructured Electrical Conductors (Carbon Nanotubes, Graphene, and Nanowires)

The principle of EGFETs is based on the creation of an electric field at the electrolyte/channel interface, which attracts or repels charge carriers within the active material. If this material is thick along the direction of the electric field, carriers will not occupy the entire thickness but only a thin layer, which means that part of the semiconductor between the source and drain remains poorly conductive, while only the top layer is conductive. An alternative is to use 1D or 2D materials such as carbon nanotubes (CNTs), nanowires such as silicon nanowires (SiNWs) or sheets such as graphene sheets. Such materials are so thin that charge carriers accumulate throughout the material, making them extremely sensitive to any electrostatic interaction. In contrast to organic semiconductors or oxides, such materials are easy to functionalize, *e.g.*, for grafting molecular probes. Some examples are given below.

2.3.2.1 Carbon Nanotubes

Carbon nanotubes, and in particular single-walled carbon nanotubes (SWCNTs), are 1D particles with a high carrier mobility of several thousand $cm^2 V^{-1} s^{-1}$. They have a large surface area which makes them very sensitive to their electrostatic environment. SWCNTs are readily deposited between the source and drain contacts of a transistor by various processes, including drop casting and even printing. The behaviour of electrolyte-gated SWCNTFETs (EG-SWCNTFETs) is very similar to that of EGFETs using organic or inorganic semiconductors, provided that the SWCNTs used are themselves semiconducting[12] (see Figure 2.12).

SWCNTs can be easily functionalized by π–π stacking, *e.g.*, with pyrene or perylene or other π-electron-rich groups. More details on EG-CNTFETs can be found in the review by Shkodra *et al.*[13]

2.3.2.2 Graphene and Its Derivatives

Among 2D materials, graphene has the best electrical conductivity, with a charge mobility of up to 5000 $cm^2 V^{-1} s^{-1}$.[14] The electronic band structure of graphene, which is a semi-metal rather than a semiconductor, is very peculiar due to its crystallinity. At some points in the reciprocal space, called Dirac points, the conduction band meets the valence band, resulting in a zero bandgap. In pristine graphene, the Fermi level is at the Dirac point, at about −4.5 eV. At this point, the charge density can be considered null (called the charge neutrality point, CNP). For the same reason, graphene is strictly ambipolar and can carry electrons or holes, the former in the valence band and the latter in the conduction band. Typically, the presence of extrinsic impurities such as ions or molecular species with a high dipolar moment in the vicinity of graphene leads to unbalanced hole/electron

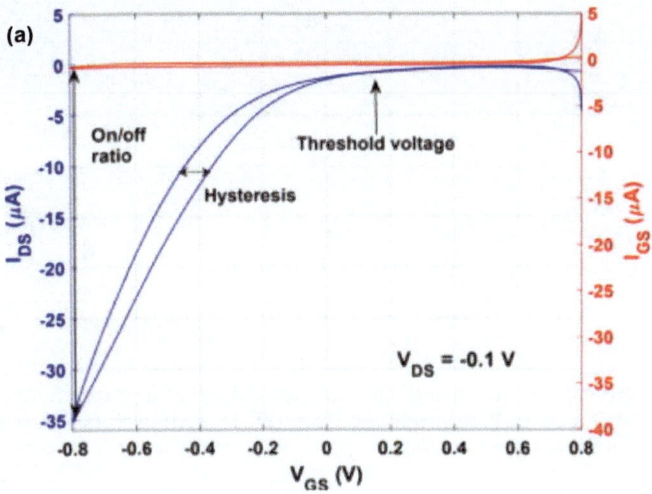

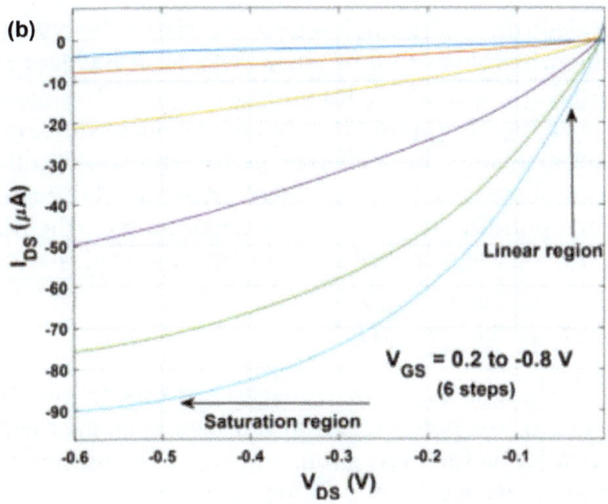

Figure 2.12 Typical (a) transfer and (b) output curves for an EG-SWCNTFET. Electrolyte: de-ionized water. Au electrode as the gate. Reproduced from ref. 13, https://doi.org/10.1063/5.0058591, under the terms of the CC BY 4.0 license, http://creativecommons.org/licenses/by/4.0/.

densities, *i.e.*, chemical doping. For this reason, graphene-based EGFETs are well suited for sensing charged biological molecules such as proteins or nucleic acids. The carrier density of graphene can also be modulated by an external electric field, typically in a field effect configuration. The highest resistivity is obtained at zero gate voltage, while a positive (or negative) voltage applied to the gate induces a conductive n-type (or p-type) channel in graphene. Typical output and transfer curves are shown in Figure 2.13. More details on EGFETs are available, for example, in the review of Vaslijevic *et al.*[15]

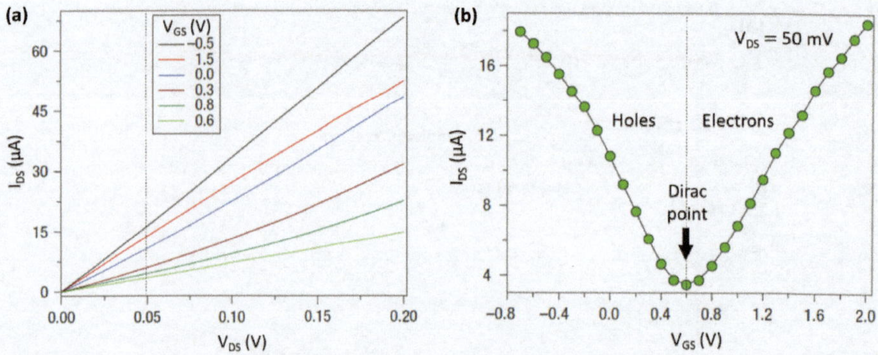

Figure 2.13 Typical (a) output and (b) transfer curves of a graphene-based electrolyte-gated FET. Reproduced from ref. 16 with permission from Springer Nature, Copyright 2008.

2.3.2.3 Silicon Nanowires or Related Materials

Silicon nanowire electrolyte-gated field-effect transistors (Si-NWEGFETs) are based on the same principle as that of other EGFETs. It involves a change in the surface potential at the solid–liquid interface of the SiNW (see eqn (2.9)). Typically, the gate voltage is applied to a reference electrode. Any change in the surface potential results in a change in the threshold voltage of the device. Figure 2.14 shows a schematic SiNW structure fabricated on a Si wafer and the corresponding transfer curves depending on the doping state of the SiNW and the charges carried by the target molecules.

2.4 Fabrication Methods

Electrolyte-gated FETs are thin-film transistors. Considering that no encapsulation is required (the active material must be in contact with the electrolyte), their fabrication is relatively simple. Only the deposition of the active material is discussed here. Several deposition techniques have been described in the literature, including spin-coating, vacuum-thermal deposition (only for small organic molecules or for making electrodes through masks) and printing techniques. If spin-coating and drop-casting are the most popular, inkjet printing and spray deposition are now being used more and more.

2.4.1 Vacuum Thermal Deposition

Vacuum thermal deposition involves placing both the substrate and the material to be deposited in a high vacuum; the material is then thermally sublimated and condensed on the substrate through a mask. Typically, gold contacts can be produced in this way. For active materials, polymers, graphene or CNTs cannot be evaporated, but only small molecules of organic semiconductors. This technique allows the formation of homogeneous, well-organised, even crystalline layers. The thickness of the film is controlled by

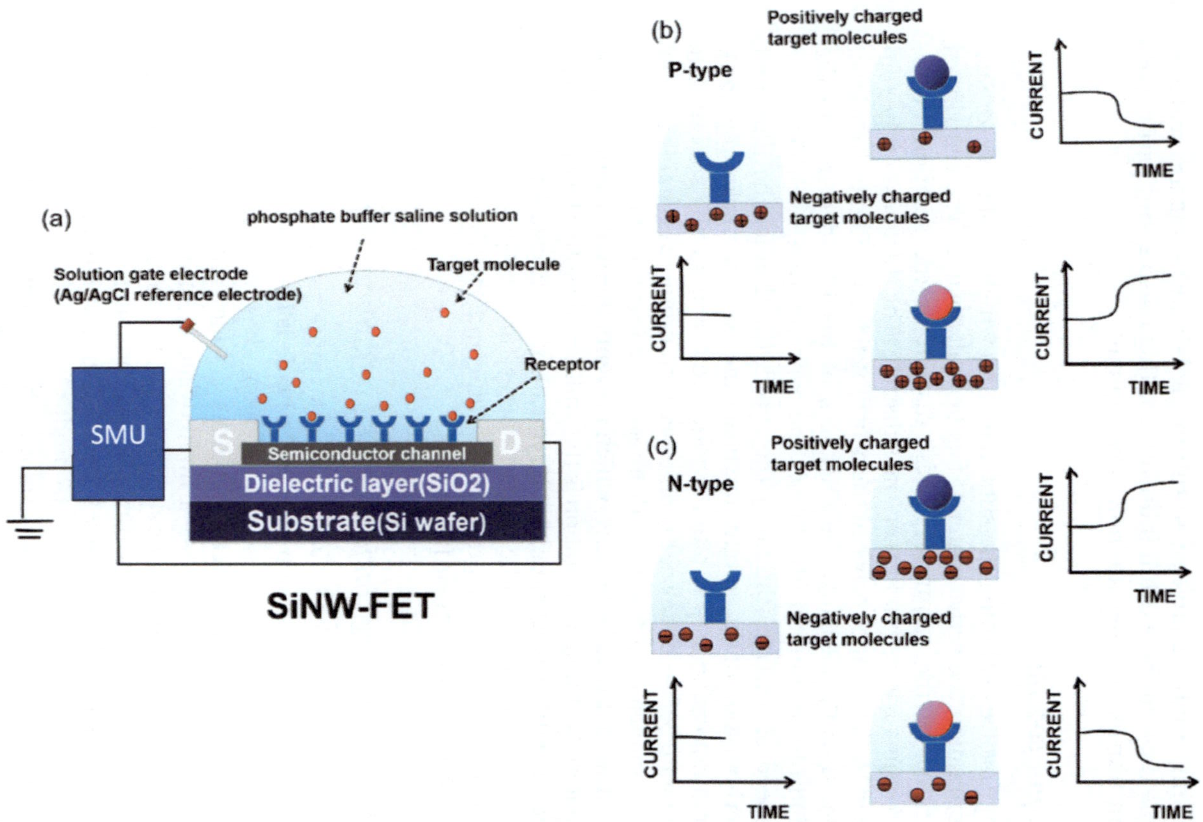

Figure 2.14 (a) Typical schematic of a SiNW-FET biosensor, showing source, drain and gate electrodes connected to a sourcemeter unit (SMU) which supplies a drain voltage V_D and a gate voltage V_G and records drain I_D and gate I_G currents. Depending on the doping state of the SiNW, the effect of a target molecule is different. For a p-type SiNW (b), a positively charged target molecule induces a decrease in drain current, while a negatively charged target molecule induces a drain current increase. (c) For an n-type SiNW, it is the opposite. Reproduced from ref. 17, https://doi.org/10.3390/s23156808, under the terms of the CC BY 4.0 license, http://creativecommons.org/licenses/by/4.0/.

the deposition time and temperature. There are many descriptions in the literature.

2.4.2 Solution-processed Deposition

Processes using solutions are easy to run in the lab and some are scalable, which means they can be used in industry. For this reason, and particularly for thin-film transistors, they are now very popular.

2.4.2.1 Spin-coating

Spin-coating involves spinning the substrate onto which a solution (or suspension) containing the active material has been deposited. The centrifugal force generated by the spinning process distributes the solution homogeneously, forming a very thin film on the substrate. The thickness of the film is determined by the balance between the centrifugal force and the viscosity of the solution. Broadly speaking, the thickness of the film depends on the concentration of the solution (the higher the concentration, the thicker the film), the speed of rotation (the faster the speed, the thinner the film) and the rate of evaporation of the solvent (the faster the rate, the thicker the film). An additional annealing step is often required to evaporate the solvent that did not evaporate before but also, in some cases, to promote crystallisation of the material. Spin-coating has the advantage over vacuum thermal evaporation that it does not require heating, which prevents thermal degradation of the active materials. A major limitation is the ability to dissolve the material or even produce a stable suspension; this is typically very delicate for graphene, for example, even if some studies report successful strategies.[18] For semiconducting oxides, they can be printed as a sufficiently stable suspension of oxide nanoparticles,[19] or of their precursors.[20]

2.4.2.2 Screen-printing, Inkjet Printing and Spray Deposition

Screen-printing is a process in which an ink (more precisely, a paste) is transferred to a substrate through a screen. A blade is used to spread the ink across the open mesh apertures of the screen; the ink then comes into contact with the substrate. It accommodates a wide range of rheological parameters for the ink, but it has a limited lateral resolution of tens of microns at best, resulting in thick patterns in the 10 μm range. Also, it does allow the production of uniform films, but it generates high ink wastage, which makes it unsuitable for costly functional inks or lab-made inks. This method has been used, for example, to produce ZnO-based EGFETs.[21]

Inkjet printing is a non-contact, solution-based technique that allows non-lithographic patterning with a lateral resolution as small as 5 μm, depending on the substrate and the rheological parameters of the ink.[22] Inkjet printing is an additive technology and is therefore much less wasteful than other solution-based processes that use subtractive patterning methods;

it generally ensures better electrical contact with the underlying electrode/material than vacuum thermal evaporation. However, it has the disadvantage of poor lateral resolution compared to lithography. A variety of printed EGFETs are described in the literature today.[23]

Spray deposition is another non-contact solution-based process. In this case, rather than using thermal or piezoelectric propulsion of the ink, the liquid is pumped and directed to the substrate by a gas stream. Similar to inkjet printing, the ink is applied to the substrate in micrometric droplets. Achieving uniform films is therefore a challenge for both techniques, particularly inkjet printing. An example of a spray-coated graphene transistor is shown below.[24] Direct ink writing is also pertinent to make tracks, contact or even print the active materials;[25] it even starts to be possible to write functional materials with a resolution of some tens of nanometres now.[26]

2.5 Biological or Non-biological Probes: Transduction Mechanisms

Detecting large biomolecules *via* field-effect transistors is challenging as the charges carried by macromolecules are distributed over a large volume or an extensive area around the molecule. Consequently, the local charge density that can be sensed at the active material/biomolecule interface is low. Additionally, based on the size of the macromolecule and the capture probe, the charges may be located far from the sensing surface. As a result, not only their density is low, but charges are screened by ions from the electrolyte. For any transduction implying charged macromolecules, it is crucial to remember the screening effect mentioned above. The electric field generated by charges carried by the biomolecules (probes and/or targets) near an interface (gate or channel) gradually disappears as ions and polarized solvent molecules situated between these charges and the surface act upon them. The effective distance of screening, known as the Debye length, denoted by λ_D, is discussed in Section 2.2.1, and given by eqn (2.5). It depends on the concentration of ions in the electrolyte, measured as the ionic strength I_S, determined by eqn (2.10):

$$I_S = \frac{1}{2}\sum_i \rho_i z_i^2 \tag{2.10}$$

where ρ_i is the density of ions i and z_i is their valence. Table 2.2 gives I_S and λ_D values for some biological buffers.

When using EGFETs to transduce the binding of a target onto a probe, it is important to consider the values presented in Table 2.2, for instance, in the presence of an antibody–antigen complex, an IgG antibody more than 10 nm tall. As such, no matter the nature of the antigen, it binds at a distance greater than 10 nm from the surface, meaning it is outside of the volume electrostatically detected by the EGFET in most electrolytes. However, if the ionic strength is sufficiently low (*e.g.*, corresponding to 100-fold diluted PBS), the antigen may be sensed, even if it is already above the Debye length.

Table 2.2 Ionic strengths and Debye lengths for citrate buffer, morpholinoethanesulfonic acid buffer (MES) and phosphate buffer saline (PBS) of different concentrations, calculated from eqn (2.10).

Buffer	Concentration/mM	Ionic strength/mM	Debye length/nm
Citrate	10	27	1.8
	100	285	0.56
MES	10	8	3.4
	100	88	1.0
PBS	0.0015	2	7.2
	0.015	19	2.2
	0.15	225	0.69

Therefore, a useful approach to increase sensitivity involves reducing the size of the capture probe.

2.5.1 DNA or RNA Aptamers

Deoxyribonucleic acid (DNA) and ribonucleic acid (RNA) exhibit the characteristics of forming pairs (adenine–thymine and guanine–cytosine). However, extended chains, such as these, can fold to produce three-dimensional structures, with particular molecular recognition (aptamers) or enzymatic activity (ribozymes). These aptamers can selectively attach to small organic molecules, cells, or large proteins. The affinity arises from a shape adjustment and a tailored distribution of weak interactions (H-bond, electrostatic interactions) between the aptamer and specific regions on the target surface. Proteins are the primary targets of aptamers, with α-thrombin, streptavidin, vascular endothelial growth factor (VEGF), hemagglutinin and adenosine-5′-triphosphate (ATP) being frequently studied. This is mainly due to proteins having dissociation constants predominantly in the nanomolar range, while small organic molecules tend to have weaker binding affinity, with dissociation constants in the micromolar range. This disparity in binding can be rationalized by the fact that there are more interaction sites between the aptamer and the target with a protein than with a small molecule. Since the aptamer may exist in "hairpin" conformations by forming cavities, it is possible for these to bind to a particular protein whose structure is also highly versatile and create a "key–lock" type complex with a significant number of linked sites. Obviously, the association between large surfaces results in a greater number of bonds compared to an association with a small molecule. This could potentially explain the increased stability of the aptamer–protein complex. Further details regarding the properties and applications of aptamers can be found in the comprehensive review article by Dunn *et al.*[27]

2.5.2 Antibodies, Nanobodies, and Peptides

When recognizing molecules, macromolecules, peptides, or proteins either in isolation or as part of a larger entity such as a virus, antibodies provide a commonly preferred option. They are typically specific to their target, termed "the antigen," even in complex matrices containing numerous other molecules or macromolecules. Additionally, antibodies boast high complexation constants, which enable the detection of very low levels of antigens. They are well-suited for detecting substances within biological fluids such as sweat, urine, serum or blood. However, they can be used in other media as well, provided that they do not interfere with the protein structure of the antibody. To ensure this, the pH should be kept around 7 and a non-zero ionic strength maintained.

Most antibodies employed in sensors are immunoglobulins (Ig). Immunoglobulins can come in several forms, with the most prevalent being IgG with a considerable molecular weight of about 150 000 g mol^{-1}. Their shape is usually symmetrical, resembling a "Y" (Figure 2.15). There are two "heavy chains" accountable for roughly 75% of the molecular weight of the antibody, and two "light chains" that account for the remaining weight. These two chains comprise a constant section, common to all IgGs, and a variable section, specific to the antigen. Therefore, the variable sections are present at both ends of the "Y" shape.

Because of its weight and bulkiness, IgG cannot be adequately arranged on the surface of a biosensor, especially on nanostructured surfaces. This creates a significant distance between the sensing surface and the antigen, as discussed earlier. For this reason, they can be more effectively substituted with smaller antibody fragments, such as traditional antibody cleavage products (Fab fragments, approximately 50 kDa), or genetic engineering products (scFv fragments – single chain variable fragments, approximately 30 kDa, or nanobodies, approximately 15 kDa) (Figure 2.15).

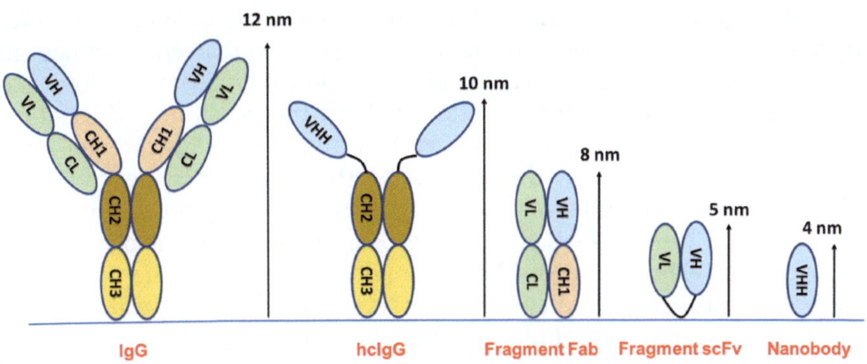

Figure 2.15 The structure of an IgG antibody consists of "constant heavy" chains (CH) and "constant light" chains (CL), forming the non-specific constant domains of the antigen. The variable domains (VH and VL) are specific and responsible for antibody–antigen binding, yielding a 12 nm height. The camelid hcIgG (heavy chain IgG) antibody, as well as the antibody fragment (Fab), single chain variable fragment (scFv) fragment, and nanobody are significantly smaller.

In conventional antibodies or their fragments, the specific part, responsible for binding with an epitope of the antigen, is composed of a part of the heavy chain (denoted by VH) and a part of the light chain (denoted by VL). However, different antibodies have been discovered in the camelid family (camel, dromedary, llama, alpaca) for which the variable part is carried by only one chain and is denoted by VHH (Figure 2.15). Nanobodies (which should be called "single domain antibodies") are precisely this VHH part isolated from the heavy chains. They are made of three sequences of 10 to 20 amino acids and are industrially produced in bioreactors from yeasts, fungi or bacteria. Thanks to their small size, nanobodies can be easily immobilized on surfaces with high density, including NPs such as quantum dots, magnetic nanoparticles, enzymes, carbon nanomaterials, *etc.*

In extreme cases where high surface densities need to be reached, the recognition zone of single-domain antibodies can be limited to a peptide, consisting of some of the amino acids present in the recognition domains. Peptides hold practical interest, as their synthesis is simpler than that of nanobodies as they can be automatically produced from the available amino acid library. However, they exhibit lower specificity and affinity compared to single-domain antibodies.

The topic of antibodies is vast, with more detailed information available in various review articles or book chapters, such as that by Karunakaran *et al.*[28] The recent publication of nanobodies is equally expansive, but the majority of the literature centres around their therapeutic use (Jovčevska *et al.*).[29]

2.5.3 Enzymes and Other Catalysts

Enzymes, which can recognize specific molecules called substrates and convert them into products, have potential applications in EGFETs. However, their extraction and purification from living organisms is often expensive, as they are typically natural proteins. Moreover, enzymes are generally unstable once removed from their natural environment and, like any protein, are prone to denaturation without the possibility of renaturation. The use of synthetic catalysts is crucial due to their heightened resilience and reproducibility. Unfortunately, these catalysts necessitate a substantial synthetic endeavor, resulting in restricted applicability of artificial enzymes. Nevertheless, nanozymes, based on nanoparticles, can be obtained with relative ease and have the potential to substitute natural enzymes. Their properties and applications have been discussed extensively in several review articles, including those by Zhou *et al.*[30] or Yang *et al.*[31] Nanoparticles, which are similar in size to enzymes, are typically composed of bimetallic nanoparticles or metal oxides to prevent the structural reorganization of monometallic nanoparticles and subsequent loss of activity. Carbon structures, such as oxide graphene sheets and occasionally carbon nanotubes, are also used. Their reactivity usually arises from a specific charge distribution on their surface. Their primary purpose is to substitute the vast majority of peroxidases and oxidases. While the catalytic activity of nanozymes can match or surpass that of natural enzymes, their selectivity does not share the same performance. Achieving selectivity is crucial for nanozyme functioning, and thus research has focused on enhancing the connection between metallic particles of oxides or carbon compounds and an organic network to form specific weak bonds with the substrate. The use of MIPs, to which catalytic NPs have been added, to mimic the selective pockets of enzymes has been shown to be highly effective.

2.5.4 Ionophores

Ion sensors (whether organic or inorganic) rely on ionophore ligands that can complex with either a family of ions or a specific ion. Abbreviations for technical terms will always be explained on first use. Typically, ionophores are designed to complex with inorganic cations due to their spherical symmetry, which allows for ligand development with equivalent symmetry – often in the form of macrocycles such as crown ethers or calixarenes. The cations interact with these cavities *via* Lewis acid–base interactions between an alkaline element of the cycle (oxygen, nitrogen, sulfur) and the cation presenting an acidic character. The size of the ligand cavity, with respect to that of the target ion, is another contributing factor to its selectivity. For instance, in the case of a K^+ ion with a diameter of 1.33 Å, the optimal crown ether is 18-crown-6-ether (see Figure 2.16). Conversely, 12-crown-4-ether excludes K^+ and demonstrates a strong affinity to the Li^+ ion (0.68 Å). Although acyclic ionophores are less commonly used, they still exhibit satisfactory affinities, although lower than that of cyclic ethers. This is because, even though they are linear, they have a folded secondary structure, which necessitates rearrangement to be able to form a complex with the cation and incurs an additional energy cost that reduces the affinity for the target. Consequently, they are flexible molecules with the ionophore ETH1001, which is specific for calcium, being an example.[32]

The situation with anions differs from that with cations. As stated earlier, since their geometry is highly diverse, their ligands – which are much less common than those used for cations – are frequently inspired by

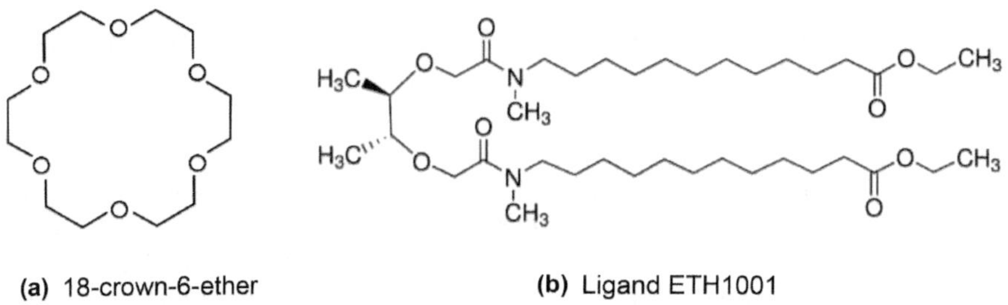

Figure 2.16 Examples of cation-specific ionophores. (a) Cyclic ether-18-crown-6, having a specific affinity for K^+; (b) the acyclic ligand ETH1001 has an affinity for Ca^{2+}.

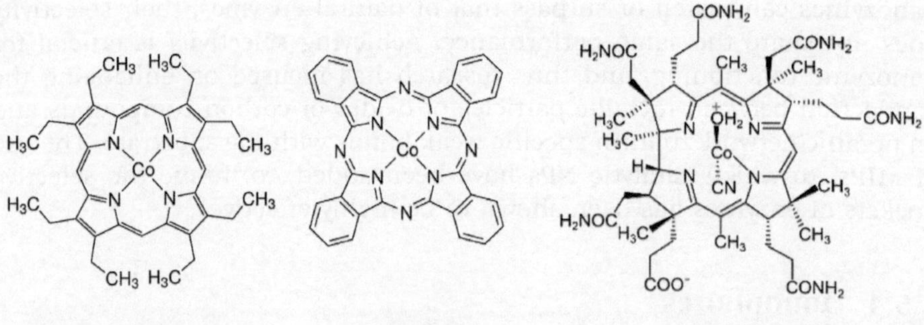

(a) 2,3,7,8,12,13,17,18-Octaethyl-21H,23H-cobalt(II) porphyrine
(b) Cobalt(II) phthalocyanine
(c) Cobyric acid

Figure 2.17 Examples of ionophores corresponding to cobalt complexes, specific for anions. Ligands (a) and (b) are selective for thiocyanate SCN^- and nitrite NO_2^- anions, while ligand (c) is selective for chloride (Cl^-) and cyanide (CN^-).

biomimicry. Therefore, we observe derivatives of cobyric acid (inspired by vitamin B12), which exhibit particular attraction towards chloride or nitrite anions, alongside phthalocyanines and porphyrins as respective ligands of thiocyanates and nitrites. All three molecules comprise a central cobalt atom, acknowledged as a hard Lewis acid, which results in a high affinity towards these anions. The ligands designated for the anions carry a centre of the acid of Lewis type (Figure 2.17) that contributes to the complexation of the anions containing Lewis bases such as the atoms Cl, P, O, N, and S present in chlorides, nitrates, sulfates, and phosphates. Similarly to cationic ligands, these ligands carry a centre of the acid of Lewis type.

Whether for cation or anion selective ionophores, these macromolecules need to be attached to the surface (gate electrode or channel material) using a spacer (typically a carbon chain). This ensures the flexibility of the ionophore and facilitates complexation.

2.5.5 Other Synthetic Ligands

Molecularly imprinted polymers (MIPs) function through a straightforward "key–lock" mechanism, initially discussed in the 1980s. This process entails the selective binding of target molecules to the polymer matrix, resulting in a high affinity and specificity towards a particular analyte. There are various techniques for producing these MIPs. The first method is the direct polymerization of a monomer in the presence of a target molecule. The second involves dissolving a polymer in the presence of the target and then precipitating the mixture. The third is a surface process also referred to as a stamping technique. All these techniques can be applied on EGFETs. Examples of MIP usage and comprehensive documentation can be found in the review article authored by BelBruno.[33]

2.6 Discussion Around Typical Applications

This section provides an overview of significant published works in the field, covering different possible geometries and materials recently used in electrolyte-gated field-effect transistors and applicable for transducing biomolecular interactions or catalysed reactions. The section also encompasses a range of possible capture probes and target molecules.

2.6.1 Nucleic Acid EGFETs

As mentioned earlier, nucleic acids are macromolecules carrying a high density of negative charge and are therefore well-suited for EGFET sensing. Relatively early in 2013, Ohno *et al.*[34] reported on a graphene-based EGFET device for DNA sensing. In their system, the DNA probe strands were immobilized onto the graphene layer forming a channel *via* non-covalent π–π interactions with a pyrene group coupled to the 5′-end of the strands (Figure 2.18(a) and (b)). If adsorption of the pyrene group (named PBASE when

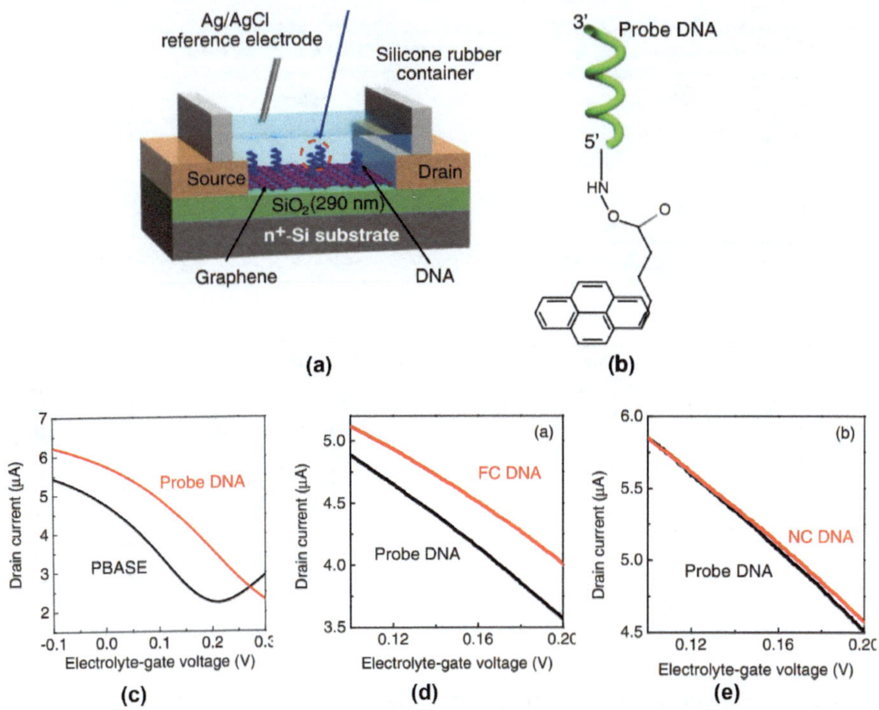

Figure 2.18 A DNA sensor based on an EGFET made of graphene (a). The probe DNA strands are fixed onto the graphene surface using a pyrene end-group (b). After adding the probe DNA (c), the transfer curve shifts towards higher gate voltages. This trend is more pronounced when a complementary target strand (d) hybridizes, but it does not take place for non-complementary strands (e). Reproduced from ref. 34 with permission from the Japan Society of Applied Physics, Copyright 2013.

it carries *N*-hydroxysuccinimide ester at the end of a spacer arm) shifted the transfer curve to the left (corresponding to n-doping), immobilization of the pyrene–DNA probe shifted the transfer curve towards more positive gate voltages, and hybridization with a complementary DNA strand amplified this shift. This was attributed to the effect of negative charges carried by the first and then the second DNA strand (Figure 2.18(c)–(e)). The authors did not discuss the sensitivity and the LOD of their sensor, but from their results, a LOD lower than that of 20 nM complementary DNA can be deduced. Cai *et al.* in 2014 confirmed this effect by using PNA (peptide nucleic acid as a probe, which is neutral and therefore does not induce any shift upon immobilization).[35] Similar behaviour was previously observed by Kergoat *et al.* on an organic EGFET.[36]

In 2015, the group of Frisbie found a reverse trend using a capacitive-coupled lateral-gate organic EGFET,[37] wherein the threshold voltage experienced a negative shift upon hybridization despite the positive shift in voltage of the DNA probe (Figure 2.19). This difference from other studies in the literature

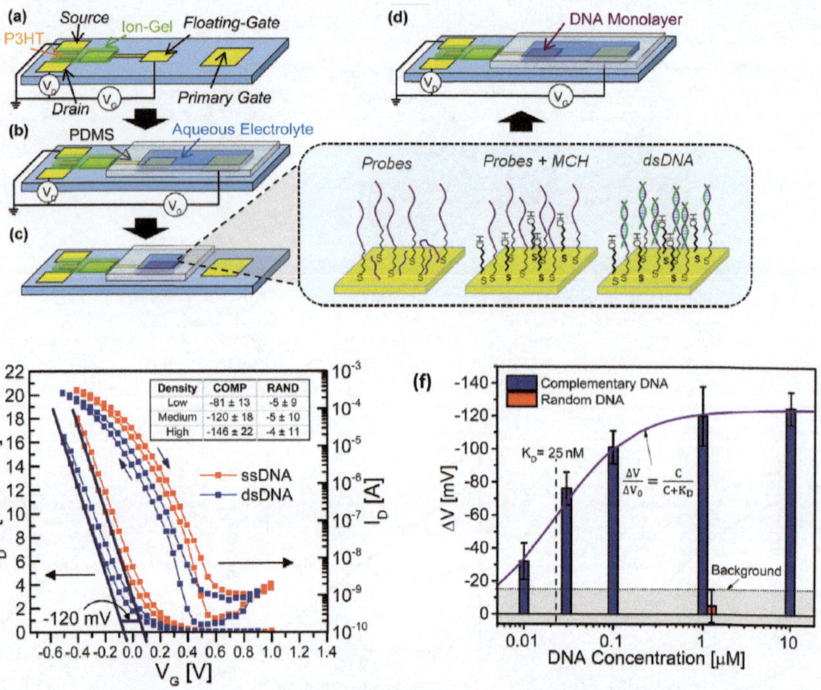

Figure 2.19 EGFET with a floating gate functionalized with DNA. The organic semiconductor P3HT was used, which was covered by an ion-gel layer, also covering the first part of the floating gate. The second part of the floating gate is functionalized by DNA probes and is covered by an electrolyte of water, which also covers a primary (command) gate (parts (a–d)). After hybridizing with the target DNA strands, it was observed that the threshold voltage shifted to the left (smaller gate voltage values) (part (e)). As shown in part (f), this voltage shift is dependent on the target DNA concentration. Reproduced from ref. 37 with permission from American Chemical Society, Copyright 2015.

is explained by the type of FET used by Frisbie and coworkers, the DNA probes here being immobilized on a floating gate driven by a capacitance-coupled command gate. This structure allows to set the transistor at its best regime so that a small change on the floating gate can be efficiently transduced into a large change in drain current for a given gate voltage.

In 2019, Campos *et al.*[38] described a graphene-based EGFET (Figure 2.20(a)) with the probe strands directly immobilized on a unique gate electrode, which showed a large Dirac point shift towards positive gate voltages upon hybridization with complementary strands over a large concentration range from 1 aM to 100 pM (Figure 2.20(b)). This trend is consistent with that observed by Ohno *et al.* when DNA probes were immobilized on the graphene channel.

Recently, in 2022 Deng *et al.*[39] described a graphene-based EGFET with DNA probe strands also immobilized by S–Au adsorption; they observed again the same trend of Dirac shift to the right upon hybridization (Figure 2.21(a)–(d)), and also with an extreme sensitivity, the LOD being down to the aM level, as in Campos *et al.* (Figure 2.21(e) and (f)).

Another innovative method to promote hybridization and try to reach even lower LODs involves producing a specific DNA-imprinted polymer, as suggested by Madj *et al.*[40] This method was recently applied to detect the cancer-associated BRCA1 gene using a MoS_2-based EGFET. The gate was modified with the probe strand, then the BRCA1 sequence was hybridized. A conducting polymer was then grown on top of the gate, and the BRCA1 sequence was dehybridized by washing in an ethanol : water (10 : 2, v/v) mixture + 0.1 M NaOH. This process leaves room for the subsequent hybridization stage, analogous to a standard molecularly imprinted polymer sensor. Due to the substantial charge density of the DNA strands, their hybridization into the imprinted holes results in a noteworthy alteration in the capacitance of the gate. Conversely to most of the related works, as those detailed above, the authors did not observe a shift in transfer curves, but rather in the output curves at a constant gate voltage.

Nucleic acids are excellent models for studying the operation mode of electrolyte-gated transistors. As shown, the extreme sensitivity of graphene to electrostatic interactions, together with its excellent stability in aqueous media, makes it one of the most promising materials for such applications.

2.6.2 Proteins, Peptides, and Pathogens

Nucleic acids are not the only charged biomolecules. Proteins, peptides and many other molecules are charged, although the charge density is usually lower than that of short DNA or RNA strands because proteins are much larger, and all amino acids are not charged at neutral pH. Protein sensing may be, for this reason, more challenging.

In 2013, Suspene *et al.*[41] synthesized copolymers of P3HT and biotin-functionalized P3HT to demonstrate that such an organic EGFET can be sensitive to the binding of avidin or streptavidin (they differ in size and in charge density). They also demonstrated that the biotin–streptavidin platform can be utilized to immobilize any type of streptavidin-labelled protein

(a) (b)

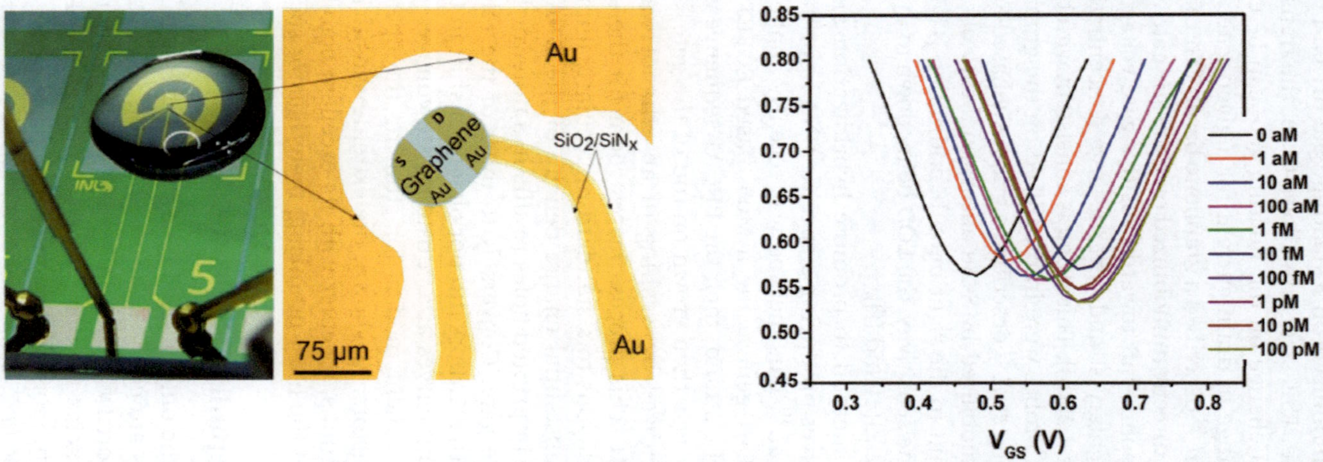

Figure 2.20 Graphene-based EGFET with DNA probes immobilized on the gate (by thiol adsorption). Details of the EGFET structure are given in (a). DNA hybridization was monitored by recording the transfer curve as a function of DNA concentration ($V_{DS} = V_{GS} = 0.1$ V), which shows a positive shift of the Dirac point with increasing DNA concentration (b). Reproduced from ref. 38 with permission from American Chemical Society, Copyright 2019.

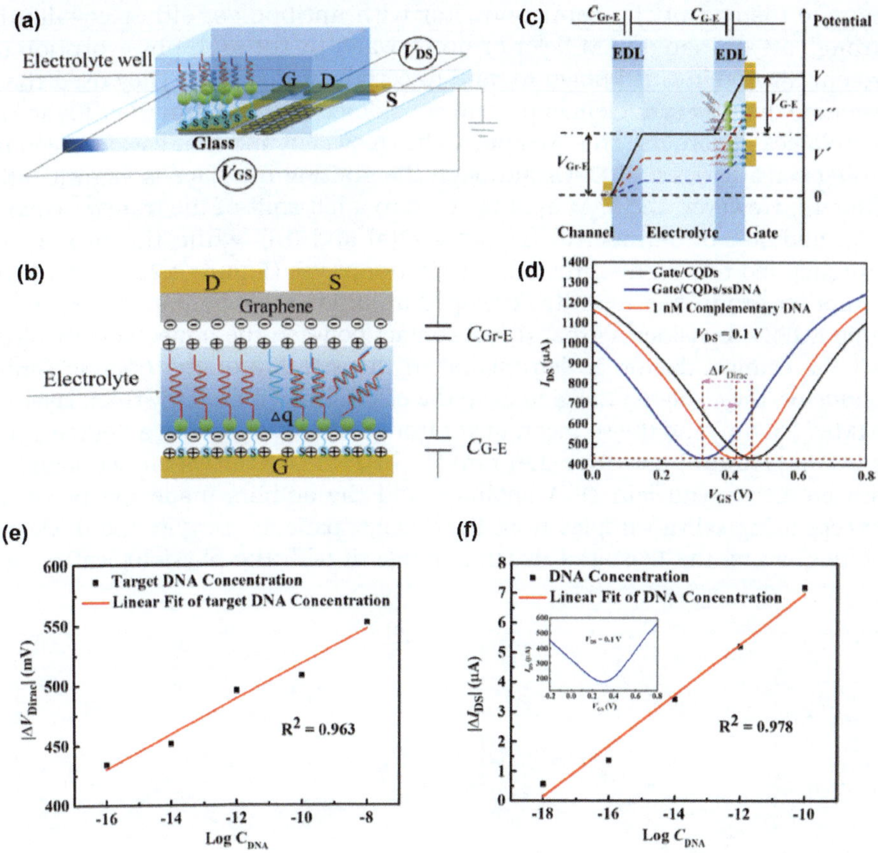

Figure 2.21 Graphene-based EGFET for which the gate was functionalized by DNA strands (a). The capacitive coupling between the gate and the graphene channel explains why any change of charge density on the gate leads to a potential shift that is transduced into a shift of the Dirac point (b–d). The device demonstrated a low LOD around the aM level when the Dirac voltage change is considered (e and f). Reproduced from ref. 39 with permission from American Chemical Society, Copyright 2022.

probe on an organic semiconductor. In 2016, Magliulo *et al.* reported a similar organic electrolyte-gated FET in which the semiconductor (P3HT) was functionalized not through affinity binding but by physisorption of anti-CRP antibodies. Upon addition of CRP (C-reactive protein) to the electrolyte, they demonstrated a change in the transfer curves that was sufficiently sensitive to allow CRP sensing from a few pM up to 2 μM.[42] It should be kept in mind that CRP has a mass of approximately 120 kDa.

Still using an organic semiconductor (actually a mixture of polystyrene for its mechanical properties and 2,8-difluoro-5,11-bis(triethylsilylethynyl)anthradithiophene (diF-TES-ADT) for its semiconducting properties), Ricci *et al.*[43] fabricated an organic EGFET in the top-gate configuration and functionalized the gate instead of the semiconductor with antibodies, either covalently through an aminated SAM layer or non-covalently through physisorption of protein G (protein G is known to bind to certain antibodies). They used their system to detect α-synuclein, a protein considered to be a biomarker for some neurological disorders. The two approaches (covalent and non-covalent) led to similar sensitivities and LODs although the grafting interface is significantly different. However, the SAM approach led to a left-shift of the transfer curves upon addition of α-synuclein (Figure 2.22(a) and (b)), while the protein G approach led to a right-shift of the transfer curves (Figure 2.22(c) and (d)).

Another group has recently developed an original configuration for an organic EGFET: considering that direct contact between the aqueous electrolyte and the semiconductor is detrimental to the stability of the organic semiconductor, they isolated the gate from the electrolyte with a fairly thick layer of PMMA (200 nm) and the semiconductor from the electrolyte with a double layer of PDMS (190 nm) and PVA (120 nm).[44] The PMMA layer was functionalized with an anti-α-synuclein DNA aptamer, and the authors made the proof of concept using saliva samples from Parkinson's patients. Despite the thick insulating layers, the LOD and dynamic range fit with the physiological range,

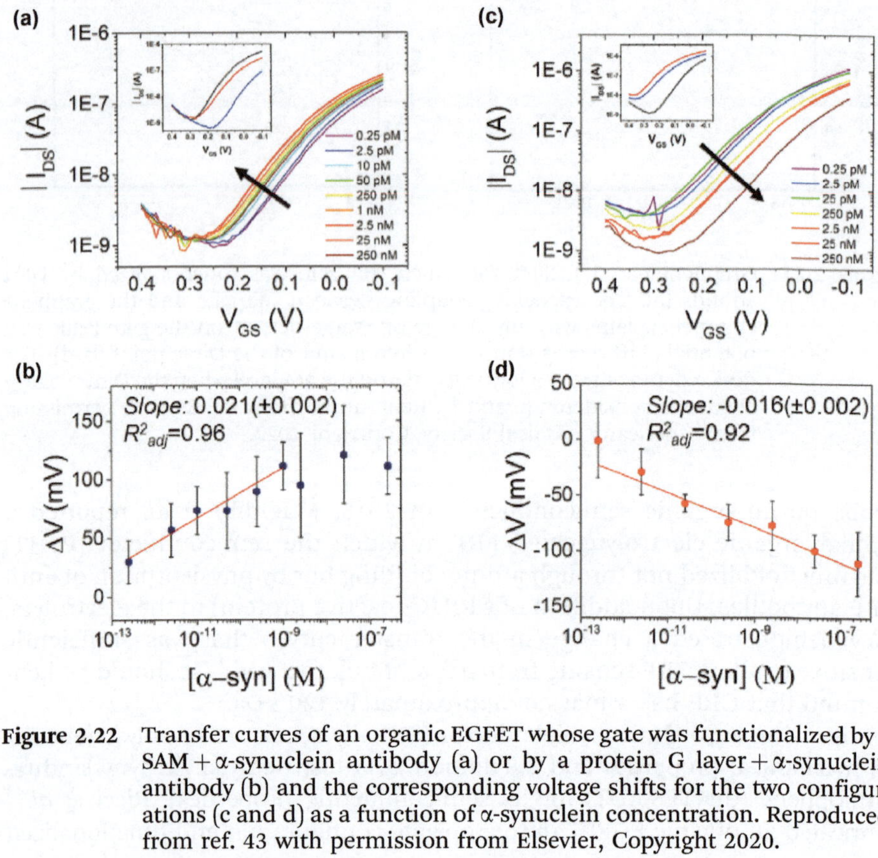

Figure 2.22 Transfer curves of an organic EGFET whose gate was functionalized by a SAM + α-synuclein antibody (a) or by a protein G layer + α-synuclein antibody (b) and the corresponding voltage shifts for the two configurations (c and d) as a function of α-synuclein concentration. Reproduced from ref. 43 with permission from Elsevier, Copyright 2020.

between 100 fg L^{-1} and 10 μg L^{-1}. The result, if confirmed, is very interesting because it solves the biggest drawback of organic EGFETs: the stability in water.

Still using metal oxides rather than organic semiconductors, Chae *et al.*[45] reported a micrometric EGFET based on IGZO (amorphous indium gallium zinc oxide) semiconducting layer included in a microfluidic chamber (Figure 2.23(a)). They used an antibody as the probe, immobilized directly on the IGZO channel. This work is interesting because the authors paid attention to the variability of the threshold voltage between transistors (see Figure 2.23(b)); they conclude in their case to a low variability, less than 10 mV, which allows V_{Th} to be considered a pertinent figure of merit (Figure 2.23(c)). They quantified this protein in the range of 1 pg mL^{-1} to 1 ng mL^{-1} with a sensitivity of about 9 mV dec^{-1}. Variability of V_{Th} is usually due to two factors: (i) dispersion among transistors, due to a mediocre fabrication process, and (ii) errors coming from the estimation method itself. Actually, many results from the literature show a quite large variability of the threshold voltage or do not deal with this problem. It must be kept in mind that the larger the ratio between the signal and the

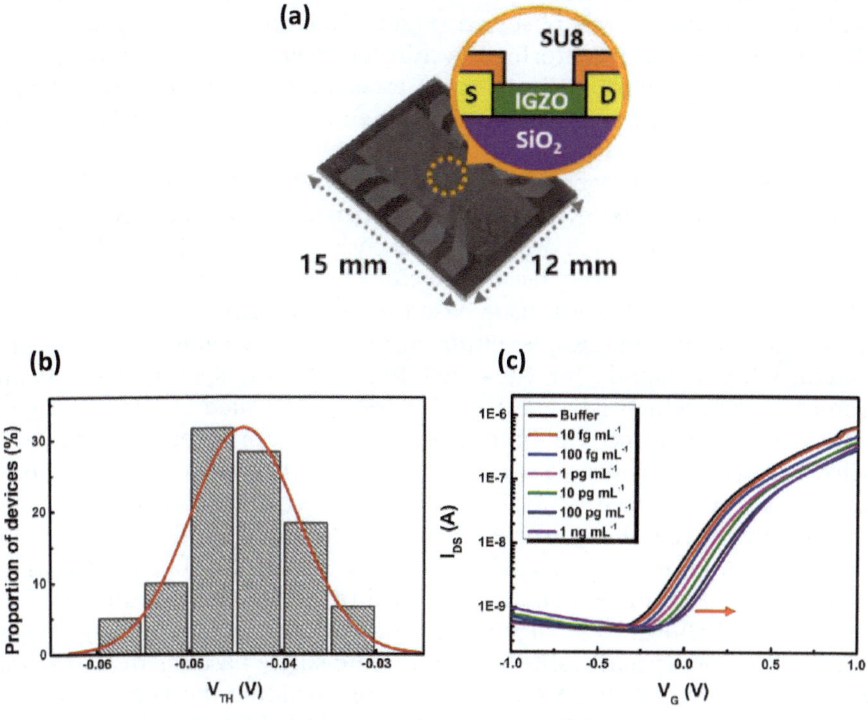

Figure 2.23 (a) Picture and sketch of an array of IGZO-based EGFETs. The IGZO layer is 50 nm thick over an active channel area of 40 × 80 μm^2. (b) Variability of V_{Th} in between devices. (c) Shift of the transfer curves as a function of the alpha-synuclein protein concentration. Reproduced from ref. 45 with permission from Elsevier, Copyright 2018.

standard deviation, the better the sensor. It must be also kept in mind that if publications focus on V_{Th}, it is not the parameter to consider if the sensor is to be applied (it would make such devices perform like potentiometric electrodes at best, which is of course not the case): it is the current change, at constant V_D and V_G, which is the pertinent figure of merit of these devices (as shown in Figure 2.20(f), for example).

The real weakness of EGFETs is induced by the effect of electrostatic screening, *i.e.*, the Debye length. Filipiak *et al.*,[46] recognizing that the sensitivity of EGFETs is constrained by the Debye length at the electrolyte/channel interface, designed a SWCNT-based EGFET that utilizes nanobodies instead of traditional antibodies (refer to the end of Section 2.5.2), enabling a reduction in the distance between the antigen and the channel. As depicted in Figure 2.24(a), the SWCNT-EGFET exhibits a good ambipolar transfer characteristic, approximately centred around 0 V *versus* the gate, comprising two reasonably symmetrical p (on the left)- and n (on the right)-branches. For this reason, the authors selected the p-branch between −0.35 V and −0.30 V, wherein the drain current I_{DS} *versus* the gate voltage displays a straight line (as illustrated in Figure 2.24(b)). After introduction of the antigen (green fluorescent protein (GFP) as a model), a slight displacement of the p-branch was observed (Figure 2.24(c)). They ensured the stability of the system by monitoring the shift for several tens of hours and got a negligible drift of 0.05 mV per hour, exceeding that of traditional SiO_2 transistors. This drift is attributed to the reference electrode rather than to SWCNTs.

With regard to pathogen detection using EGFETs, the most striking examples relate to the recent COVID-19 RNA virus. A review has recently been published that gives a good overview of recently published approaches.[47] It is interesting to focus on a very recent and efficient approach called CRISPR/Cas.

Instead of developing a sensor based on simple DNA/RNA or RNA/RNA hybridization, the CRISPR/Cas approach relies on the catalytic activity of a protein. CRISPR stands for Clustered Regularly Interspaced Short Palindromic Repeat, which is a specific RNA sequence called crRNA. This RNA sequence has the property of binding to a protein, more precisely a nuclease called Cas, which is activated when the correct RNA sequence (crRNA) binds. Cas13a has a very specific binding affinity for its crRNA and combines this advantage with the ability to cleave the RNA sequence independently, not only the RNA bound to the crRNA, but also untargeted (but neighboring) collateral single-stranded RNA. Cas13-based DNA detection has already been described and shown to be highly efficient, particularly in viral diagnostics, because it is very specific (able to discriminate single base mismatches) and works at very low target DNA concentrations, avoiding the need for amplification steps (such as PCR). Most, if not all, publications to date have focused on optical detection, typically fluorescence. However, Yu *et al.*[48] have recently proposed an approach using an EGFET, summarized in Figure 2.25 (the use of CRISPR/Cas has already been described by Hajian *et al.* in 2019, also on a graphene FET, but not for virus detection).[49]

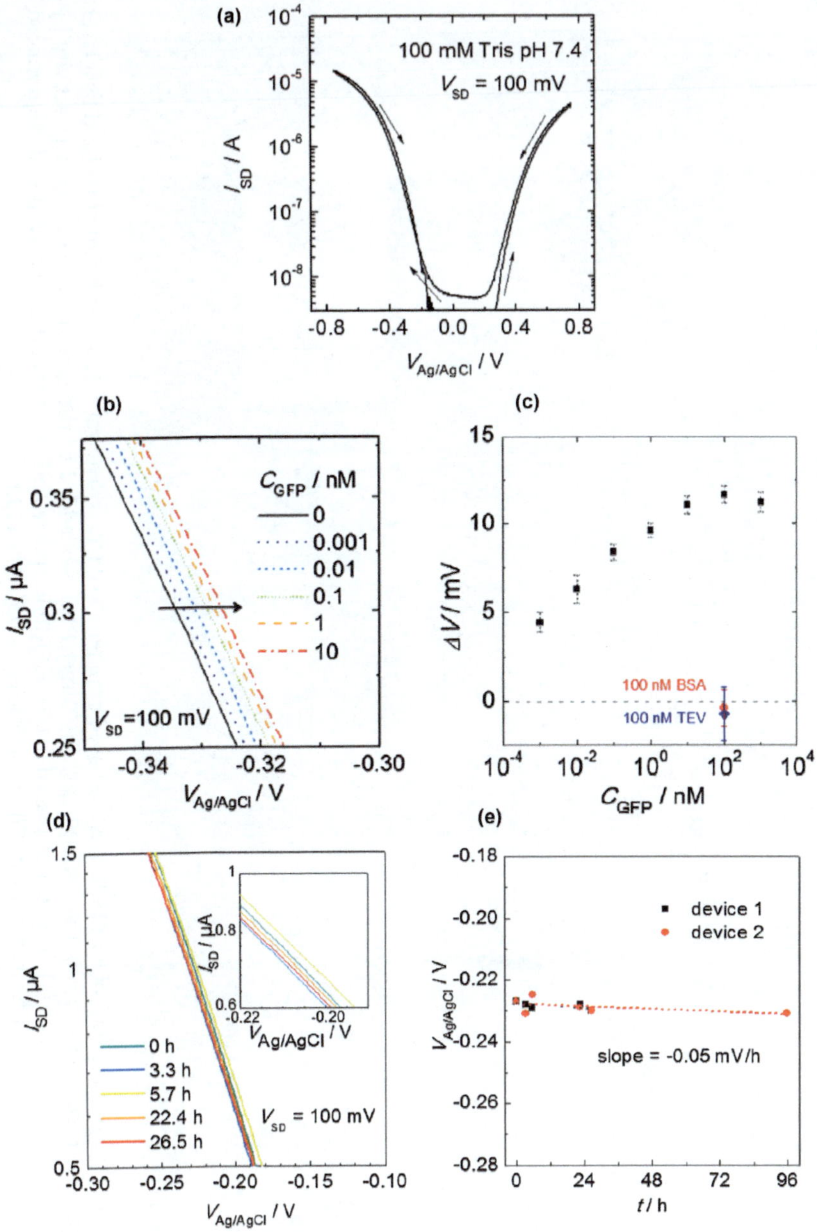

Figure 2.24 (a) Transfer curve of a SWCNT EGFET, showcasing the ambipolar characteristic of the SWCNT. Any given alteration in the electrostatic environment can shift the curve to the left or right, depending on the polarity of the charge, as it works for graphene. Nanobodies are immobilized on the channel, and a green fluorescent protein (GFP) is used as an antigen to induce negative charges. A sensitivity of roughly 4 $mV\,dec^{-1}$ (c) is observed for negative charges, leading to a right shift (b), whereas a minor drift is noticed over time (d and e). Reproduced from ref. 46 with permission from Elsevier, Copyright 2018.

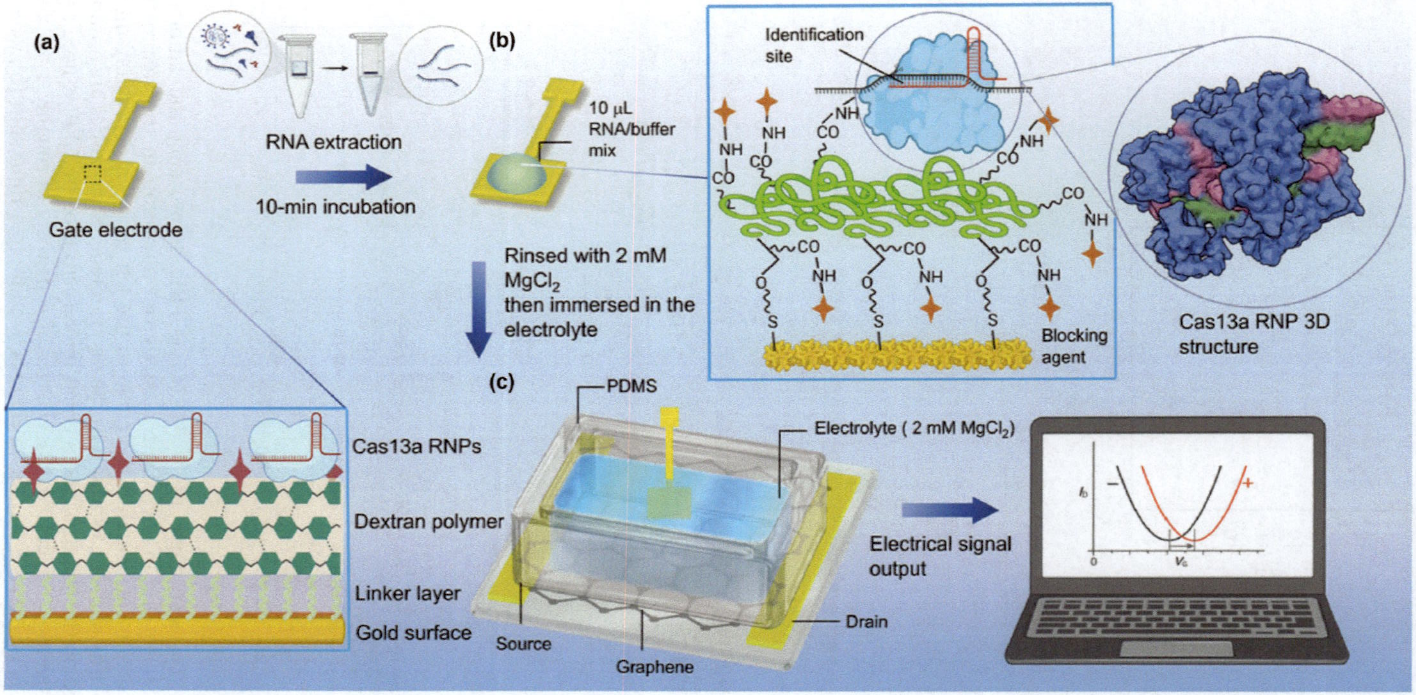

Figure 2.25 Illustration of the process flow of a CRISPR/Cas13a-based graphene EGFET. (a) The gate electrode is functionalized by a layer of dextran, which covalently binds the CRISPR/Cas. (b) Target RNA was added to the gate to activate cleavage activity by hybridizing to the crRNA, then (c) the gate was placed in the EGFET electrolyte and transfer curves were recorded. Reproduced from ref. 48 with permission from American Chemical Society, Copyright 2022.

In their configuration, the gate electrode is used as a sensing surface. It is first functionalized by a layer of biopolymer on which the CRISPR/Cas13a is immobilized. RNA extracts in buffer (which acts as an electrolyte in the EGFET) are added directly to the gate. The transistor is a conventional 3-terminal transistor (gate, source, and drain) with the gate at the top. When the target RNA strand binds to the crRNA inside the Cas13a, the CRISPR/Cas starts its cleavage activity. The authors showed that RNA cleavage shifts the transfer curve of the transistor to the right (more positive gate voltages) and attributed this change not to a change in the gate capacitance C, but to a change in the amount of charges q carried by the target RNA (cleaved RNA is shorter, so the amount of charges carried decreases), so that $V_{\mathrm{Dirac}} = q/C$ and consequently $\Delta V_{\mathrm{Dirac}} = \Delta q/C$ (Figure 2.26).

Another very important application of EGFETs for pathogen detection was reported by White *et al.*[50] The device is an organic EGFET, but consists of a control gate (functionalized) and a floating gate, the latter divided into two parts (marked R and L in Figure 2.27(a)). The control gate and the R-side of the floating gate are in contact with an aqueous electrolyte into which the sample is introduced. The L-side of the floating gate is in contact with an ion gel which also covers the organic semiconductor (P3HT). The R-side of the floating gate is functionalized with the capture probe (in this case a DNA aptamer), while the control gate, of a larger area, is not and serves as a pseudo-reference to which an external gate voltage is applied. The other originality is that the transistor is used in a NOT (inverter) gate, as shown in Figure 2.27(d): it is in series with a resistor. This circuit works like a voltage divider, with a constant resistor (R_{L}) and a variable resistor

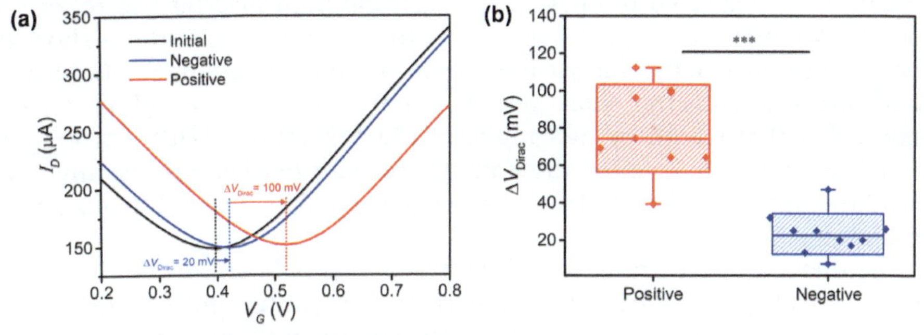

Figure 2.26 Analytical results obtained on RNAs obtained from patient swabs, after an RT-PCR step. (a) Transfer curves obtained from the CRISPR/Cas13a-based graphene EGFET described above, before the addition of the target RNA, and then incubation of the gate with an extract from a positive patient (an RNA concentration of 10^{-13} M is estimated) (red curve). The same procedure was performed with extracts from patients known to be COVID negative (blue curve). (b) The voltage shift of the Dirac point obtained for 15 samples (7 positive and 8 negative) is shown on an error rectangle plot, showing two distinct areas. Reproduced from ref. 48 with permission from American Chemical Society, Copyright 2022.

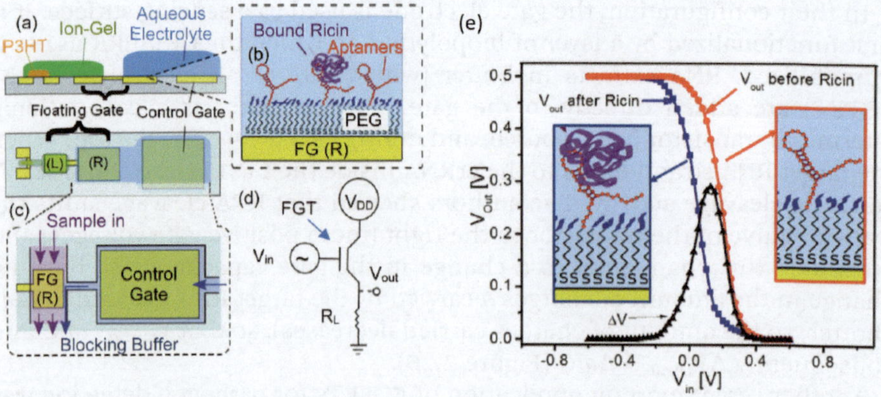

Figure 2.27 Floating gate EGFET developed for ricin protein detection. (a) Side view of the device showing a control gate and the R-side of a floating gate immersed in the sample electrolyte, the L-side of the floating gate covered by an ion gel which also covers the semiconductor (P3HT). (b) It is the R-side of the floating gate which is functionalized by an aptamer specific for the ricin protein. (c) While the control gate is rinsed with a blank buffer, the R-side of the floating gate is rinsed with the sample solution. (d) The whole device is mounted as an inverter. (e) Inverter characteristics, before and after ricin addition; the interpretable signal is obtained by subtraction of the two curves (ΔV). Reproduced from ref. 50 with permission from the Authors, Copyright 2016.

(the transistor). The output signal (V_{out}) is measured between the transistor and the resistor, while the input signal (V_{in}) is applied to the control gate. The device is expected to operate as a conventional inverter, *i.e.* to give a low V_{out} voltage for a high V_{in} voltage and *vice versa*. The authors showed that the presence of the analyte (here ricin, a protein of about 60 kDa) leads to a shift in the inverter characteristic (Figure 2.27(e)). It would have been interesting if the authors had replaced the constant R_L with a reference transistor functionalized by the aptamer but not receiving the sample, to avoid the impractical step of subtracting the signal as they did (black curve in Figure 2.27(e)).

The final originality of this work is that the authors also worked with alternating voltages. Figure 2.28(a) shows the ΔV_{out} values for different ricin concentrations during one period, *i.e.* for V_{in} varying from −0.3 V to 0.3 V and back. It is interesting to note the asymmetry of the output signal and, clearly, the increasing amplitude with increasing ricin concentration. Figure 2.28(b) shows plots of ΔV_{out} as a function of ricin concentration for different matrices (water, milk, and orange juice) and for a blank. The instrument is able to detect ricin at concentrations greater than 0.1 nM. Curiously, this approach has not been used by other researchers. However, this excellent work opens the way to AC characterization of EGFETs and their use in active electronic components such as logic gates, the simplest of which is the NOT gate, to make complex decision-making chemical logic gates.

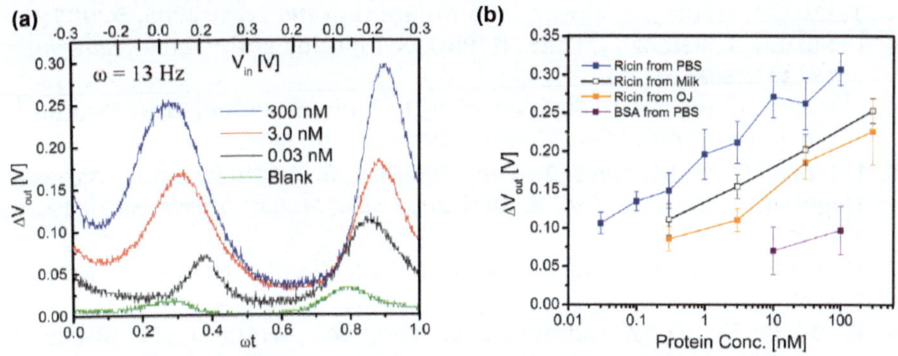

Figure 2.28 Results obtained for ricin detection on the floating-gate EGFET described above, under the AC regime. (a) $V_{in}(t) = 0.3\ \sin(\omega t)$ (volts), while $\Delta V_{out}(t)$ shows a complex shape, the amplitude of which depends on ricin concentration. (b) Plots of the maximum value of ΔV_{out} for various ricin concentrations. Reproduced from ref. 50 with permission from the Authors, Copyright 2016.

2.7 Conclusion

Soon after the first description of liquid electrolyte-gated field-effect transistors, researchers understood the potential of such devices for sensing and, in particular, for biosensing applications, since an aqueous electrolyte is an ideal medium for all the biochemical processes on which biosensors are usually based. Due to the great variety of active materials (1D or 2D metallic conductors, semi-metals, inorganic semiconductors, and organic semiconductors) and the great variety of biomolecules to be used as a capture probe or target, the operation of EGFETs has not yet been fully rationalized, which is one of the weaknesses of these systems today. Another weakness, derived from a strength, is their great sensitivity to their environment, which means that all chemical parameters must be well controlled when carrying out an experiment. Otherwise, a small change in pH, temperature, ionic strength or the presence of light can significantly affect the result and make it unreliable. Nevertheless, electrolyte-gated transistors appear to be promising systems, not only because of their ability to amplify small potential changes into large current variations, but also because of their simplicity: they are easy to fabricate as thin films, including by additive technologies such as printing, and also easy to incorporate into more advanced electronic circuits such as logic gates, paving the way for transistor-based chemical logics.

References

1. J. Wu, *Chem. Rev.*, 2022, **122**, 10821.
2. L. Herlogsson, X. Crispin, N. D. Robinson, M. Sandberg, O. J. Hagel, G. Gustafsson and M. Berggren, *Adv. Mater.*, 2007, **19**, 97.

3. T. Ozel, A. Gaur, J. A. Rogers and M. Shim, *Nano Lett.*, 2005, **5**, 905.
4. L. Fillaud, T. Petenzi, J. Pallu, B. Piro, G. Mattana and V. Noel, *Langmuir*, 2018, **34**, 3686.
5. J. Pallu, M. Avci-Adali, P. Mackeben, L. Mohammadnejad, G. Mattana, V. Noël and B. Piro, *Org. Electron.*, 2019, **75**, 105402.
6. J. Jeong, G. C. Marques, X. Feng, D. Boll, S. A. Singaraju, J. Aghassi-Hagmann, H. Hahn and B. Breitung, *Adv. Mater. Interfaces*, 2019, **6**, 1901074.
7. G. Y. Wang, K. Lian and T. Y. Chu, *IEEE J. Electron Devices Soc.*, 2021, **9**, 939.
8. F. Torricelli, D. Z. Adrahtas, Z. Bao, M. Berggren, F. Biscarini, A. Bonfiglio, C. A. Bortolotti, C. D. Frisbie, E. Macchia, G. G. Malliaras, I. McCulloch, M. Moser, T. Q. Nguyen, R. M. Owens, A. Salleo, A. Spanu and L. Torsi, *Nat. Rev. Methods Primers*, 2021, **1**, 66.
9. M. Parmeggiani, A. Verna, A. Ballesio, M. Cocuzza, E. Piatti, V. Fra, C. F. Pirri and S. L. Marasso, *Sensors*, 2019, **19**, 4497.
10. F. A. Viola, F. Melloni, A. Molazemhosseini, F. Modena, M. Sassi, L. Beverina and M. Caironi, *Adv. Electron. Mater.*, 2023, **9**, 2200573.
11. M. S. Chae, J. H. Park, H. W. Son, K. S. Hwang and T. G. Kim, *Sens. Actuators, B*, 2018, **262**, 876.
12. T. W. Odom, J. L. Huang, P. Kim and C. M. Lieber, *Nature*, 1998, **391**, 62.
13. B. Shkodra, M. Petrelli, M. A. Costa Angeli, D. Garoli, N. Nakatsuka, P. Lugli and L. Petti, *Appl. Phys. Rev.*, 2021, **8**, 041325.
14. K. S. Novoselov, D. Jiang, F. Schedin, T. J. Booth, V. V. Khotkevich, S. V. Morozov and A. K. Geim, *Proc. Natl. Acad. Sci. U. S. A.*, 2005, **102**, 10451.
15. S. Vasilijević, R. Boukraa, N. Battaglini and B. Piro, *Synth. Met.*, 2023, **295**, 117355.
16. A. Das, S. Pisana, B. Chakraborty, S. Piscanec, S. K. Saha, U. V. Waghmare, K. S. Novoselov, H. R. Krishnamurthy, A. K. Geim, A. C. Ferrari and A. K. Sood, *Nat. Nanotechnol.*, 2008, **3**, 210.
17. H. Li, D. Li, H. Chen, X. Yue, K. Fan, L. Dong and G. Wang, *Sensors*, 2023, **23**, 6808.
18. S. Lu, B. N. Smith, H. Meikle, M. J. Therien and A. D. Franklin, *Nano Lett.*, 2023, **23**, 2100.
19. J. S. Gebauer, V. Mackert, S. Ognjanović and M. Winterer, *J. Colloid Interface Sci.*, 2018, **526**, 400.
20. X. Liu, T. J. Tarn, F. Huang and J. Fan, *Particuology*, 2005, **19**, 1.
21. J. T. Carvalho, V. Dubceac, P. Grey, I. Cunha, E. Fortunato, R. Martins, A. Clausner, E. Zschech and L. Pereira, *Nanomaterials*, 2019, **9**, 169.
22. J. Lemarchand, N. Bridonneau, N. Battaglini, F. Carn, G. Mattana, B. Piro, S. Zrig and V. Noël, *Angew. Chem., Int. Ed.*, 2022, **61**, e202200166.
23. K. Chennit, N. Delavari, S. Mekhmoukhen, R. Boukraa, L. Fillaud, S. Zrig, N. Battaglini, B. Piro, V. Noël, I. Zozoulenko and G. Mattana, *Adv. Mater. Technol.*, 2023, **8**, 2200300.

24. B. Fenech-Salerno, M. Holicky, C. Yao, A. E. Cass and F. Torrisi, *Nanoscale*, 2023, **15**, 3243.
25. K. S. Pandya, S. S. Shindalkar and B. Kandasubramanian, *Prog. Addit. Manuf.*, 2023, 1.
26. M. Pascual, N. Bigan, A. M'Barki, R. Mental, I. Allegro and U. Lemmer, *Advanced Fabrication Technologies for Micro/Nano Optics and Photonics XVI*, 2023, pp. 124330E.
27. M. Dunn, R. Jimenez and J. Chaput, *Nat. Rev. Chem.*, 2017, **1**, 76.
28. C. Karunakaran, K. Bhargava and R. Benjamin, *Biosensors and Bioelectronics*, Elsevier, 1st edn, 2015.
29. I. Jovčevska and S. Muyldermans, *BioDrugs*, 2020, **34**, 11.
30. Y. Zhou, B. Liu, R. Yang and J. Liu, *Bioconjugate Chem.*, 2017, **28**, 2903.
31. W. Yang, X. Yang, L. Zhu, H. Chu, X. Li and W. Xu, *Coord. Chem. Rev.*, 2021, **448**, 214170.
32. R. D. Johnson and L. G. Bachas, *Anal. Bioanal. Chem.*, 2003, **376**, 328.
33. J. J. BelBruno, *Chem. Rev.*, 2019, **119**, 94.
34. Y. Ohno, S. Okamoto, K. Maehashi and K. Matsumoto, *Jpn. J. Appl. Phys.*, 2013, **52**, 110107.
35. B. Cai, S. Wang, L. Huang, Y. Ning, Z. Zhang and G. J. Zhang, *ACS Nano*, 2014, **8**, 2632.
36. L. Kergoat, B. Piro, M. Berggren, M. C. Pham, A. Yassar and G. Horowitz, *Org. Electron.*, 2012, **13**, 1.
37. S. P. White, K. D. Dorfman and C. D. Frisbie, *Anal. Chem.*, 2015, **87**, 1861.
38. R. Campos, J. Borme, J. R. Guerreiro, G. Machado, M. F. G. Cerqueira, D. Y. Petrovykh and P. Alpuim, *ACS Sens.*, 2019, **4**, 286.
39. M. Deng, J. Li, B. Xiao, Z. Ren, Z. Li, H. Yu, J. Li, J. Wang, Z. Chen and X. Wang, *Anal. Chem.*, 2022, **94**, 3320.
40. S. Mansouri Majd, F. Mirzapour, M. Shamsipur, I. Manouchehri, E. Babaee, A. Pashabadi and R. Moradian, *Talanta*, 2023, **257**, 124394.
41. C. Suspène, B. Piro, S. Reisberg, M. C. Pham, H. Toss, M. Berggren, A. Yassar and G. Horowitz, *J. Mater. Chem. B*, 2013, **1**, 2090.
42. M. Magliulo, D. De Tullio, I. Vikholm-Lundin, W. M. Albers, T. Munter, K. Manoli, G. Palazzo and L. Torsi, *Anal. Bioanal. Chem.*, 2016, **408**, 3943.
43. S. Ricci, S. Casalini, V. Parkula, M. Selvaraj, G. D. Saygin, P. Greco, F. Biscarini and M. Mas-Torrent, *Biosens. Bioelectron.*, 2020, **167**, 112433.
44. R. S. Massey, E. M. McConnell, D. Chan, M. R. Holahan, M. C. DeRosa and R. Prakash, *ACS Sens.*, 2023, **8**, 3116.
45. M. S. Chae, J. H. Park, H. W. Son, K. S. Hwang and T. G. Kim, *Sens. Actuators, B*, 2018, **262**, 876.
46. M. S. Filipiak, M. Rother, N. M. Andoy, A. C. Knudsen, S. Grimm, C. Bachran, L. K. Swee, J. Zaumseil and A. Tarasov, *Sens. Actuators, B*, 2018, **255**, 1507.
47. N. H. Al-Hardan, M. Firdaus-Raih, M. A. A. Hamid and A. Jalar, *Crit. Rev. Solid State Mater. Sci.*, 2023, 1–27.

48. H. Yu, H. Zhang, J. Li, Z. Zhao, M. Deng, Z. Ren, Z. Li, C. Xue, M. G. Li and Z. Chen, *ACS Sens.*, 2022, 7, 3923.
49. R. Hajian, S. Balderston, T. Tran, T. deBoer, J. Etienne, M. Sandhu, N. A. Wauford, J. Y. Chung, J. Nokes, M. Athaiya, J. Paredes, R. Peytavi, B. Goldsmith, N. Murthy, I. M. Conboy and K. Aran, *Nat. Biomed. Eng.*, 2019, **3**, 427.
50. S. P. White, S. Sreevatsan, C. D. Frisbie and K. D. Dorfman, *ACS Sens.*, 2016, **1**, 1213.

CHAPTER 3

Challenges in the Detection of Emerging Novel Pathogens and Evolving Known Pathogens

N. A. SHAMSUSAH,[a] M. E. I. MOHD FADLI,[b] R. EMRIZAL,[c] S. A. HANIFAH[b] AND M. FIRDAUS-RAIH*[c]

[a] Universiti Kebangsaan Malaysia, Department of Earth Sciences and Environment, Bangi, 43600, Malaysia; [b] Universiti Kebangsaan Malaysia, Department of Chemical Sciences, Bangi, 43600, Malaysia; [c] Universiti Kebangsaan Malaysia, Department of Applied Physics, Bangi, 43600, Malaysia
*Email: m.firdaus@ukm.edu.my

3.1 Novel Pathogens and Evolving Known Pathogens

Within the span of slightly over a century, many epidemics had escaped regional containment to become pandemics. At the time of writing, the most recent pandemic, COVID-19, spanned from 2020 to 2023 and was caused by the SARS-CoV-2 virus.[1] Like COVID-19, the overwhelming majority of pandemics in the 20th and early 21st centuries were caused by viruses. Many, if not the majority, of those viruses had been previously described and are known to science; this category includes novel strains of those known pathogens.[2] Nevertheless, there have also been newly emerged pathogens such as Ebola,[3] HIV,[4] avian influenza,[5] Nipah,[6] SARS,[7] swine influenza[8] and of course SARS-CoV-2,[9] being among the more notable ones. The need to be able to rapidly and accurately identify newly emerged pathogens is obvious. Such a capacity is an important tool in controlling the spread of infections

Detection Science Series No. 24
Field-effect Transistor Biosensors for Rapid Pathogen Detection
Edited by Naif H. M. Al-Hardan, Azman Jalar, Muhammad Azmi Abdul Hamid and Mohd Firdaus Raih

Published by the Royal Society of Chemistry, www.rsc.org

and thus allowing for prevention of epidemics and their subsequent expansion into pandemics.

While the emergence of new pathogens is a more frequent and increasing occurrence, the threat of evolving but previously described pathogens cannot be overlooked. The capacity of microorganisms to evolve, adapt and develop a capacity for resistance against antimicrobial agents had been predicted by the discoverer of antibiotics, Sir Alexander Fleming.[10] In such situations, it is therefore imperative that there is the capability to distinguish strains that are susceptible to antimicrobial agents from those that are resistant. This will allow for the correct treatment to be administered in a timely manner. Such resistant organisms are expected to be genetically very similar to those that were previously susceptible, thus making it even more challenging to differentiate them.

Through the years, the standards, accuracy and speed of pathogen identification have been cumulatively improved on. The identification of pathogens by culturing and microscopy techniques is clearly slow and not the best means of preventing the rapid spread of infections. The speed of identification is significantly faster and the accuracy is much higher when immunological methods or genetics-based methods are used. While immunological methods may be fast and easy to execute, they can suffer from a lack of specificity when the differences are not detectable immunologically. This chapter discusses the use of biologically sensitive field-effect transistors (Bio-FETs) for the accurate and rapid detection of pathogens with a specific focus on the biorecognition elements (BREs) that provide the analyte specificity.

3.2 A Contextual Overview of Bio-FETs

The development and use of electrical, electrochemical, optical, photonic, magnetic, piezoelectric, and other forms of biosensors have been reviewed by Abbas *et al.*[11] Field-effect transistor (FET) biosensors proved to be one of the most promising solutions among a large range of electrical sensing devices due to advantages such as speed, low cost, and simplicity.[12] Since the invention of the first FET biosensor fifty years ago by Bergveld,[13] this technique has been evolving in many forms since 1970 and is an ideal strategy for the rapid and precise detection of diverse analytes. In general, a FET is a solid-state device in which the electroconductivity of the semiconductor between the source and drain terminals is controlled by a third gate electrode *via* an insulator.[14] Because the majority of these devices are built using standard methods, they require essentially little post-processing; however, to change their surface and boost their sensitivity, other materials such as nanostructures are synthesized and added to their structure.[14,15]

Materials such as SiO_2, Si_3N_4 and Ta_2O_5 are commonly used as the structure of the FET biosensing device when designing a successful FET biosensor.[16–18] Nanotechnology integration paves the way for the application of nano-transducers such as nanowires, carbon nanotubes, nanoparticles, nanorods, and so on in the creation of novel biosensors.[19] When it comes to sensitive, selective, label-free, fast, and cost-effective devices, several nanostructures, such as silicon nanowires, graphene, and carbon nanotubes, are among the most popular and useful options.[20] The mechanical strength, high surface-to-volume ratio, extraordinary chemical and thermal firmness, and superior semiconductivity of these materials make them excellent candidates for immobilizing biorecognition elements.[21] Numerous pioneering investigations have leveraged these principles to develop advanced devices for POC diagnosis of various pathogenic bacteria or viruses, which are comprehensively detailed in the following sections.

To recognize specific analytes, biological receptors such as antibodies, nucleic acids, aptamers, enzymes, cells, microorganisms, or synthetic biomaterials are mounted on sensing channels that are connected to the source and drain electrodes.[22] The transducer system turns biochemical changes into detectable signals after exposing the biosensor to target analytes and producing specialized biological complexes such as antigen–antibody complexes, enzyme–substrate complexes, DNA structures, and so on. Attaching charged biomolecules to the surface of a gate dielectric is analogous to applying a voltage with a gate electrode, resulting in threshold voltage changes. As a result, the strategy behind FET biosensors is the dependency of conductance on adsorbed species.[23] These biosensors harness the inherent sensitivity of FETs to minute changes in the electrostatic field of the target molecules in the form of biorecognition elements such as antigens, nucleic acids, membrane proteins, glycolipids and extracellular vesicles (Table 3.1).

3.3 Pathogen Components Amenable to Detection

In this section, we review the different components of the pathogen that would be amenable to detection. We limit our discussion to only bacterial and viral pathogens and omit parasites or fungal pathogens. Bacterial pathogens are directly detectable in the form of the whole cell or the components that are fragmented from the cell including analytes that are secreted by the cell into the surrounding environment. Whole cell detection would require the sensor to have the capacity to detect analytes present on the outside of the cell; this typically includes proteins, lipids and/or carbohydrate molecules that make up the outer membrane. For viruses, the capacity to detect the whole virus particle or virion is similar to that of a whole cell albeit less complex; this would still require a similar set of analytes to be detected, namely proteins, lipids and carbohydrate molecules. The ability to detect such whole cells would allow development of specific sensors that can distinguish pathogenic species from non-pathogens or discriminate a known pathogen *versus* a newly emerged strain while at the same time not depending on the prior need for extensive studies to characterize target analytes or the specific biology of the pathogen.

The detection of whole cell pathogens can be one of the most selective and sensitive methods when detecting a specific strain or species of pathogens.

Table 3.1 Macromolecules that have served as targets for Bio-FET analysis.

Analyte (macromolecule)	Bio-FET type	Bioreceptor used	Limit of detection	Linear range	Ref.
Whole cell (Ebola virus)	Antibody	EBOV whole virion and equine-origin antibody	2.4 pg mL^{-1}	2.4×10^{-12}–1.2×10^{-7} g mL^{-1}	24
Whole cell (*Mycobacterium tuberculosis*)	Antibody	Anti-Ag85B antibody	10×10^{-4} ng mL^{-1}	10×10^{-4}–1 ng mL^{-1}	25
Protein (SARS-CoV-2)	Antibody	SARS-CoV-2 spike antibody	1.6×10 PFU per mL	1.6×10–1.6×10^{4} PFU per mL	26 and 27
Protein (SARS-CoV-2)	Antibody	SARS-CoV-2 S1 antibody	4.12 fg mL^{-1}	0.1 to 5×10^{3} fg mL^{-1}	28
Protein (SARS-CoV-2)	Antibody	SARS-CoV-2 spike antibody	25 fg μL^{-1}	2.5×10–10×10^{6} fg μL^{-1}	29
Protein (Ebola virus)	Antibody	Anti-Ebola antibody	1 ng mL^{-1}	1–444 ng mL^{-1}	30
Protein (Zika virus)	Aptamer	Aptamer	0.01 pg mL^{-1}	0.01–1000 pg mL^{-1}	31
Protein (avian influenza virus)	Aptamer	Aptamer	5.9×10^{-3} nM	10×10^{-3}–10 nM	32
Deoxyribonucleic acid (influenza A virus)	DNA	DNA	10×10^{-4} nM	10×10^{-4}–10 nM	33
Deoxyribonucleic acid (hepatitis B virus)	DNA	DNA	1×10^{-12} μM	5×10^{-7}–50 μM	34
Whole cell (*E. coli*)	Antibody	Anti-*E. coli* antibody (ab30522)	5×10^{3} CFU per mL	5×10^{3}–5×10^{5} CFU per mL	35
Cell membrane (Gram-positive bacteria)	Antibody	Vancomycin	50 CFU per mL	50–10^{6} CFU per mL	36
Protein (*E. coli* strain PKL1162)	Glycosylated surfaces	Mannose-specific type 1 fimbriae	1.9×10^{5} CFU per mL	1.9×10^{5}–4.2×10^{9} CFU per mL	37
Toxin (*Staphylococcus aureus*)	Cell membrane	Red blood cell membrane	1.43×10^{2} CFU per mL	0.6×10^{2}–2.23×10^{2} CFU per mL	38
Cell membrane (Gram positive and negative bacteria)	Antibody	Vancomycin and magainin I	1 CFU per mL	1–9 CFU per mL	39
Cell membrane (Gram positive and negative bacteria)	—	—	10^{3} CFU per mL	10–10^{5} CFU per mL	40

Each bacterium or virus has specific mutations that result in their own distinguishing characteristics. In the formation of a working FET biosensor for the detection of a whole cell pathogen, antibodies or aptamers are usually used as the bioreceptor due to their characteristic of being highly selective toward their target. The successful detection of the Ebola virus in a study published by Jin *et al.*[24] is one such example. The study used a FET modified with reduced graphene oxide that made use of the EBOV whole virion and equine-origin antibody to detect the Ebola virus. The resulting FET biosensor has successfully managed to detect Ebola virus particles at concentrations as low as 2.4 pg mL^{-1} and is highly selective toward the Ebola virus when tested with other common viruses.

Although the use of analytes that are secreted by pathogens, using unique combinations of proteins, or various macromolecules that are made accessible upon cell rupture is possible with Bio-FETs, they are also detectable *via* various other established methods that utilize immunological or molecular biology approaches. Many of such methods are already the gold standard for the detection of many pathogens. Examples of such analytes are specific proteins that can be detected by immunological means or nucleic acid fragments that could be amplified by polymerase chain reactions and then either visualized or sequenced. Although the use of the above-mentioned analytes is either a part of, or can be added to, the existing sensor capability, their use has specific drawbacks. In general, using such macromolecules requires prior extensive characterization and, as a result, this may make them unsuitable as a means of rapidly responding to a newly emerged pathogen or novel strain. In the next section, we delve into the biological and structural features of external pathogen components amenable to detection in order for developers of Bio-FET sensors to assess the feasibility of their use as bio-recognition elements.

3.4 Bacterial Membrane Associated Components

Gram-negative bacteria exhibit a highly variable outer membrane (OM) covered with molecules that can activate immune responses, known as antigens. The serotyping of bacteria is based on unique combinations of these antigens.[41] The main types of antigens present on the cell surface encompass O- (somatic), K- (capsular), H- (flagellar) antigens and exopolysaccharides (EPS).[42,43] These antigens play diverse roles, including aiding in movement (H-antigen), providing protection against harsh conditions (K- and O-antigens), engaging with the environment (K- and O-antigens) and enhancing the structural integrity of the cell's external membrane (O-antigen).[44]

The predominant composition of the outer layer of the OM consists of lipopolysaccharide (LPS), which includes lipid A, the core polysaccharide and the O-antigen.[45] Within Gram-negative bacteria, the O-antigen has gained significance as a critical factor in virulence. Positioned distally, it forms the highly immunogenic element of LPS. The O-antigen is a variable carbohydrate chain, varying in length from two to seven residues, composed of diverse sugar types and their derivatives (such as pentose, heptose, mannose, rhamnose, galactose and glucose), which results in specificity for different serotypes.[46]

The genes essential for the synthesis of the O-antigen are grouped together within a specific chromosomal region known as the O-antigen biosynthesis gene cluster (O-AGC) in numerous bacterial species.[47] Within these O-AGCs, there are two distinct genes, namely *wzx* (encoding the O-antigen flippase) and *wzy* (encoding the O-antigen polymerase), which are mostly unique and have been identified as key targets for molecular O serogroup classification.[48] Often, the O-antigen is the primary contributor to the diversity of cell surfaces in Gram-negative bacteria, offering a selective advantage within its specific environment.[49] Moreover, it functions as a pivotal virulence factor linked to bacterial pathogenesis.[50] This configuration not only aids in the molecular classification of bacterial strains but also facilitates exploration of bacterial evolution.[51]

Capsules, known as K-antigens, are attached to the cell surface of bacteria *via* covalent bonds to phospholipid or lipid A molecules.[52] In the context of pathogenic *Escherichia coli*, K-antigens play a significant role in determining virulence. These surface carbohydrates create a protective layer external to various other surface structures, such as LPS, providing a shield against various environmental stresses and acting as a barrier against the host's immune responses.[53–56] The presence of specific polysaccharide capsules on the surfaces of disease-causing bacterial strains renders these capsules valuable serological targets for monitoring and preventing infections.

Surface carbohydrates play essential roles in the ability of *Acinetobacter baumannii* to adjust to different surroundings and rapidly develop resistance to multiple drugs as well as enhance its virulence.[53,57,58] K-antigens have demonstrated their efficacy as the foundation for successful vaccines targeting both Gram-negative and Gram-positive human pathogens.[59] For instance, *Francisella tularensis*, a bacterium causing tularaemia, has been regarded as possessing a capsule for more than four decades. This conclusion is drawn from observations of changes in colony appearance from smooth to rough phenotypes and the identification of conserved polysaccharide structures within the organism.[60]

In contrast, the H-antigen is a protein-based antigen associated with the structure of flagella.[61] In classical bacteriology, the H-antigen, also known as *E. coli* flagella, represents elongated, slender tubular formations composed of flagellin subunits forming the walls. The distinct serological characteristics of H-antigens are established by specific epitopes exclusively present on the external surface of flagella. These epitopes are identifiable through the conventional agglutination method using entire bacterial cells bearing flagella.[62] On the other hand, EPS are large molecules released externally, forming either tightly bound capsules or loosely attached slime layers in microorganisms.[42] They primarily function in guarding against desiccation, phagocytosis, cell identification, defence against phage attacks, resistance to antibiotics or toxins and managing osmotic stress.[63]

The Enterobacteriaceae family members, including *Klebsiella*, *Proteus*, *Shigella*, *Yersinia* and *Salmonella*, which are responsible for a range of infectious conditions, encompassing intestinal and hospital-acquired infections, share a distinctive glycolipid known as the enterobacterial common antigen (ECA) on their cell surfaces.[64–68] While Enterobacterales express various antigens like K-, O-, and H-antigens, ECA stands out as unique because it is exclusive to this order and remains consistent, enabling cross-reactivity among Enterobacterales members.[69,70] The ECA comprises repetitive linear units of a trisaccharide structure consisting of 4-acetamide-4,6-dideoxy-D-galactose (Fuc4NAc), *N*-acetyl-D-mannosaminuronic acid (ManNAcA), and *N*-acetyl-D-glucosamine (GlcNAc). Ranked as the second most notable immunogenic element, it holds a position closely following the lipopolysaccharide (LPS) O-antigen.[71] Several ECA variants have been identified since its discovery in *E. coli*, including ECA-PG, ECA-LPS, and ECA-CYC.[68]

The predominant form, ECA-PG, is linked to diacylglycerol and exclusively present in *Salmonella*.[72] ECA-LPS is tethered to the lipid A core region of LPS,[73] while ECA-CYC is a water-soluble cyclic form with four or six trisaccharide repeat units, found in the periplasmic space rather than on the bacterial surface.[74] The significance of the ECA in virulence within the Enterobacteriaceae family is emphasized by its influence on attributes like bile resistance and motility.[75] Similar to the O-antigen, ECAs operate as polysaccharide antigens attached to the bacterial cell surface, triggering the generation of anti-ECA antibodies in mice. However, these antibodies directed against the ECA provide limited passive immune defence against Enterobacteriaceae strains, primarily due to their uniqueness in comparison to other virulence factors and surface polysaccharides.

Bacterial cells can also release extracellular vesicles (EVs), and there are distinct variations, particularly between Gram-positive and Gram-negative cells that have been extensively reviewed by McMillan and Kuehn.[76] Extracellular vesicles originating from Gram-negative bacteria have been extensively investigated and are commonly known as outer membrane vesicles (OMVs). These OMVs are generated through the budding process from the outer membrane. Initially, it was believed that Gram-positive cells, with their robust cell wall, were incapable of releasing EVs. However, an increasing number of studies have demonstrated the existence of EVs from Gram-positive cells.[77–79] Gram-positive EVs emerge from the inner membrane and traverse the cell wall, also referred to as the peptidoglycan layer, before being released.

Some investigations have proposed that a weakening of the peptidoglycan layer could facilitate and enhance EV release.[77–79] EVs serve various functions, including the secretion of cellular components that contribute to pathogenesis, resilience under challenging conditions and the regulation of microbial interactions within bacterial communities.[80] OMVs encompass surface proteins, intracellular proteins, nucleic acids, fragments of peptidoglycan, and other cellular materials. Their outer membrane bears resemblance to that of the parent cell, featuring a bilayer structure with

LPS positioned on the outer leaflet and an inner leaflet composed of phospholipids.[80,81]

In *Mycoplasma*, a variety of lipid-associated membrane proteins (LPPs) have the capability to modulate immune responses.[82] Several noteworthy LPPs have been identified in different *Mycoplasma* species. For instance, LppQ in *M. mycoides*,[83] P60 in *M. capricolum*[84] and P30 in *M. pneumoniae*[85] have been described. These *Mycoplasma* LPPs play significant roles as virulence factors and targets for growth inhibitory antibodies. They can impact various functions, including apoptosis,[86] antigenic variation,[87] molecular transport,[88] nuclease activity[89] and adhesion.[90] In the context of *M. agalactiae*, a prominent pathogen causing Agalactia syndrome in sheep and goats, membrane proteins such as P30,[91] cytadhesin P40,[92] P48,[93] lipoprotein MAG_5040,[89] pyruvate dehydrogenase,[76] Vpmas,[94] MAG_1560 and MAG_6130[95] have been identified.

3.5 Virion Surface Associated Components

As for viruses, many of them exhibit diverse glycoproteins on their surfaces. For instance, the hepatitis C virus possesses two envelope glycoproteins named E1 and E2.[96,97] These proteins are integral to viral infectivity and hold promise as potential subunit vaccine candidates.[97] Similarly, the Ebola virus features glycoproteins GP1 and GP2, which facilitate cell attachment and fusion, making them the primary targets of host antibodies. Interestingly, the Ebola virus employs an RNA editing mechanism to regulate the expression of these GP1 and GP2 genes. High levels of their expression can disrupt normal cell physiology.[98]

In the case of HIV-1, the GP120 protein is responsible for initiating viral entry into CD4 cells.[99] Recent research has revealed that different types of glycans in HIV-1 result in varying levels of infectivity. Viruses with more oligomannose and less structural complexity exhibit higher efficiency in infection. This underscores the pivotal role of mature oligosaccharide structures in envelope glycans during the HIV-1 infection process. The GP120 protein of the envelope is produced through N-linked glycosylation of the GP160 protein in the endoplasmic reticulum, followed by further folding and cleavage in the Golgi complex.[100]

In addition, the viral membrane serves a spectrum of crucial functions that contribute to the processes of viral replication and transmission.[101] This membrane consists of a complex organization of lipids, lipid aggregates, glycoproteins and compounds combining lipids and glycoproteins.[102] The orchestrated interaction between the envelope glycoproteins and the structural configuration of the envelope lipids facilitates the fusion of the virus with the plasma or endosomal membranes of the host, thereby enabling the virus to enter the host cell. The intricate glycoprotein covering of the viral particle arises from the mutual evolution of the virus and its host, resulting in its captivating characteristics.[103] Additionally, it aids in the exit of replicated virus from the host cell through the process of budding and fission, often involving the host membrane.[104] Furthermore, the organizational structure of the envelope lipids acts as a safeguard for the viral RNA content against the harsh extracellular environment.[105] The fundamental constituents of the envelope, which define its structure and function, are lipids typically derived from the membranes of host cells. The assembly of the viral envelope within the host cell enhances the diversity of structural formations of lipid aggregates, leading to a broad array of specific interactions between proteins and lipids.[106]

Significantly, the capability of a virus to cause infection is intricately linked to the presence of its glycoproteins, rather than its nucleic acid content. Among enveloped viruses, there is a prevailing reliance on membrane proteins, critical for establishing specific interactions with ligands on host cells. These specialized proteins also play a pivotal role in the initial stages of fusing with cell membranes and initiating the internalization of the virus. It is worth noting that these envelope proteins are frequently subject to glycosylation.[107] This glycosylation process takes place within the secretory pathway of the endoplasmic reticulum and Golgi complex, orchestrated by the enzymatic activities of host cell-produced glycosyltransferase enzymes. The process is essential to ensure compatibility between the virion and its host, which is vital for the ability of the virus to cause disease. The glycans present on the envelope proteins act as immunological barriers, preventing evasion by the host's immune system.[108]

In line with this intricate interplay between viral glycoproteins and infection, severe acute respiratory syndrome coronavirus 2 (SARS-CoV-2) possesses four primary structural proteins, namely spike (S), membrane (M), envelope (E) and nucleocapsid (N) proteins.[109] The S protein comprises two distinct subunits, S1 and S2, which play crucial roles in virus attachment, fusion and entry.[110] Specifically, the S1 subunit contains a receptor-binding domain that interacts with host cell receptors, particularly human angiotensin-converting enzyme 2 (Figure 3.1).[110] On the other hand, the S2 subunit facilitates the fusion of viral and cellular membranes, enabling the virus to enter the host cell. The N proteins are responsible for binding to the viral RNA genome, forming the nucleocapsid structure, while M proteins serve dual functions by releasing nutrients through the transmembrane and contributing to the formation of the viral envelope.[110] The E proteins are relatively small polypeptides that orchestrate the assembly and release of the virus. [110] Among the four major viral structural proteins (S, E, M, and N), the two most abundant proteins S and N (NP) demonstrated sufficient specificity.[111,112] Consequently, these two proteins are frequently employed in practical serological testing.

The N-antigen, being highly prevalent in the viral particle, is expected to induce the highest levels of antibody titres during infection. Therefore, serological tests utilizing the N-antigen are anticipated to exhibit the highest sensitivity. Okba *et al.*[113] found that ELISA tests utilizing the S1-antigen were more effective in distinguishing between coronavirus disease 2019 (COVID-19) patients and groups affected by human coronavirus (HcoV) or Middle East

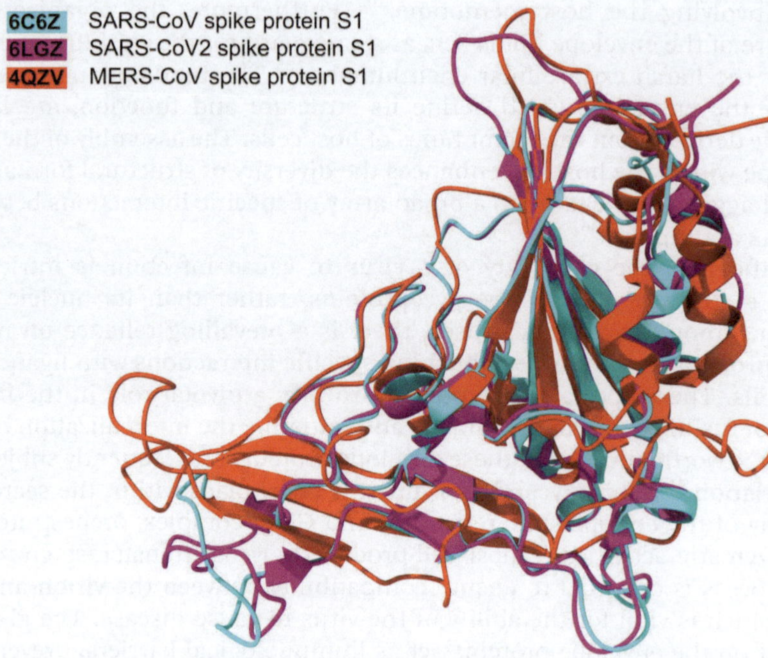

Figure 3.1 Superposition of spike protein S1 subunit structures from SARS-CoV (PDB ID: 6C6Z), SARS-CoV-2 (PDB ID: 6LGZ) and MERS-CoV (PDB ID: 4QZV) shows the conserved regions where the structures overlap and the variable regions where the structures do not overlap and can thus provide the molecular level distinction for detection. The coordinates used to generate these images were retrieved from the Protein Data Bank database.

respiratory syndrome coronavirus (MERS-CoV). Conversely, ELISA tests employing the NP-antigen demonstrated greater sensitivity in identifying seropositivity among patients with mild COVID-19 cases. In a direct comparison, the S-antigen yielded slightly higher ELISA values compared to the NP-antigen for COVID-19 patients.[30] Shifting focus to a different viral context, the hemagglutinin (HA) and neuraminidase (NA) proteins of the influenza A virus exhibit antagonistic activities upon interacting with sialic acid (SA), a critical receptor for virus attachment. The HA binds to SA through its receptor-binding sites, while NA is a receptor-destroying enzyme by removing SAs.[114]

Another viral infection, dengue which is transmitted by *Aedes* mosquitoes, is caused by any of the four dengue virus serotypes (DENV1 to DENV4).[115] The genetic material of DENV comprises a single-stranded RNA with a positive polarity and carries instructions for a polyprotein synthesis. This polyprotein is processed by both host and viral proteases, leading to the formation of three structural proteins (C, E, and prM/M) and seven non-structural proteins (NS1, NS2A, NS2B, NS3, NS4A, NS4B and NS5) through proteolytic cleavage.[116] The NS proteins play a pivotal role in viral genome

Table 3.2 Components of pathogens that have been employed in Bio-FETs.

Cell component	Type	Organism	Ref.
Whole virion		Ebola virus	24
Whole bacteria		*E. coli* K12	119
Surface glycoprotein	GP1	Ebola virus	120 and 121
	GP120	HIV-1	122
Spike protein	S1	SARS-CoV-2	26–28
Lipoprotein	LipL32	*Leptospira interrogans*	123
Antigen		Influenza A (H1N1) virus	124
	N-antigen	SARS-CoV-2	125
	NS1-antigen	Dengue virus	126
Hemagglutinin		Influenza A (H1N1) virus	127
Toxin		Cyanobacteria	128
Extracellular vesicles		*E. coli*	129 and 130
Capsid protein	p24	HIV-1	131

replication within the host cell. The NS1-antigen is discharged into the bloodstream and can be identified during the initial phases of dengue infection, often preceding the development of antibodies against the virus. This distinctive attribute renders the NS1-antigen a valuable asset for the early detection of dengue fever. Moreover, levels of the dengue NS1-antigen have been demonstrated to correlate with the severity of the disease, with higher concentrations observed in individuals afflicted by dengue haemorrhagic fever compared to those with less severe cases of dengue fever.[117] In human immunodeficiency virus (HIV), the p24-antigen holds the distinction of being the most prevalent protein within the HIV and plays a crucial role in the formation of the protective capsid that surrounds the genetic material of the virus. The clinical application of detecting the HIV p24-antigen is employed for the diagnosis of early-stage HIV infection, a phase during which the presence of HIV-specific antibodies remains undetectable (Table 3.2).[118]

3.6 Challenges for Bio-FET Detection

The utilization of FETs for discriminative sensing of diverse biomolecular analytes represents a significant advancement in the field of biosensing for specific and rapid pathogen detection. As reviewed in the earlier sections and in other chapters, many examples of Bio-FETs in use for pathogen detection already exist. Current Bio-FETs are able to detect charge differences due to multiple variations of amino acids that make up the proteins from different pathogen species (Figure 3.2). The challenge remains as the capability to identify the analytes optimally amenable to detection followed by the capacity to detect charge variations that differentiate molecular level changes as little as a single amino acid mutation. This would allow detection of newly emerged mutants. To accomplish this would require the technological means to rapidly prototype new compositions of analytes or variations of one target analyte in order to respond to new pathogens or the

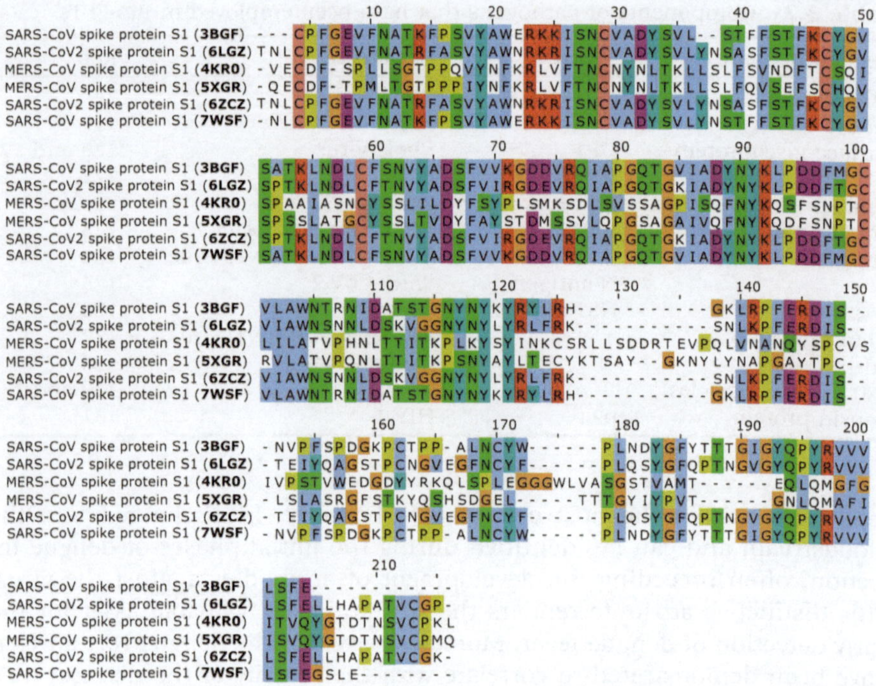

Figure 3.2 Multiple sequence alignment that exhibits the presence of single amino acid mutation between spike protein S1 subunits of different origins (SARS-CoV, SARS-CoV-2, or MERS-CoV). The variability at the molecular level of amino acid sequences even for very similar viruses results in structural elements such as the spike proteins having the potential to be an analyte for Bio-FETs.

emergence of new strains. By leveraging on our understanding of specific interactions between the biomolecules and FET surfaces, we can advance our ability to identify pathogens rapidly and accurately, thereby enhancing disease diagnosis, environmental monitoring and public health outcomes.

References

1. I. Polatoğlu, T. Oncu-Oner, I. Dalman and S. Ozdogan, *MedComm*, 2023, **4**, e228.
2. R. E. Baker, A. S. Mahmud, I. F. Miller, M. Rajeev, F. Rasambainarivo, B. L. Rice, S. Takahashi, A. J. Tatem, C. E. Wagner, L.-F. Wang, A. Wesolowski and C. J. E. Metcalf, *Nat. Rev. Microbiol.*, 2022, **20**, 193–205.
3. K. B. Laupland and L. Valiquette, *Can. J. Infect. Dis. Med. Microbiol.*, 2014, **25**, 128–129.
4. D. Fajardo-Ortiz, M. Lopez-Cervantes, L. Duran, M. Dumontier, M. Lara, H. Ochoa and V. M. Castano, *PLoS One*, 2017, **12**, e0178293.

5. E. Imperia, L. Bazzani, F. Scarpa, A. Borsetti, N. Petrosillo, M. Giovanetti and M. Ciccozzi, *Microbiol. Res.*, 2023, **14**, 635–645.
6. D. D. Kulkarni, C. Tosh, G. Venkatesh and D. Senthil Kumar, *Indian J. Virol.*, 2013, **24**, 398–408.
7. P. Vijayanand, E. Wilkins and M. Woodhead, *Clin. Med.*, 2004, **4**, 152–160.
8. W. Ma, *Virus Res.*, 2020, **288**, 198118.
9. Y.-C. Liu, R.-L. Kuo and S.-R. Shih, *Biomed. J.*, 2020, **43**, 328–333.
10. A. Fleming, *Noble Lecture*, 1945, pp. 83–93.
11. S. Abbas, A. Ahmed, I. M. S. Al-kadmy and A. A. Sattar, *TrAC, Trends Anal. Chem.*, 20202020, 133.
12. L. Syedmoradi, A. Ahmadi, M. L. Norton and K. Omidfar, *Mikrochim. Acta*, 2019, **186**, 11.
13. J. C. Dutta, P. K. Sharma and H. R. Thakur, *Adv. Intell. Syst.*, 2017, **553**, 687–697.
14. R. K. Ratnesh, A. Goel, G. Kaushik, H. Garg, Chandan, M. Singh and B. Prasad, *Mater. Sci. Semicond. Process.*, 2021, **134**, 106002.
15. N. P. Shetti, S. D. Bukkitgar, K. R. Reddy, C. V. Reddy and T. M. Aminabhavi, *Colloids Surf., B*, 2019, **178**, 385–394.
16. T. Wadhera, D. Kakkar, G. Wadhwa and B. Raj, *J. Electron. Mater.*, 2019, **48**, 7635–7646.
17. X. Ma, R. Peng, W. Mao, Y. Lin and H. Yu, *Electrochem. Sci. Adv.*, 2023, **3**, 1–20.
18. B. Shkodra, M. Petrelli, M. A. Costa Angeli, D. Garoli, N. Nakatsuka, P. Lugli and L. Petti, *Appl. Phys. Rev.*, 2021, **8**, 4.
19. M. Rabbani, M. E. Hoque and Z. Bin Mahbub, *Nanosensors in biomedical and environmental applications: Perspectives and prospects*, Elsevier Inc., 2020.
20. D. Sadighbayan, M. Hasanzadeh and E. Ghafar-zadeh, *TrAC, Trends Anal. Chem.*, 2020, **133**, 116067.
21. R. Ahmad, T. Mahmoudi, M. S. Ahn and Y. B. Hahn, *Biosens. Bioelectron.*, 2018, **100**, 312–325.
22. L. Meng, A. P. F. Turner and W. C. Mak, *Biotechnol. Adv.*, 2020, **39**, 107398.
23. H. S. Magar, R. Y. A. Hassan and A. Mulchandani, *Sensors*, 2021, **21**, 19.
24. X. Jin, H. Zhang, Y. T. Li, M. M. Xiao, Z. L. Zhang, D. W. Pang, G. Wong, Z. Y. Zhang and G. J. Zhang, *Microchim. Acta*, 2019, **186**, 1–9.
25. X. Xie, J. Ma, H. Wang, Z. Cheng, T. Li, S. Chen, Y. Du, J. Wu, C. Wang and X. Xu, *Lab Chip*, 2022, **22**, 1702–1713.
26. T.-R. Cui, Y.-C. Qiao, J.-W. Gao, C.-H. Wang, Y. Zhang, L. Han, Y. Yang and T.-L. Ren, *Molecules*, 2021, **26**, 6947.
27. G. Seo, G. Lee, M. J. Kim, S.-H. Baek, M. Choi, K. B. Ku, C.-S. Lee, S. Jun, D. Park, H. G. Kim, S.-J. Kim, J.-O. Lee, B. T. Kim, E. C. Park and S. Il Kim, *ACS Nano*, 2020, **14**, 12257–12258.
28. M. A. Zamzami, G. Rabbani, A. Ahmad, A. A. Basalah, W. H. Al-Sabban, S. Nate Ahn and H. Choudhry, *Bioelectrochemistry*, 2022, **143**, 107982.
29. P. Fathi-Hafshejani, N. Azam, L. Wang, M. A. Kuroda, M. C. Hamilton, S. Hasim and M. Mahjouri-Samani, *ACS Nano*, 2021, **15**, 11461–11469.

30. Y. Chen, X. Tong, Y. Li, B. Gu, J. Yan, Y. Liu, H. Shen, R. Huang and C. Wu, *PLoS Pathog.*, 2020, **16**, e1008796.
31. N. B. F. Almeida, T. A. S. L. Sousa, V. C. F. Santos, C. M. S. Lacerda, T. G. Silva, R. F. Q. Grenfell, F. Plentz and A. S. R. Andrade, *Beilstein J. Nanotechnol.*, 2022, **13**, 873–881.
32. J. Kwon, Y. Lee, T. Lee and J. H. Ahn, *Anal. Chem.*, 2020, **92**, 5524–5531.
33. T. L. Tran, T. T. Nguyen, T. T. Huyen Tran, V. T. Chu, Q. Thinh Tran and A. Tuan Mai, *Phys. E*, 2017, **93**, 83–86.
34. M. Shariati, M. Vaezjalali and M. Sadeghi, *Anal. Chim. Acta*, 2021, **1156**, 338360.
35. G. Wu, M. Meyyappan and K. W. Chiu Lai, *2016 IEEE 16th International Conference on Nanotechnology (IEEE-NANO)*, 2016, pp. 25–28.
36. A. Moudgil, S. Singh, N. Mishra, P. Mishra and S. Das, *Adv. Mater. Technol.*, 2020, **5**, 1–11.
37. N. Formisano, N. Bhalla, M. Heeran, J. Reyes Martinez, A. Sarkar, M. Laabei, P. Jolly, C. R. Bowen, J. T. Taylor, S. Flitsch and P. Estrela, *Biosens. Bioelectron.*, 2016, **85**, 103–109.
38. H. Gong, F. Chen, Z. Huang, Y. Gu, Q. Zhang, Y. Chen, Y. Zhang, J. Zhuang, Y. K. Cho, R. H. Fang, W. Gao, S. Xu and L. Zhang, *ACS Nano*, 2019, **13**, 3714–3722.
39. K. H. Kim, S. J. Park, C. S. Park, S. E. Seo, J. Lee, J. Kim, S. H. Lee, S. Lee, J. S. Kim, C. M. Ryu, D. Yong, H. Yoon, H. S. Song, S. H. Lee and O. S. Kwon, *Biosens. Bioelectron.*, 2020, **167**, 112514.
40. A. Dey, A. Singh, D. Dutta, S. S. Ghosh and P. K. Iyer, *J. Mater. Chem. A*, 2019, **7**, 18330–18337.
41. A. L. van Wuijckhuijse and B. L. M. van Baar, in *Principles of Bacterial Detection: Biosensors, Recognition Receptors and Microsystems*, Springer New York, New York, NY, 2008, pp. 929–954.
42. P. Cescutti, in *Microbial Glycobiology*, Elsevier, 2010, pp. 93–108.
43. R. M. Robins-Browne and E. L. Hartland, *J. Gastroenterol. Hepatol.*, 2002, **17**, 467–475.
44. E. R. Rojas, G. Billings, P. D. Odermatt, G. K. Auer, L. Zhu, A. Miguel, F. Chang, D. B. Weibel, J. A. Theriot and K. C. Huang, *Nature*, 2018, **559**, 617–621.
45. T. J. Silhavy, D. Kahne and S. Walker, *Cold Spring Harbor Perspect. Biol.*, 2010, **2**, a000414.
46. S. Merino, V. Gonzalez and J. M. Tomás, *Future Microbiol.*, 2016, **11**, 903–918.
47. H. Wang, H. Zheng, Q. Li, Y. Xu, J. Wang, P. Du, X. Li, X. Liu, L. Zhang, N. Zou, G. Yan, Z. Zhang, H. Jing, J. Xu and Y. Xiong, *Front. Microbiol.*, 2017, **8**, 1857.
48. C. DebRoy, P. M. Fratamico, X. Yan, G. Baranzoni, Y. Liu, D. S. Needleman, R. Tebbs, C. D. O'Connell, A. Allred, M. Swimley, M. Mwangi, V. Kapur, J. A. Raygoza Garay, E. L. Roberts and R. Katani, *PLoS One*, 2016, **11**, e0147434.

49. L. Wang, Q. Wang and P. R. Reeves, *Subcellular Biochemistry*, 2010, vol. 53, pp. 123–152.
50. M. Caboni, T. Pédron, O. Rossi, D. Goulding, D. Pickard, F. Citiulo, C. A. MacLennan, G. Dougan, N. R. Thomson, A. Saul, P. J. Sansonetti and C. Gerke, *PLoS Pathog.*, 2015, **11**, e1004749.
51. P. D. Kuchur, A. I. Churkina, A. A. Rybina and A. S. Komissarov, *bioRxiv*, 2022, preprint, DOI: 10.1101/2022.04.05.486866.
52. B. F. Cress, J. A. Englaender, W. He, D. Kasper, R. J. Linhardt and M. A. G. Koffas, *FEMS Microbiol. Rev.*, 2014, **38**, 660–697.
53. E. Geisinger and R. R. Isberg, *PLoS Pathog.*, 2015, **11**, e1004691.
54. A. Moran, *Microbial Glycobiology: Structures, Relevance and Applications*, Academic Press, 1st edn, 2009.
55. J. A. Iwashkiw, A. Seper, B. S. Weber, N. E. Scott, E. Vinogradov, C. Stratilo, B. Reiz, S. J. Cordwell, R. Whittal, S. Schild and M. F. Feldman, *PLoS Pathog.*, 2012, **8**, e1002758.
56. T. A. Russo, N. R. Luke, J. M. Beanan, R. Olson, S. L. Sauberan, U. MacDonald, L. W. Schultz, T. C. Umland and A. A. Campagnari, *Infect. Immun.*, 2010, **78**, 3993–4000.
57. B. S. Weber, C. M. Harding and M. F. Feldman, *J. Bacteriol.*, 2016, **198**, 880–887.
58. K. Wieland, P. Chhatwal and R.-P. Vonberg, *Am. J. Infect. Control*, 2018, **46**, 643–648.
59. T. V. Adamkiewicz, B. J. Silk, J. Howgate, W. Baughman, G. Strayhorn, K. Sullivan and M. M. Farley, *Pediatrics*, 2008, **121**, 562–569.
60. H. N. Carlisle, V. Hinchliffe and S. Saslaw, *J. Immunol.*, 1962, **89**, 638–644.
61. F. Ørskov and I. Ørskov, *Methods in Microbiology*, 1978, vol. 11, pp. 1–77.
62. Y. A. Ratiner, S. Salmenlinna, M. Eklund, M. Keskimäki and A. Siitonen, *J. Clin. Microbiol.*, 2003, **41**, 1033–1040.
63. J. Angelin and M. Kavitha, *Int. J. Biol. Macromol.*, 2020, **162**, 853–865.
64. E. C. Böttger, M. Jürs, T. Barrett, K. Wachsmuth, S. Metzger and D. Bitter-Suermann, *J. Clin. Microbiol.*, 1987, **25**, 377–382.
65. H.-M. Kuhn, U. Meier-Dieter and H. Mayer, *FEMS Microbiol. Lett.*, 1988, **54**, 195–222.
66. A. K. Rai and A. M. Mitchell, *mBio*, 2020, **11**, 4.
67. S. Ramia, D. Amsterdam, H. Mayer and E. Neter, *Immunol. Commun.*, 1982, **11**, 491–494.
68. P. D. Rick, H. Mayer, B. A. Neumeyer, S. Wolski and D. Bitter-Suermann, *J. Bacteriol.*, 1985, **162**, 494–503.
69. K. Jann and O. Westphal, *Microbial polysaccharides: The antigens*, Elsevier, 1975.
70. C. M. Kunin, M. V. Beard and N. E. Halmagyi, *Exp. Biol. Med.*, 1962, **111**, 160–166.
71. T. K. Gozdziewicz, C. Lugowski and J. Lukasiewicz, *J. Biol. Chem.*, 2014, **289**, 2745–2754.
72. C. Lugowski and E. Romanowska, *Eur. J. Biochem.*, 1978, **91**, 89–97.

73. P. Kiss, J. Rinno, G. Schmidt and H. Mayer, *Eur. J. Biochem.*, 1978, **88**, 211–218.
74. P. J. A. Erbel, K. Barr, N. Gao, G. J. Gerwig, P. D. Rick and K. H. Gardner, *J. Bacteriol.*, 2003, **185**, 1995–2004.
75. T. Inoue, R. Shingaki, S. Hirose, K. Waki, H. Mori and K. Fukui, *J. Bacteriol.*, 2007, **189**, 950–957.
76. H. M. McMillan and M. J. Kuehn, *EMBO J.*, 2021, **40**, 21.
77. S. Bose, S. Aggarwal, D. V. Singh and N. Acharya, *Microb. Cell*, 2020, **7**, 312–322.
78. P. Briaud and R. K. Carroll, *Infect. Immun.*, 2020, **88**, 12.
79. L. Brown, J. M. Wolf, R. Prados-Rosales and A. Casadevall, *Nat. Rev. Microbiol.*, 2015, **13**, 620–630.
80. C. Schwechheimer and M. J. Kuehn, *Nat. Rev. Microbiol.*, 2015, **13**, 605–619.
81. A. T. Jan, *Front. Microbiol.*, 2017, **8**, 1053.
82. R. P. Dellinger, M. M. Levy, A. Rhodes, D. Annane, H. Gerlach, S. M. Opal, J. E. Sevransky, C. L. Sprung, I. S. Douglas, R. Jaeschke, T. M. Osborn, M. E. Nunnally, S. R. Townsend, K. Reinhart, R. M. Kleinpell, D. C. Angus, C. S. Deutschman, F. R. Machado, G. D. Rubenfeld, S. Webb, R. J. Beale, J.-L. Vincent and R. Moreno, *Intensive Care Med.*, 2013, **39**, 165–228.
83. D. C. Angus, W. T. Linde-Zwirble, J. Lidicker, G. Clermont, J. Carcillo and M. R. Pinsky, *Crit. Care Med.*, 2001, **29**, 1303–1310.
84. J. C. Mira, L. F. Gentile, B. J. Mathias, P. A. Efron, S. C. Brakenridge, A. M. Mohr, F. A. Moore and L. L. Moldawer, *Crit. Care Med.*, 2017, **45**, 253–262.
85. T. Evans, *Clin. Med.*, 2018, **18**, 146–149.
86. J.-L. Vincent, *PLoS Med.*, 2016, **13**, e1002022.
87. L. Evans, A. Rhodes, W. Alhazzani, M. Antonelli, C. M. Coopersmith, C. French, F. R. Machado, L. Mcintyre, M. Ostermann, H. C. Prescott, C. Schorr, S. Simpson, W. J. Wiersinga, F. Alshamsi, D. C. Angus, Y. Arabi, L. Azevedo, R. Beale, G. Beilman, E. Belley-Cote, L. Burry, M. Cecconi, J. Centofanti, A. Coz Yataco, J. De Waele, R. P. Dellinger, K. Doi, B. Du, E. Estenssoro, R. Ferrer, C. Gomersall, C. Hodgson, M. H. Møller, T. Iwashyna, S. Jacob, R. Kleinpell, M. Klompas, Y. Koh, A. Kumar, A. Kwizera, S. Lobo, H. Masur, S. McGloughlin, S. Mehta, Y. Mehta, M. Mer, M. Nunnally, S. Oczkowski, T. Osborn, E. Papathanassoglou, A. Perner, M. Puskarich, J. Roberts, W. Schweickert, M. Seckel, J. Sevransky, C. L. Sprung, T. Welte, J. Zimmerman and M. Levy, *Intensive Care Med.*, 2021, **47**, 1181–1247.
88. S. M. Opal and X. Wittebole, *Crit. Care Clin.*, 2020, **36**, 11–22.
89. I. Grondman, A. Pirvu, A. Riza, M. Ioana and M. G. Netea, *Biochem. Soc. Trans.*, 2020, **48**, 1–14.
90. C. Pierrakos and J.-L. Vincent, *Crit. Care*, 2010, **14**, R15.
91. M. Mierzchała-Pasierb and M. Lipińska-Gediga, *Anaesthesiol. Intensive Ther.*, 2019, **51**, 299–305.

92. J. Lee and D. Banerjee, *Crit. Care Clin.*, 2020, **36**, 105–113.
93. L. Doyle and M. Wang, *Cells*, 2019, **8**, 727.
94. C. Théry, K. W. Witwer, E. Aikawa, M. J. Alcaraz, J. D. Anderson, R. Andriantsitohaina, A. Antoniou, T. Arab, F. Archer, G. K. Atkin-Smith, D. C. Ayre, J. Bach, D. Bachurski, H. Baharvand, L. Balaj, S. Baldacchino, N. N. Bauer, A. A. Baxter, M. Bebawy, C. Beckham, A. Bedina Zavec, A. Benmoussa, A. C. Berardi, P. Bergese, E. Bielska, C. Blenkiron, S. Bobis-Wozowicz, E. Boilard, W. Boireau, A. Bongiovanni, F. E. Borràs, S. Bosch, C. M. Boulanger, X. Breakefield, A. M. Breglio, M. Á. Brennan, D. R. Brigstock, A. Brisson, M. L. Broekman, J. F. Bromberg, P. Bryl-Górecka, S. Buch, A. H. Buck, D. Burger, S. Busatto, D. Buschmann, B. Bussolati, E. I. Buzás, J. B. Byrd, G. Camussi, D. R. Carter, S. Caruso, L. W. Chamley, Y. Chang, C. Chen, S. Chen, L. Cheng, A. R. Chin, A. Clayton, S. P. Clerici, A. Cocks, E. Cocucci, R. J. Coffey, A. Cordeiro-da-Silva, Y. Couch, F. A. Coumans, B. Coyle, R. Crescitelli, M. F. Criado, C. D'Souza-Schorey, S. Das, A. Datta Chaudhuri, P. de Candia, E. F. De Santana, O. De Wever, H. A. del Portillo, T. Demaret, S. Deville, A. Devitt, B. Dhondt, D. Di Vizio, L. C. Dieterich, V. Dolo, A. P. Dominguez Rubio, M. Dominici, M. R. Dourado, T. A. Driedonks, F. V Duarte, H. M. Duncan, R. M. Eichenberger, K. Ekström, S. EL Andaloussi, C. Elie-Caille, U. Erdbrügger, J. M. Falcón-Pérez, F. Fatima, J. E. Fish, M. Flores-Bellver, A. Försönits, A. Frelet-Barrand, F. Fricke, G. Fuhrmann, S. Gabrielsson, A. Gámez-Valero, C. Gardiner, K. Gärtner, R. Gaudin, Y. S. Gho, B. Giebel, C. Gilbert, M. Gimona, I. Giusti, D. C. Goberdhan, A. Görgens, S. M. Gorski, D. W. Greening, J. C. Gross, A. Gualerzi, G. N. Gupta, D. Gustafson, A. Handberg, R. A. Haraszti, P. Harrison, H. Hegyesi, A. Hendrix, A. F. Hill, F. H. Hochberg, K. F. Hoffmann, B. Holder, H. Holthofer, B. Hosseinkhani, G. Hu, Y. Huang, V. Huber, S. Hunt, A. G. Ibrahim, T. Ikezu, J. M. Inal, M. Isin, A. Ivanova, H. K. Jackson, S. Jacobsen, S. M. Jay, M. Jayachandran, G. Jenster, L. Jiang, S. M. Johnson, J. C. Jones, A. Jong, T. Jovanovic-Talisman, S. Jung, R. Kalluri, S. Kano, S. Kaur, Y. Kawamura, E. T. Keller, D. Khamari, E. Khomyakova, A. Khvorova, P. Kierulf, K. P. Kim, T. Kislinger, M. Klingeborn, D. J. Klinke, M. Kornek, M. M. Kosanović, Á. F. Kovács, E. Krämer-Albers, S. Krasemann, M. Krause, I. V Kurochkin, G. D. Kusuma, S. Kuypers, S. Laitinen, S. M. Langevin, L. R. Languino, J. Lannigan, C. Lässer, L. C. Laurent, G. Lavieu, E. Lázaro-Ibáñez, S. Le Lay, M. Lee, Y. X. F. Lee, D. S. Lemos, M. Lenassi, A. Leszczynska, I. T. Li, K. Liao, S. F. Libregts, E. Ligeti, R. Lim, S. K. Lim, A. Linē, K. Linnemannstöns, A. Llorente, C. A. Lombard, M. J. Lorenowicz, Á. M. Lörincz, J. Lötvall, J. Lovett, M. C. Lowry, X. Loyer, Q. Lu, B. Lukomska, T. R. Lunavat, S. L. Maas, H. Malhi, A. Marcilla, J. Mariani, J. Mariscal, E. S. Martens-Uzunova, L. Martin-Jaular, M. C. Martinez, V. R. Martins, M. Mathieu, S. Mathivanan, M. Maugeri, L. K. McGinnis, M. J. McVey, D. G. Meckes, K. L. Meehan, I. Mertens,

V. R. Minciacchi, A. Möller, M. Møller Jørgensen, A. Morales-Kastresana, J. Morhayim, F. Mullier, M. Muraca, L. Musante, V. Mussack, D. C. Muth, K. H. Myburgh, T. Najrana, M. Nawaz, I. Nazarenko, P. Nejsum, C. Neri, T. Neri, R. Nieuwland, L. Nimrichter, J. P. Nolan, E. N. Noltet Hoen, N. Noren Hooten, L. O'Driscoll, T. O'Grady, A. O'Loghlen, T. Ochiya, M. Olivier, A. Ortiz, L. A. Ortiz, X. Osteikoetxea, O. Østergaard, M. Ostrowski, J. Park, D. M. Pegtel, H. Peinado, F. Perut, M. W. Pfaffl, D. G. Phinney, B. C. Pieters, R. C. Pink, D. S. Pisetsky, E. Pogge von Strandmann, I. Polakovicova, I. K. Poon, B. H. Powell, I. Prada, L. Pulliam, P. Quesenberry, A. Radeghieri, R. L. Raffai, S. Raimondo, J. Rak, M. I. Ramirez, G. Raposo, M. S. Rayyan, N. Regev-Rudzki, F. L. Ricklefs, P. D. Robbins, D. D. Roberts, S. C. Rodrigues, E. Rohde, S. Rome, K. M. Rouschop, A. Rughetti, A. E. Russell, P. Saá, S. Sahoo, E. Salas-Huenuleo, C. Sánchez, J. A. Saugstad, M. J. Saul, R. M. Schiffelers, R. Schneider, T. H. Schøyen, A. Scott, E. Shahaj, S. Sharma, O. Shatnyeva, F. Shekari, G. V. Shelke, A. K. Shetty, K. Shiba, P. R. Siljander, A. M. Silva, A. Skowronek, O. L. Snyder, R. P. Soares, B. W. Sódar, C. Soekmadji, J. Sotillo, P. D. Stahl, W. Stoorvogel, S. L. Stott, E. F. Strasser, S. Swift, H. Tahara, M. Tewari, K. Timms, S. Tiwari, R. Tixeira, M. Tkach, W. S. Toh, R. Tomasini, A. C. Torrecilhas, J. P. Tosar, V. Toxavidis, L. Urbanelli, P. Vader, B. W. van Balkom, S. G. van der Grein, J. Van Deun, M. J. van Herwijnen, K. Van Keuren-Jensen, G. van Niel, M. E. van Royen, A. J. van Wijnen, M. H. Vasconcelos, I. J. Vechetti, T. D. Veit, L. J. Vella, É. Velot, F. J. Verweij, B. Vestad, J. L. Viñas, T. Visnovitz, K. V Vukman, J. Wahlgren, D. C. Watson, M. H. Wauben, A. Weaver, J. P. Webber, V. Weber, A. M. Wehman, D. J. Weiss, J. A. Welsh, S. Wendt, A. M. Wheelock, Z. Wiener, L. Witte, J. Wolfram, A. Xagorari, P. Xander, J. Xu, X. Yan, M. Yáñez-Mó, H. Yin, Y. Yuana, V. Zappulli, J. Zarubova, V. Žėkas, J. Zhang, Z. Zhao, L. Zheng, A. R. Zheutlin, A. M. Zickler, P. Zimmermann, A. M. Zivkovic, D. Zocco and E. K. Zuba-Surma, *J. Extracell. Vesicles*, 2018, 7, 1.

95. L. V. Michel and T. Gaborski, *J. Biol. Chem.*, 2022, **298**, 102483.
96. R. J. Kuhn, W. Zhang, M. G. Rossmann, S. V. Pletnev, J. Corver, E. Lenches, C. T. Jones, S. Mukhopadhyay, P. R. Chipman, E. G. Strauss, T. S. Baker and J. H. Strauss, *Cell*, 2002, **108**, 717–725.
97. A. Varki, R. D. Cummings and J. D. Esko, *Essentials of Glycobiology*, Cold Spring Harbor Laboratory Press, 2009.
98. O. Dolnik, V. A. Volchkova, B. Escudero-Perez, P. Lawrence, H.-D. Klenk and V. E. Volchkov, *J. Infect. Dis.*, 2015, **212**, S322–S328.
99. N. Chirmule and S. Pahwa, *Microbiol. Rev.*, 1996, **60**, 386–406.
100. A. Malaspina, B. S. Collins, A. Dell, G. Alter and T. M. Onami, *AIDS Res. Hum. Retroviruses*, 2013, **29**, 1407–1417.
101. F. A. Rey and S.-M. Lok, *Cell*, 2018, **172**, 1319–1334.
102. J. Zhang, A. Pekosz and R. A. Lamb, *J. Virol.*, 2000, **74**, 4634–4644.
103. WHO, Dengue: Guidelines for Diagnosis, Treatment, Prevention and Control: New Edition, World Health Organization, Geneva, 2009.

104. S. C. Harrison, *Nat. Struct. Mol. Biol.*, 2008, **15**, 690–698.
105. R. B. Chan, L. Tanner and M. R. Wenk, *Chem. Phys. Lipids*, 2010, **163**, 449–459.
106. M. Lorizate and H.-G. Krausslich, *Cold Spring Harbor Perspect. Biol.*, 2011, **3**, a004820.
107. CDC, Chikugunya Virus. http://www.cdc.gov/chikungunya/.
108. C. A. M. de Haan, H. Vennema and P. J. M. Rottier, *J. Virol.*, 2000, **74**, 4967–4978.
109. S. Satarker and M. Nampoothiri, *Arch. Med. Res.*, 2020, **51**, 482–491.
110. W. Y. Lim, B. L. Lan and N. Ramakrishnan, *Biosensors*, 2021, **11**, 434.
111. C. Dobaño, R. Santano, A. Jiménez, M. Vidal, J. Chi, N. Rodrigo Melero, M. Popovic, R. López-Aladid, L. Fernández-Barat, M. Tortajada, F. Carmona-Torre, G. Reina, A. Torres, A. Mayor, C. Carolis, A. L. García-Basteiro, R. Aguilar, G. Moncunill and L. Izquierdo, *Transl. Res.*, 2021, **232**, 60–74.
112. B. Tilocca, A. Soggiu, M. Sanguinetti, V. Musella, D. Britti, L. Bonizzi, A. Urbani and P. Roncada, *Microbes Infect.*, 2020, **22**, 188–194.
113. N. M. A. Okba, M. A. Müller, W. Li, C. Wang, C. H. GeurtsvanKessel, V. M. Corman, M. M. Lamers, R. S. Sikkema, E. de Bruin, F. D. Chandler, Y. Yazdanpanah, Q. Le Hingrat, D. Descamps, N. Houhou-Fidouh, C. B. E. M. Reusken, B.-J. Bosch, C. Drosten, M. P. G. Koopmans and B. L. Haagmans, *Emerging Infect. Dis.*, 2020, **26**, 1478–1488.
114. R. Du, Q. Cui and L. Rong, *Viruses*, 2019, **11**, 458.
115. M. G. Guzman and E. Harris, *Lancet*, 2015, **385**, 453–465.
116. E. G. Acosta, A. Kumar and R. Bartenschlager, *Advances in Virus Research*, 2014, vol. 88, pp. 1–109.
117. D. H. Libraty, P. R. Young, D. Pickering, T. P. Endy, S. Kalayanarooj, S. Green, D. W. Vaughn, A. Nisalak, F. A. Ennis and A. L. Rothman, *J. Infect. Dis.*, 2002, **186**, 1165–1168.
118. A. M. Anderson, W. R. Tyor, M. J. Mulligan, D. Waldrop-Valverde, J. L. Lennox and S. L. Letendre, *Clin. Infect. Dis.*, 2018, **67**, 137–140.
119. L. Könemund, L. Neumann, F. Hirschberg, R. Biedendieck, D. Jahn, H.-H. Johannes and W. Kowalsky, *Sci. Rep.*, 2022, **12**, 4397.
120. A. Maity, X. Sui, B. Jin, H. Pu, K. J. Bottum, X. Huang, J. Chang, G. Zhou, G. Lu and J. Chen, *Anal. Chem.*, 2018, **90**, 14230–14238.
121. A. Poghossian, M. Jablonski, D. Molinnus, C. Wege and M. J. Schöning, *Front. Plant Sci.*, 2020, **11**, 598103.
122. A. G. Leija-Montoya, M. L. Benítez-Hess and L. M. Alvarez-Salas, in *Nucleic Acids – From Basic Aspects to Laboratory Tools*, InTech, 2016.
123. R. Nagraik, A. Kaushal, S. Gupta, S. Sethi, A. Sharma and D. Kumar, *J. Biosci. Bioeng.*, 2020, **129**, 441–446.
124. C. Park, W. Choi, D. Kim, B. Jin and J.-S. Lee, *IEEE Sens. J.*, 2019, **19**, 10985–10990.
125. Q. Peng, W. Huang, D. Chen, Z. Gao, Y. Yang and Q. Yuan, *Chin. J. Chem.*, 2023, **41**, 2253–2260.

126. D. Wasik, A. Mulchandani and M. V. Yates, *Anal. Chem.*, 2018, **90**, 679–684.
127. S. Hideshima, H. Hinou, D. Ebihara, R. Sato, S. Kuroiwa, T. Nakanishi, S.-I. Nishimura and T. Osaka, *Anal. Chem.*, 2013, **85**, 5641–5644.
128. F. Tan, N. M. Saucedo, P. Ramnani and A. Mulchandani, *Environ. Sci. Technol.*, 2015, **49**, 9256–9263.
129. A. Pandey, Y. Gurbuz, V. Ozguz, J. H. Niazi and A. Qureshi, *Biosens. Bioelectron.*, 2017, **91**, 225–231.
130. M. Kazemzadeh, C. L. Hisey, P. Dauros-Singorenko, S. Swift, K. Zargar-Shoshtari, W. Xu and N. G. R. Broderick, *IEEE Sens. J.*, 2022, **22**, 1128–1137.
131. E. Macchia, L. Sarcina, R. A. Picca, K. Manoli, C. Di Franco, G. Scamarcio and L. Torsi, *Anal. Bioanal. Chem.*, 2020, **412**, 811–818.

Part II: Real-time Detection – the Employment of BioFETs

CHAPTER 4

Rapid Detection of Microorganisms Based on FET Devices

SUSHANT SHARMA,[a] MAYUR JAGDISHBHAI PATEL,[a] RAMKRISHNA DAS ADHIKARI,[b] MIZANUR ALAM,[b] HIMANGSHU BAISHYA,[b] DEEPAK YADAV,[a] MANAB KALITA,[a] DHEERAJ KHUBCHANDANI[a] AND PARAMESWAR KRISHNAN IYER*[a,b]

[a] Department of Chemistry, Indian Institute of Technology Guwahati, Guwahati-781039, Assam, India; [b] Centre for Nanotechnology and School for Health Science and Technology, Indian Institute of Technology Guwahati, Guwahati-781039, Assam, India
*Email: pki@iitg.ac.in

4.1 Introduction

The development of point-of-care (PoC) detection kits has become increasingly important in the modern age of medical diagnosis systems due to the high demand for early detection of bacteria and viruses with high selectivity and sensitivity that are frequently considered to be pathogenic organisms for concerns related to human health, the environment, and the food industry.[1,2] Therefore, diagnosing and treating illnesses are crucial to maintaining human health. Traditional detection technologies for pathogens, such as polymerase chain reaction (PCR) and enzyme-linked

Detection Science Series No. 24
Field-effect Transistor Biosensors for Rapid Pathogen Detection
Edited by Naif H. M. Al-Hardan, Azman Jalar, Muhammad Azmi Abdul Hamid and Mohd Firdaus Raih

Published by the Royal Society of Chemistry, www.rsc.org

immunosorbent assay (ELISA), require skilled personnel and specialized equipment. Also, optical-based readout methods, particularly those that employ fluorescent indicators and others that make use of sophisticated spectroscopic apparatus, have long dominated this area. Once more, observational or physical methods typically have lower precision, and in particular situations, more sophisticated imaging or spectroscopic techniques are required. However, all these techniques are time-consuming, labour-intensive, and expensive, have low throughput, and occasionally have poor sensitivities. The development of semiconductor-integrated circuit technology has played an inevitable role in the introduction of effective, efficient, and affordable detection tools as a result of the growing demands for rapid detection mechanisms for microorganisms. Biosensors have proven to be a better option for diagnosing diseases over the past ten years. Biosensors are analytical tools that convert a biochemical reaction brought on by a particular analyte or collection of analytes into a quantifiable signal proportional to the target analytes. Two essential components are required for developing a functional biosensor: (i) the bio-recognition element, which may exist in the form of enzymes, antibodies, aptamers, or ion-selective membranes, and (ii) the bio-transduction device, which transforms the input signal into a quantifiable output signal based on interactions between the bio-recognition element (receptor) and the analytes (host).[3] Biosensors are classified into electrochemical, fluorescent, colorimetric, field effect transistor (FET), surface plasmon resonance (SPR), and piezoelectric biosensors, among others, based on their mode of sensing.

FET biosensors have garnered the most interest among them due to their superior electronic properties, inherent signal amplification, and direct interaction of the channel with the environment, which improves control over the surface charge. This suggests that surface FET-based biosensors may be more sensitive to biological events that occur at the channel surface. These events cause a change in the semiconductor channel surface potential, which in turn modulates the conductance of the channel and enables the device to detect a variety of biomolecules with high sensitivity. Apart from these, their small size, inexpensive manufacturing, and simple handling characteristics make them a potential tool for the quick detection of microorganisms.[4] To date, a wide range of semiconductor materials, including metal oxides, carbon nanotubes, compounds based on polymers, inorganic materials, and different composites, have been used as the channel in FET biosensors. FET devices are classified into inorganic-FET, organic-FET, 2D-FET, and other categories based on the semiconducting channel materials used.[5]

Based on the active semiconducting materials used and their characteristic parameters, this chapter provides a basic overview of the various types of FET biosensor devices and their use in the quick, label-free detection of bacteria and viruses that pose a risk to human health and the ecosystem.

4.2 Working Principle

The operation of a field-effect transistor (FET) is based on the principle that the third electrode (the gate) controls the flow of current through the semiconducting channel connecting the two conductive (source and drain) terminals, allowing current variations over a wide range of orders of magnitude. The electrical characteristics of FETs are utilized by FET-based biosensors, a class of sensing technology devices, to identify biological substances. The gate–insulator surface of a FET is frequently functionalized with bioreceptors like enzymes, antibodies, or single-stranded DNA (ssDNA) probes in order to specifically recognize and capture a particular target analyte (or host).[6] Protein-based bio-recognition components such as enzymes and antibodies are frequently used because of their high sensitivity (detection down to the nM range) and extensive commercial availability. For a biosensor to function well, the right immobilization method for receptor binding is essential. Physical adsorption, trapping, covalent binding, cross-linking, and layer-by-layer technique are some of the commonly utilized immobilization techniques.[7] The formation of a space-charge region in the semiconductor channel by means of an electric field perpendicular to the gate surface is a common operating mechanism, even though each FET biosensor has a unique structure. Biochemical reactions or biomolecular interactions can cause any form of charge or potential change, which biosensors are able to detect at or close to the gate insulator/electrolyte (analyte) interface. By analogy, with the effect of an additional gate voltage, the binding of charged target molecules of the analyte will modulate the effective charge on the gate surface. As a result, the semiconductor–insulator interface undergoes equal variations of opposite charge density. This will cause a shift in the device's conductance or current and capacitance–voltage behaviours, with respect to the original input properties.[8] Figure 4.1 shows the schematic of the device architecture and detection mechanism of a FET-based biosensor.

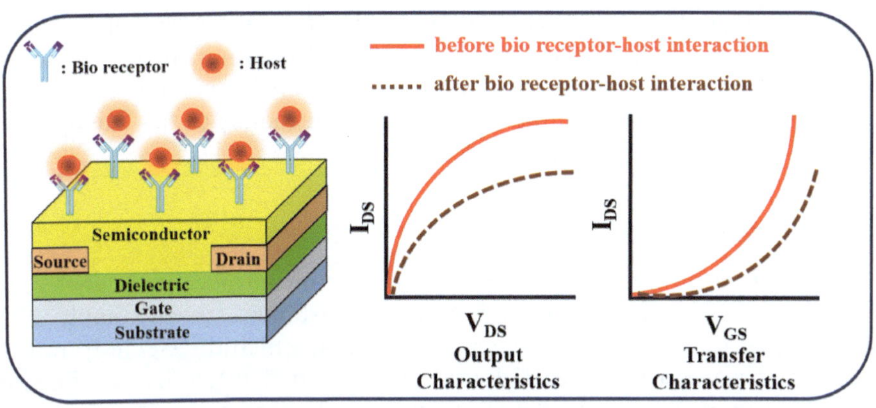

Figure 4.1 FET biosensor device architecture and detection mechanism.

4.3 Characteristic Parameters

Every biosensor possesses certain static and dynamic characteristics. The performance of a biosensor is influenced by the optimization of these features. The characteristic parameters[9–12] of a FET-based biosensor are discussed below.

4.3.1 Selectivity

The most significant feature of a biosensor is its selectivity. It refers to the features of a sensor that determine whether it can respond selectively to a group of analytes or even to a single analyte.

The interaction of an antigen and an antibody is the best illustration of selectivity. Antibodies often serve as bioreceptors and are immobilized on the surface of the transducer. The antigen-containing solution is then exposed to the transducer, where antibodies only interact with the antigens. Selectivity is the primary factor to be taken into account when choosing bioreceptors for a biosensor.

4.3.2 Reproducibility

This is also one of the most important features in biosensing and refers to the ability of a biosensor to produce an identical response under identical testing conditions.

Reproducibility is demarcated as the precision and accuracy of the transducer and electronics in a biosensor. If a sample is tested more than once, then the accuracy relates to the ability of the sensor to produce a mean value that is close to the true value, while precision refers to the ability of the sensor to produce identical results each time.

4.3.3 Sensitivity

This is considered the most essential feature of a biosensor. The relationship between the change in analyte concentration and the intensity of the signal produced by the transducer is known as the sensitivity of a biosensor. A biosensor should ideally produce a signal in response to tiny changes in the concentration of the target analyte.

4.3.4 Limit of Detection (LOD)

The limit of detection is defined as the lowest concentration of analyte that the sensor can detect under particular circumstances, most notably at a given temperature. It is also the concentration at which the signal from the analytical process can be confidently distinguished from the baseline or control solution. It can be calculated from the linear fit obtained from the response curve. The response curve is the plot of %R *versus* the concentration

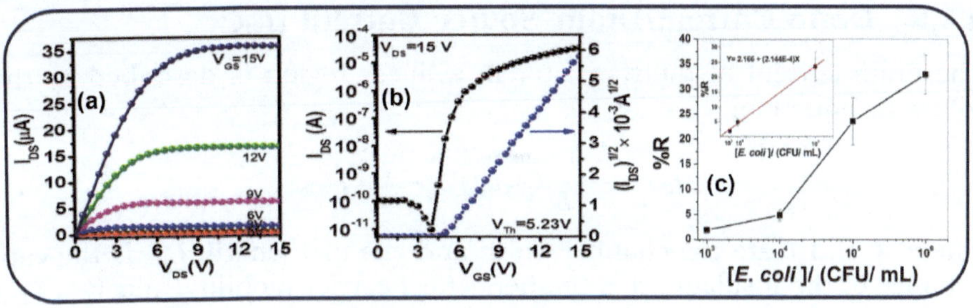

Figure 4.2 (a) Output characteristics and (b) transfer characteristics of FET-based biosensors, and (c) response curve. Reproduced from ref. 19 with permission from Elsevier, Copyright 2018.

of the bacterial/virus solution where $\%R = [(\text{Resistance}_{\text{final}} - \text{Resistance}_{\text{initial}})/\text{Resistance}_{\text{initial}}] \times 100$.

Figure 4.2(c) depicts the response curve of the anti-*E. coli*/AuNPs/Al_2O_3/rGO biosensor to *E. coli* addition. The inset of Figure 4.2(c) shows the linear fitting from 10^3 CFU per mL to 10^6 CFU per mL of the bacterial *E. coli* concentration with 10^3 CFU per mL as the minimum LOD value.

4.3.5 Stability

In particular, for biosensors used for continuous monitoring, the stability of the biosensor is an important feature. The ability of a biosensor device to withstand performance changes over time in response to interruptions caused by external factors is determined by this property. These could be expressed as variables like temperature, humidity, or other environmental factors.

4.3.6 Response Time

The time required for a sensor to respond to a step concentration change from zero to a specific concentration value is referred to as response time.

4.3.7 Range or Linearity

The relative deviation of an experimentally obtained calibration graph from an ideal straight line is referred to as linearity.

Linearity is mathematically denoted by the equation $y = mc$, where c is the analyte concentration, y is the signal output, and m is the sensitivity of the biosensor. The range of analyte concentrations being tested as well as the biosensor's resolution can both affect the sensor's linearity. The smallest change in the concentration of an analyte necessary to cause a change in the biosensor response is referred to as the resolution of the biosensor.

4.3.8 Drain Current/Drain–Source Current (I_{DS})

The drain current of the transistor in a linear region is described by the following equation:

$$I_{DS} = -\frac{W}{L} C_{ox}\mu[(V_{GS} - V_{Th})V_{DS} \quad (4.1)$$

where W and L are the channel width and channel length, C_{ox} is the capacitance of the insulator, μ is the field-effect carrier mobility, and V_{GS}, V_{DS}, and V_{Th} are the gate–source, drain–source, and threshold voltages, respectively. The plot of I_{DS} *versus* drain–source voltage (V_{DS}) gives output characteristics, while the plot of I_{DS} *versus* gate–source voltage (V_{GS}) results in transfer characteristics (Figure 4.2(a) and (b)).

4.3.9 Field Effect Mobility (μ)

The field effect mobility measures the average charge carrier drift velocity per unit electric field.

4.3.10 Current On/Off Ratio (I_{ON}/I_{OFF})

I_{ON}/I_{OFF} is the ratio of the maximum I_{DS} ("on" current) value to the minimum I_{DS} ("off" current) value, obtained from transfer characteristics plotted on a logarithmic scale. The ratio characterizes the ability of the device to switch a signal from "on" to "off".

4.3.11 Threshold Voltage (V_{Th})

The threshold voltage can be defined as the gate voltage at which the conduction channel starts to form. V_{Th} can be extracted by plotting I_{DS} *vs.* V_{GS} and extrapolating to $I_{DS} = 0$ as shown in Figure 4.2(b).

4.3.12 Transconductance

Transconductance (g_m) is the change of I_{DS} with respect to V_{GS} for a low drain voltage V_{DS} in the linear region.

$$g_m = -\frac{\partial I_{DS}}{\partial V_{GS}} = \frac{W\mu C_{ox}}{L} V_{DS} \quad (4.2)$$

4.4 Active Materials for FETs

4.4.1 Carbon Nanotubes

A carbon nanotube (CNT) is a 1-dimensional (1D) allotrope of carbon having a ribbon of sp^2 hybridized graphene sheet with a hexagonal lattice completely coiled into a cylindrical tube. Single-walled carbon nanotubes (SWCNTs) have gained immense utilization in biosensors due to their quasi-ballistic transport at low voltage, higher transconductance, high sensitivity, high selectivity, fast response speed, short recovery time, low cost, low detection limit, and ease of fabrication. In addition, the absence of bulk C-atoms and lowered dimensionality offer a high surface-to-volume ratio, which allows for efficient identification of low concentrations of viruses which bind to the nanotube surface.

4.4.2 Nanowires

Nanowires (NWs) are highly functional 1D structures with diameter in the range of 1–100 nm and offer unique electrical properties. The benefits of nanowires are that they may respond quickly to an electric field and have a large specific surface area and can be composed of a range of biological receptors (such as peptide nucleic acids, aptamers, antibodies, *etc.*). Examples of commonly used nanowires are silicon (Si), zinc oxide (ZnO), organic nanowires (PPy-NWs), and indium oxide (In_2O_3). The nanowire-based biosensor can detect very low concentration markers in a tiny sample size due to a high surface-to-volume ratio.

4.4.3 2D Materials

Graphene has a honeycomb structure made of sp^2 hybridized carbon atoms and is one atom thick. It has attracted a lot of interest for use as a channel material in FET sensors because of its remarkable electrical properties, high flexibility and biocompatibility, enormous specific surface area, and straightforward chemical functionalization. After the success of graphene, a new family of 2D materials known as transition metal dichalcogenides (TMDCs) such as MoS_2 and WSe_2 has been explored for the channel material in order to overcome the issue of no bandgap in graphene materials (electrons can flow any energy level). TMDCs are composed of 2D stacked layers of covalently bonded transition metal and dichalcogenide atoms arranged in a hexagonal lattice and held together by weak van der Waals interactions. They displayed outstanding electrical characteristics and ambient temperature, making them promising FET materials.

4.4.4 π-Conjugated Organic Molecules

These π-conjugated organic molecules have a core carbon backbone of alternating single and double bonds along the molecules giving rise to a partially delocalized π-molecular orbital that significantly defines its electronic properties. Organic semiconductors based on molecules of low molecular weight have attracted significant research interest because of their significant advantages such as defined molecular structure, good film-forming

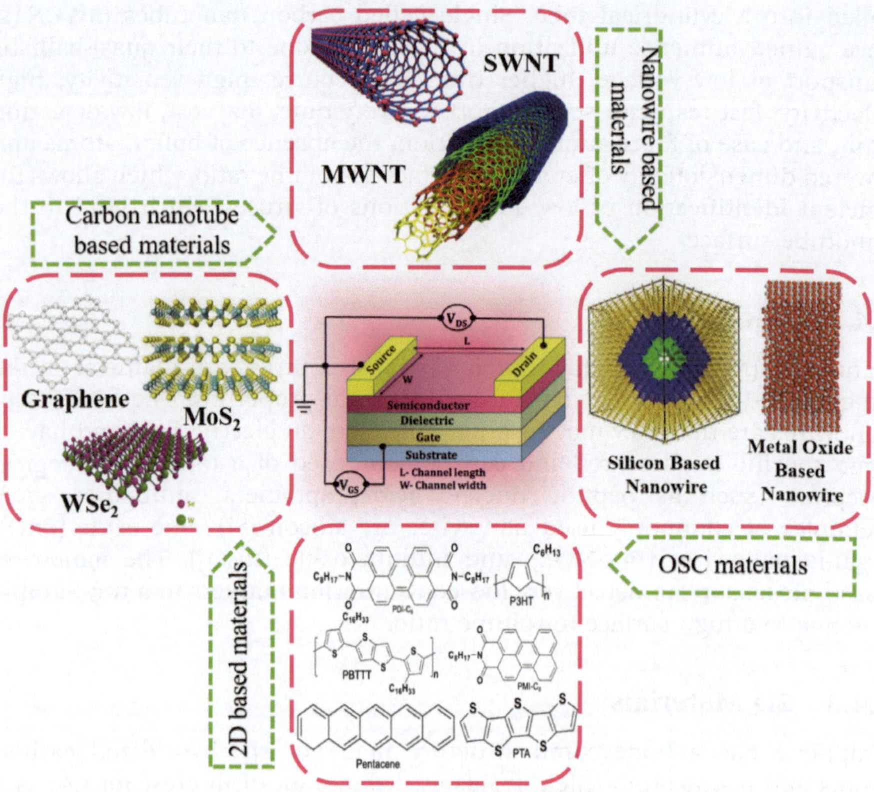

Figure 4.3 Active materials for FET based biosensors.

ability, and high purity without variation between batches of synthesis over semiconductor materials. Figure 4.3 shows the active materials used in FET-based biosensor devices.

4.5 Microorganisms

This chapter deals with the detection of a number of bacterium and virus types. The detection/sensing technique of microorganisms is discussed in Sections 4.6 and 4.7. This section provides a brief outline of the type and microorganisms and their effect on life forms.

Bacteria are an essential part of our ecosystem with some large and diverse groups being harmless and commonly classified as either Gram-positive or Gram-negative bacteria. However, some types of bacteria are extremely harmful to human health. The Shiga toxins by the enterohemorrhagic strains of *E. coli* bacteria are extremely lethal. *E. coli* O157:H7 is the worst type of *E. coli* borne through food and water. This bacterium causes diarrhea, anemia, kidney failure, and hemolytic uremic syndrome. *Borrelia burgdorferi*, commonly referred to as "Lyme", cause Lyme disease affecting people worldwide with symptoms such as fever, headache, and erythema migrans (rash skin condition). *Salmonella infantis*, a bacterium from the Salmonella family, causes gastrointestinal illness and fever called salmonellosis and is the most deadly pathogenic bacterium for human food-borne diseases due to its antibiotic resistance.

Viruses are infectious sub-microscopic agents constituting nucleic acid segments (DNA or RNA) surrounded by a protein coating. They impact life forms by invading a host cell and replicating themselves utilizing the components of the host cell. Influenza viruses are enclosed viruses with genomes made up of short, negative-sense RNA segments. Under appropriate humidity and temperature conditions, they can sustain for several hours. They cause mild to severe illness, dry cough, dizziness, and sometimes death as well. Hepadnaviridae family viruses, particularly, hepatitis B and hepatitis C, are fatal and cause chronic infection and liver inflammation. Hepatitis B is a DNA virus replicating *via* an RNA intermediate, while hepatitis C is an RNA virus replicating on the cellular membrane by RNA replication. Hepatitis B is spread through blood and bodily fluids, but hepatitis C is usually transmitted only through blood. The hemorrhagic fever virus (Ebola virus) is another lethal pathogen that disrupts the body's clotting system, resulting in internal bleeding when blood leaks from small blood vessels. The dominant viral disease dengue caused by the dengue virus enters the human body through mosquito bites, specifically *Aedes aegypti*. Dengue viruses are classified into DEN-1, DEN-2, DEN-3, and DEN-4 with the DEN-2 virus being more severe than the others. The dengue virus causes extremely high fever called hemorrhagic fever which is life-threatening. Coronaviruses (CoVs), a type of enveloped RNA viruses, typically respond to respiratory and intestinal infections in animals and humans. The 2019-nCoV, also known as severe acute respiratory syndrome coronavirus 2 (SARS-CoV-2), belongs to this family of CoVs. The envelope (E), membrane (M), spike (S), and nucleocapsid (N) proteins comprise the coronavirus. Of particular significance, N and S are the two most critical proteins among these four. The N protein aids the virus in properly forming the capsid and the overall viral structure and the S protein aids in the virus attachment to the host cell causing its wide replication, acute respiratory distress syndrome (ARDS), and respiratory failure.

4.6 Bacterial Detection by FETs

4.6.1 Carbon Nanotube FETs (CNT-FETs)

The *E. coli* bacterial screening technique is based on aptamer-functionalized single-walled carbon-nanotube field-effect transistor (SWNT-FET) arrays combined with the most probable number (MPN) technique. Instead of an incubating dish, an RNA-based *E. coli* aptamer functionalized SWNT-FET is utilized. CDI-Tween 20 molecules and streptavidin immobilized the biotinylated aptamers to the sidewalls of the SWNTs through biotin and avidin noncovalent interactions. A schematic diagram of the bacterial detection

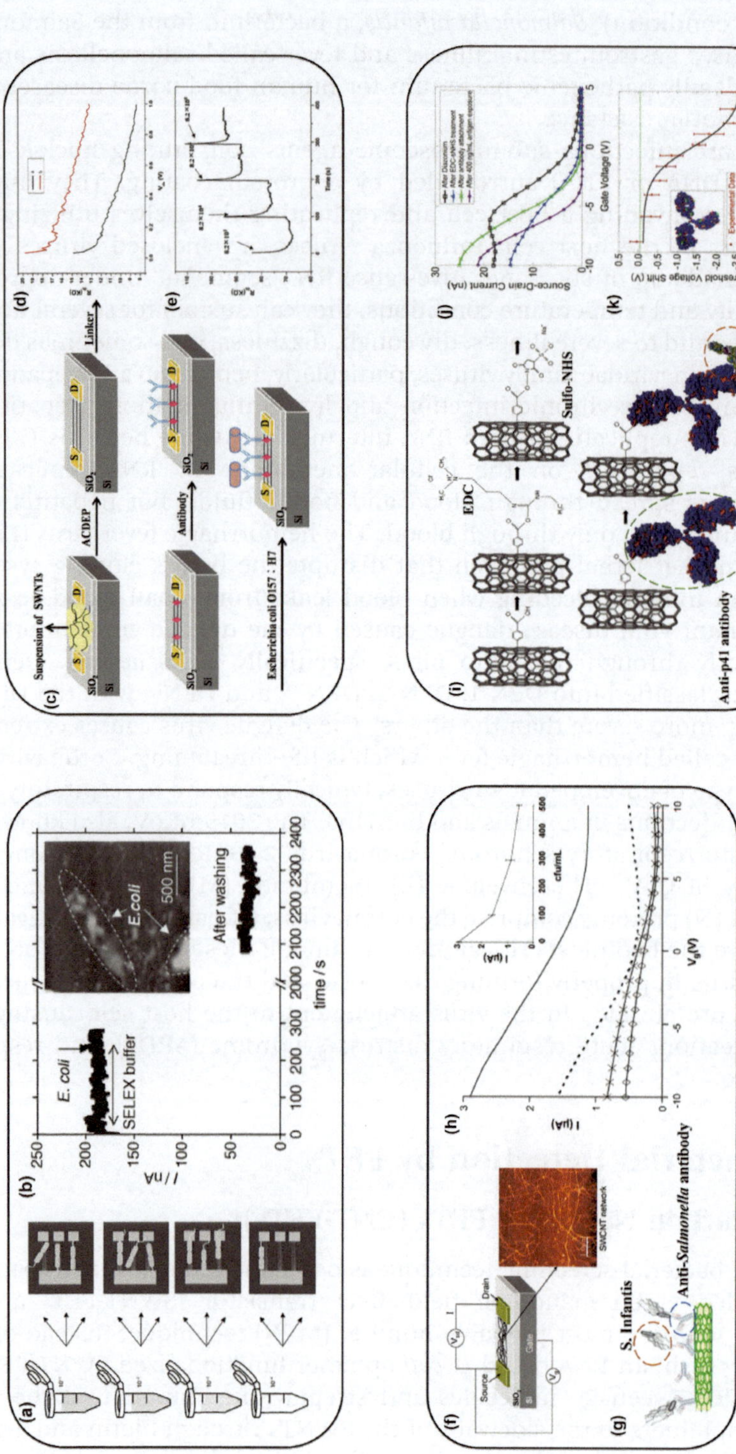
(a)
(b)
SELEX buffer
E. coli
After washing
500 nm
time / s
I / nA
(c)
Suspension of SWNTs
ACDEP
Linker
(d)
(e)
Antibody
Escherichia coli O157 : H7
(f)
(g)
S. Infantis
Anti-Salmonella antibody
SWCNT network
(h)
(i)
(j)
(k)
EDC
Sulfo-NHS
Anti-p41 antibody
Flagellar antigen
Gate Voltage (V)
Source-Drain Current (nA)
FLA Concentration (ng/mL)
Threshold Voltage Shift (V)

technique is shown in Figure 4.4(a). After binding of 10^7 CFU per mL of *E. coli* to the aptamer functionalized SWNT-FET, the electrical conductivity of the SWNT-FET decreased drastically by 50% (Figure 4.4(b)) and displayed a discriminating nature with respect to *Salmonella infantis*.[13] Another strategy to detect the lethal *E. coli* variant "*E. coli* O157:H7" is to employ a solution of the pyrene moiety and two succinimidyl ester functional groups of PASE (1-pyrene butanoic acid succinimidyl ester) as a linker molecule and the $-NH_2$ group of anti-*E-coli* antibodies to functionalize the SWCNT channel through π–π stacking and covalent bonding. Due to *E. coli* O157:H7 interaction with the antibody, a significant decrease in the source–drain resistance occurred (Figure 4.4(d)) with a limit of detection as low as 8.2×10^2 CFU per mL. Also, the resistance changed from 84 kΩ to 110 kΩ along with the increase of the concentration of *E. coli* O157:H7 solution (Figure 4.4(e)), which indicates the possibility of the CNT-FET biosensor to detect *E. coli* O157:H7. The fabrication and detection processes of the *E. coli* O157:H7 bacteria are illustrated in Figure 4.4(c).[14]

Salmonella infantis bacteria can also be recognized and detected by CNT-FETs. Anti-Salmonella antibodies are immobilized onto the side walls of CNTs by the adsorption approach. A linear aliphatic chain surfactant, Tween 20, binds to the surfaces of CNTs through hydrophobic interactions to protect them from the non-specific binding (NSB) of other bacteria or proteins. The bacteria are linked to immobilized CNTs through the interaction of anti-Salmonella antibodies with the somatic O antigen (O-antigen) and the flagellar H antigen (H-antigen) in the outer membrane and the flagellums of the *S. infantis* bacterium, respectively (Figure 4.4(f) and (g)). Detection is achieved when charge transfer occurs from bacterial protein

Figure 4.4 (a) Detection of *E. coli* bacteria utilizing an array of three aptamer-functionalized SWNT-FETs paired with an MPN table and (b) electrical characteristics of *E. coli* detection on the SWNT-FET device. Reproduced from ref. 13 with permission from John Wiley & Sons, Copyright 2008 Wiley-VCH Verlag GmbH & Co. KGaA, Weinheim. (c) Schematic illustration of the fabrication of the CNT-FET biosensor, (d) resistance *vs.* voltage curve of the introduction of O157:H7 into antibody–CNT-FET with PSE (a: PASE–CNT-FET, b: pristine CNT-FET), and (e) the time dependence of CNT-FET resistance after the introduction of *E. coli* O157:H7 at various concentrations into the antibody-modified CNT-FET. Reproduced from ref. 14 with permission from Scientific Research, Copyright 2012. (f) Experimental setup for detecting *Salmonella infantis*, (g) schematic of the antigen–antibody interaction of *S. infantis* with an SWCNT and (h) V_g *vs.* I_{DS} curve of the typical CNT-FET before and after exposure to *S. infantis*. Reproduced from ref. 15 with permission from Elsevier, Copyright 2008. (i) Functionalization chemistry for the Lyme antibody and binding of the flagellar antigen, (j) I_{DS} *vs.* V_g characteristics at subsequent stages of nanotube functionalization, and (k) threshold voltage variation as a function of antigen concentration. Reproduced from ref. 16 with permission from Elsevier, Copyright 2013.

amino groups to CNTs upon the increase of bacterial concentration. This causes a decrease in source–drain current at a negative gate voltage (Figure 4.4(h)). The obtained LOD is 100 CFU per mL of *S. infantis* bacteria within a one-hour time period.[15]

Borrelia burgdorferi, commonly referred to as "Lyme", which causes Lyme disease, is manifested by insects, and its early detection is very necessary. An active SWCNT on Si/SiO_2 support-based FET device is functionalized with the *Borrelia burgdorferi* flagellar antibody for Lyme detection. The nanotubes are treated with diazonium salts to establish sp^3 hybridization sites that result in carboxylic acid moieties. The carboxylic acid is activated with EDC followed by sulfo-NHS stabilization. The sulfo-NHS is displaced by the primary amine in a surface lysine residue (depicted in red) on the anti-p41 antibody (Figure 4.4(i)). During the exposure stage, the flagellar antigen binds to the epitope on the antibody which resulted in a 50–90% drop in I_{ON} as well as a 3–4 V decrease in transistor threshold voltage towards a negative value with increasing concentration (Figure 4.4(j) and (k)). This threshold change is indicative of the nanotube FET being locally gated by the residues of flagellar protein bound to the antibody. This sensor reliably detected antigens with a lower concentration value of 1 ng mL^{-1} which correlates its importance to levels of clinical interest. The effectiveness of this FET sensor as a point-of-care diagnostic tool is suggested by its rapidity and ease of use, which are superior to conventional immunoassays.[16]

4.6.2 Nanowire FETs (NW-FETs)

This is a new approach based on filtration to detect five food-poisoning microorganisms. The first phase of the immunoassay process involves immobilizing the antibody on a polystyrene microbead and filtering it through a pore-size-appropriate filter. Target bacterial analytes using microbeads cannot pass through syringe filters if they develop a complex structure. As a result, there was no observable change in the pH of the filtrate solution. Unbound microbeads, on the other hand, were able to pass through the filter, and the immobilized enzyme (urease) reacted with the urea to produce ammonia. This caused a pH shift of the filtrate solution that modulates the channel current of the SiNW FET (Figure 4.5(a)). The microbead size and pore size are adjusted for effective bacterial detection. Five antibody bacteria (anti-*E. coli* 0157 antibodies, anti-*S. typhimurium* antibody, anti-*L. monocytogenes* antibody, anti-*B. cereus* antibody, and anti-*S. aureus* antibody) collected from horse serum were subjected to a selectivity test using a one-step immunoassay. Due to a significant difference between the drain current and threshold voltage measured in relation to pH shift, greater sensitivity can be attained. Figure 4.5(b) depicts the selectivity for each bacterium. Figure 4.5(b) depicts the selectivity for each bacterium. Through a one-step immunoassay method, very low LOD values of 124 cells per mL for an anti-*S. aureus* antibody, 80 cells per mL

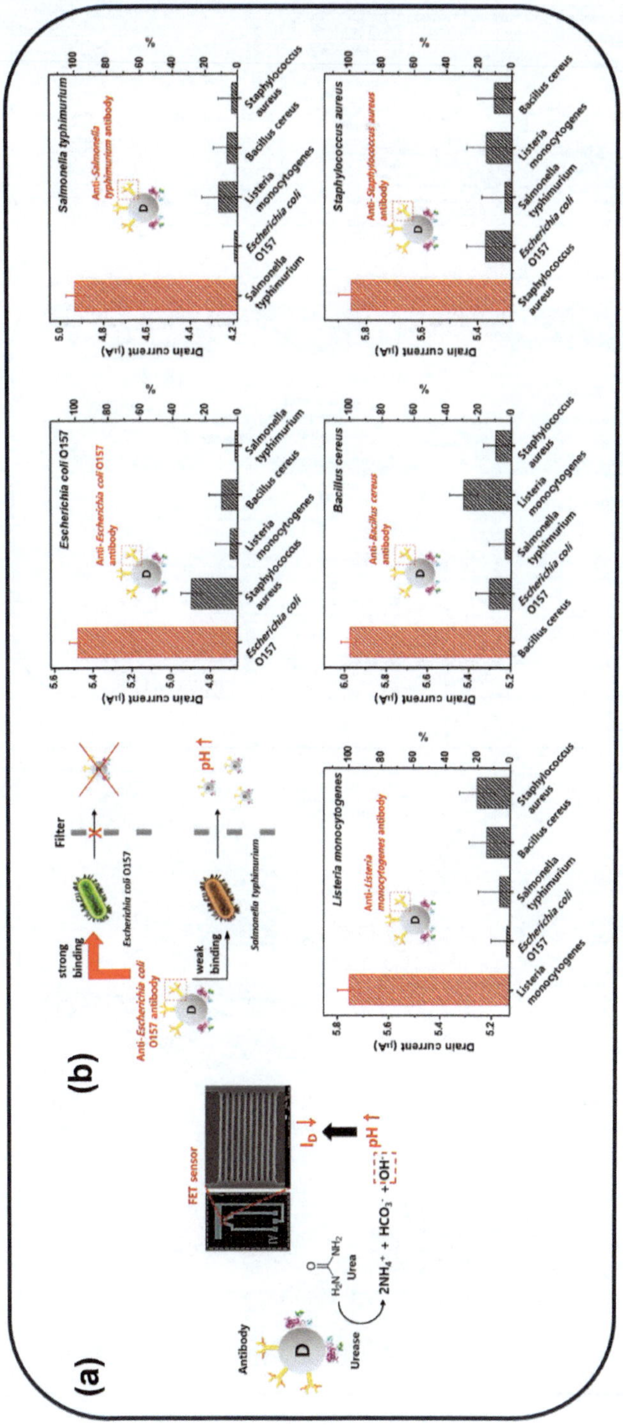

Figure 4.5 (a) Selectivity comparison for five bacteria and (b) one-step immunoassay process with no wash. Reproduced from ref. 17 with permission from Elsevier, Copyright 2023.

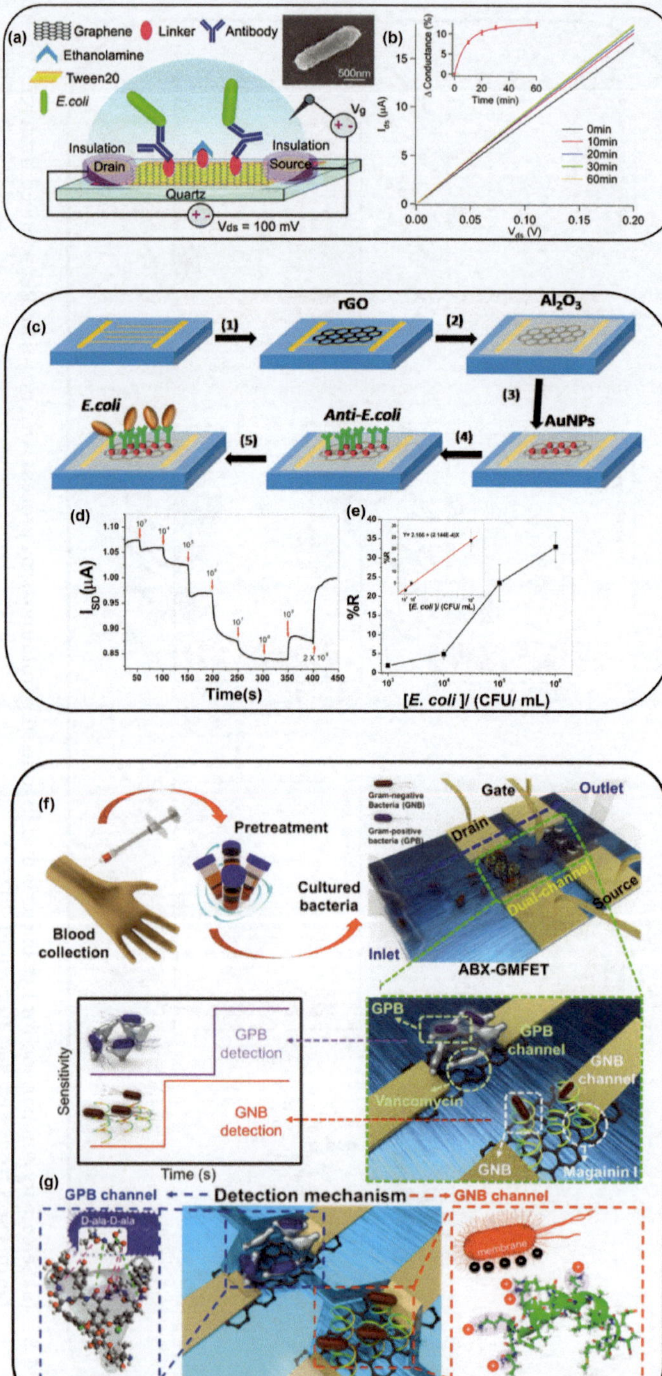

(a)
Graphene
Linker
Antibody
Ethanolamine
Tween20
E.coli
500nm
Vg
Insulation
Insulation
Drain
Source
Quartz
Vds = 100 mV
(b)
Δ Conductance (%)
Time (min)
Ids (μA)
Vds (V)
0min
10min
20min
30min
60min
(c)
rGO
Al2O3
(1)
(2)
(3)
AuNPs
(4)
(5)
E.coli
Anti-E.coli
(d)
ISD (μA)
Time(s)
2 X 10^8
(e)
%R
[E. coli]/ (CFU/ mL)
(f)
Pretreatment
Cultured bacteria
Blood collection
Gram-negative Bacteria (GNB)
Gram-positive bacteria (GPB)
Gate
Outlet
Drain
Dual-channel
Source
Inlet
ABX-GMFET
Sensitivity
GPB detection
GNB detection
Time (s)
GPB
GPB channel
GNB channel
Vancomycin
GNB
Magainin I
(g)
GPB channel
Detection mechanism
GNB channel
D-ala-D-ala
membrane

for an anti-*E. coli* 0157 antibody, 20 cells per mL for an anti-S. typhimurium antibody, 8 cells per mL for an anti-*B. cereus* antibody, and 2 cells per mL for an anti-*L. monocytogenes* antibody were found.[17]

4.6.3 Graphene FETs (G-FETs)

Graphene also stood as a promising material for *E. coli* detection. Anti-*E. coli* antibodies were immobilized onto a graphene film developed by the CVD technique to specifically detect *E. coli* bacteria using linker molecules (1-pyrenebutanoic acid succinimidyl ester) whose pyrene group interacted with the graphene surface through strong π–π interaction and the succinimidyl ester group that reacted covalently with the amino group on the antibody. Tween 20 passivates the uncoated graphene area to prevent non-specific binding (Figure 4.6(a)). The device conductance increases significantly after exposure to *E. coli* bacteria at concentrations as low as 10 CFU per mL (Figure 4.6(b)). The increase in graphene conductance with respect to time is due to an increase in hole density driven by the highly negatively charged bacterial wall since the graphene FET is operated in the p-type zone ($V_g = 0$ V).[18]

Drinking water contaminated with *Escherichia coli* can be harmful to humans, especially if medical assistance is delayed after the development of the first symptoms. *E. coli* strains produce Shiga toxins that are lethal to human life; therefore, early detection and real-time surveillance are critical. Current approaches, such as PCR and colony counting, lack the sensitivity required to detect bacteria at low concentrations. A reduced graphene oxide (rGO) FET modified with *E. coli*-specific antibodies with the device architecture anti-*E. coli*/AuNPs/Al_2O_3/rGO/FET recognize specific bacteria in river water. rGO is coated with Al_2O_3 to passivate the defects in rGO and then tagged with AuNPs that serve as anchoring sites as gold is generally considered an ideal element for immobilization of bio-recognition molecules

Figure 4.6 (a) Anti-*E. coli* antibody-functionalized graphene-FET for the detection of *E. coli* (inset: SEM image of an *E. coli* attached to antibody-functionalized graphene) and (b) output curve with time variation (inset: conductance variation with time). Reproduced from ref. 18 with permission from the Royal Society of Chemistry. (c) A schematic demonstrating the various phases in the production of an anti-*E. coli*/AuNPs/Al_2O_3/rGO FET sensor, (d) detection of *E. coli* (CFU per mL) in water in real time, and (e) biosensor response curve to *E. coli* addition. Reproduced from ref. 19 with permission from Elsevier, Copyright 2018. (f) Schematic of the entire ABX-GMFET system with dual channels for separating GPB and GNB in pre-treated cultured samples and (g) diagram illustrating the ABX-GMFET in dual-channel microfluidics. The enlarged diagrams illustrate the detection mechanisms for GPB (left) and GNB (right) in buffer solution. Reproduced from ref. 20 with permission from Elsevier, Copyright 2020.

(Figure 4.6(c)). When exposed to a liquid sample containing *E. coli*, the conductivity of the rGO channel in the FET is modulated which accounts for *E. coli* detection. A linear decrease in source–drain current with bacterial addition is observed, which is due to the ions moving quickly from the liquid drop on the device surface to the contact area. The sensor can detect the species in the range of 10^3 to 10^5 CFU per mL (Figure 4.6(d) and (e)).[19]

A POCT device using a hybrid microfluidic with an active graphene material is a good alternative for the early detection of *E. coli*. This portable and real-time monitoring device employs a graphene micro-patterned FET coupled with dual antibiotic bioprobes. It achieved a low LOD of 10^0 CFU per mL. A complete ABX-GMFET system with dual channels for differentiating GPB (Gram-positive bacteria) and GNB (Gram-negative bacteria) in pre-treatment cultured samples is presented in Figure 4.6(f). Electrostatic gating effects explain the sensing mechanisms: (i) in the GNB channel, an interaction between the negative charge of the LPS layer in the outer membrane and the positive charge of the cationic AMPs (right figure), and (ii) in the Gram-positive channel, an interaction between the D-alanyl-D-alanine moiety at the lipid peptidoglycan terminus and vancomycin (left figure). The FETs are created using chemical vapor deposition (CVD) of graphene, and a polydimethylsiloxane microfluidic chip was subsequently linked to the bioprobe-conjugated GFET. *E. coli*, *Salmonella enterica*, *Staphylococcus aureus*, and *Enterococcus faecium* were the four bacterial species examined. *E. coli* and *S. enterica* were recognized using specialized vancomycin bioprobes for quick detection, whereas *S. aureus* and *E. faecium* were detected using magainin I bioprobes. A simplified detection mechanism is shown in Figure 4.6(g).[20]

4.6.4 Transition-metal Dichalcogenide FETs (TMDC-FETs)

A highly sensitive and selective 2D MoS_2 FET detects streptavidin bacteria. This FET is designed using a mechanically exfoliated MoS_2 adhered on top of Si/SiO_2 with an ionic liquid as the source of the gate voltage. A dielectric layer of hafnium oxide (HfO_2) serves the gating purposes (Figure 4.7(a)). The device is pH dependent for selective sensing and achieves sensitivity values as high as 713 for a pH change from 4 to 5. The dielectric layer HfO_2 is functionalized with biotin and a molecule of 3-aminopropyl(triethoxy)silane (APTES) in order to detect the streptavidin bacterium. The output characteristics of the device are shown in Figure 4.7(b). The observation follows that the addition of streptavidin solution (10 μM in 0.01 PBS) leads to a decrease in current (red curve) due to the negative charge of the protein, as the pH of the solution is more than the isoelectric point (pI) of streptavidin. A similar current in pure buffer and streptavidin solution was observed in an unfunctionalized device (with only APTES), confirming there are no false signals. In the subthreshold region, the sensitivity of the MoS_2-based FET biosensor exhibits a LOD of 100 fM of streptavidin at 196 μA.[21]

Employing MoS_2, a very sensitive and dependable biosensor on a suspended MoS_2 channel FET can identify *E. coli* bacteria. A single atomic layer of MoS_2

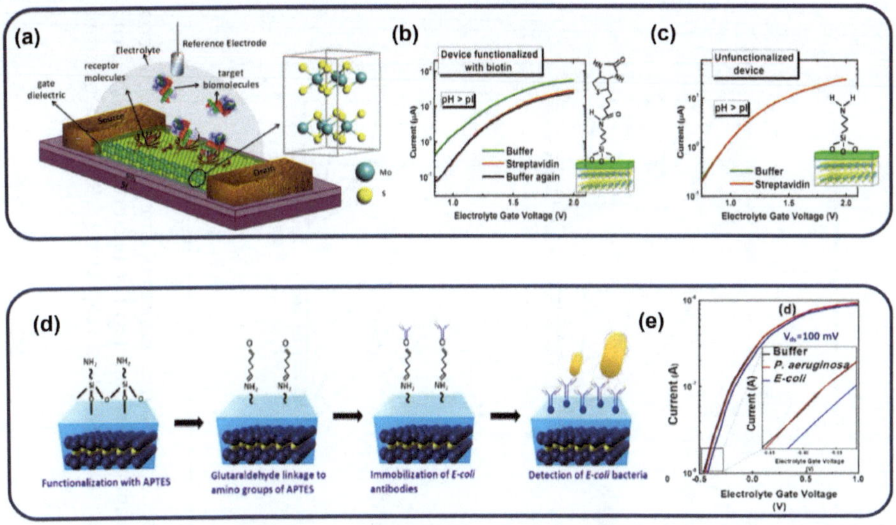

Figure 4.7 (a) Device architecture of the MoS_2 bio-FET, electrical characteristics of (b) biotin-functionalized and (c) unfunctionalized devices. Reproduced from ref. 21 with permission from American Chemical Society, Copyright 2014. (d) Schematic of *E. coli* bacterial detection and (e) I_D–V_{LG} characterization of *E. coli* and *P. aeruginosa* (100 CFU per mL) compared with PBS buffer solution. Reproduced from ref. 22 with permission from Elsevier, Copyright 2021.

flakes is deposited on top of a nano-spaced electrode using a unique dry stamping process. PMMA serves as a sacrificial layer to deposit the 2D material onto the nanogap electrode which is then removed by soaking the device in acetone that dissolves the PMMA without harming the transferred 2D. A dielectric layer of HfO_2 is then deposited on top of the suspended device to passivate the metal electrodes as well as prevent direct contact with the electrolyte or biomolecules. For *E. coli* antibody immobilization on the HfO_2 surface, the suspended device was absorbed into APTES for 2 hours which ensured APTES binding to the dielectric surface by ether (–O–) linkage. Subsequently, the HfO_2–APTES substrate was then immersed in a glutaraldehyde solution for approximately 1 hour for anti-*E. coli* immobilization. An ultralow power (V_{DS} = 100 mV) suspended MoS_2 FET biosensor then selectively detects *E. coli* bacteria at concentrations as low as 10 CFU per mL by altering the conductance of the channel by 9%. The current in the MoS_2 channel decreases due to an increased hole concentration on the dielectric surface caused by the strongly negatively charged bacterial wall (Figure 4.7(d) and (e)).[22]

4.6.5 Organic FETs (OFETs)

A highly sensitive π-conjugated organic molecule based FET (OFET) detects bacteria without the need for surface modification or immobilization. An OFET device consists simply of a tri-layer dielectric system of aluminium

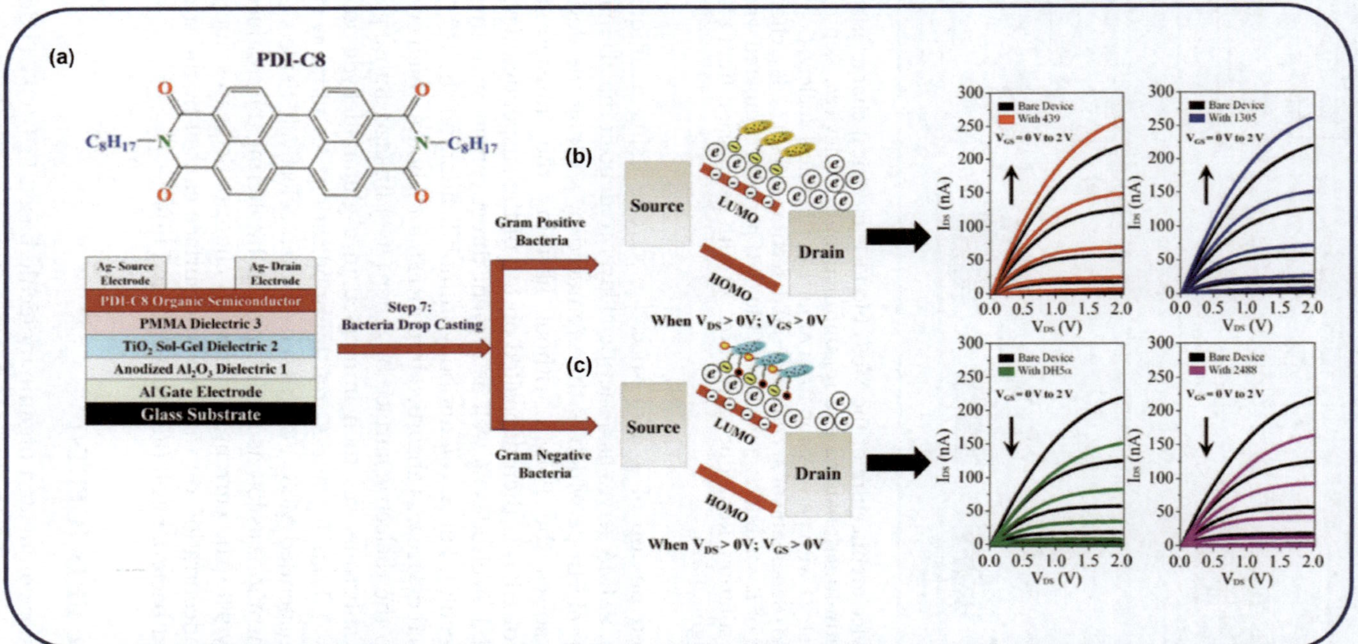

Figure 4.8 (a) Device architecture with PDI-C8 as the active semiconducting material, (b) detection mechanism of Gram-positive bacteria (*Enterococcus faecalis* (439) and *Bacillus cereus* (1305)), and (c) detection mechanism of Gram-negative bacteria (*Escherichia coli* (DH5α) and *Pseudomonas aeruginosa* (2488)) with the corresponding output characteristics, respectively. Reproduced from ref. 23 with permission from the Royal Society of Chemistry.

oxide (Al_2O_3), titanium oxide (TiO_2), poly(methylmethacrylate) (PMMA) polymer, and the n-type *N,N*-dioctyl-3,4,9,10-perylenedicarboximide (PDI-C8) organic semiconducting material (OSC). The Al_2O_3 dielectric thin film is formed by the anodization technique, the TiO_2 film (coated using a sol–gel solution of TiO_2), and PMMA (coated by dissolving in anisole solution) by a simple spin coating technique and PDI-C8 by the thermal vacuum evaporation technique over a glass substrate. This OFET device is operated at an ultra-low voltage of ~2 V (Figure 4.8(a)) to detect Gram-positive (*Enterococcus faecalis* (439) and *Bacillus cereus* (1305) and Gram-negative (*Escherichia coli* (DH5α) and *Pseudomonas aeruginosa* (2488)) bacteria. The mechanism followed that at $V_{DS} > 0$ and $V_{GS} = 0$, no current (I_{DS}) flowed in the bare device. With increasing V_{GS}, the current I_{DS} accumulated and flowed through the channel/dielectric interface. When the Gram-positive bacterium-containing solution was drop-cast on the OSC, a channel formed through the interface under the conditions of $V_{DS} > 0$ and $V_{GS} = 0$ (in the absence of an external field), which resulted in a small current I_{DS} in the system. With further V_{GS} increase, a stronger channel formed and the current I_{DS} increased due to the combined effect of the external gate bias and Gram-positive bacteria (Figure 4.8(b)). Teichoic acid, a negatively charged substance found in the cell walls of Gram-positive bacteria, was responsible for the extra channel that was created (in the absence of an electric field). However, the presence of Gram-negative bacteria created a resistive channel that decreased the current I_{DS}. The current I_{DS} further decreased with increasing V_{GS} (Figure 4.8(c)). This striking difference in the current behaviour of the OFET device was because the Gram-negative bacteria have an outer membrane that contains biological components (lipids, proteins, lipopolysaccharides, *etc.*), while Gram-positive bacteria are devoid of the outer membrane with only teichoic acid highly organized in their cell wall in comparison to Gram-negative bacteria. Due to this structural difference and asymmetry in the cell wall of bacteria, the I_{ON}/I_{OFF}, mobility, and V_{th} show striking dissimilarities and account for their rapid, label-free, and easy detection with a LOD value of 10^3 CFU per mL. Table 4.1 summarises the changes in device parameter values on exposure to the bacterial solution.[23]

Table 4.1 Summary of the changes in OFET device parameter values on exposure to bacteria.[a]

Device	Name of bacteria	V_{Th} (V)	μ ($cm^2 V^{-1} s^{-1}$)	I_{ON}/I_{OFF}
Bare	—	0.20 (±0.03)	0.30 (±0.02)	10^4
With Gram-positive bacteria	*Enterococcus faecalis* (439)	0.50 (±0.04)	0.50 (±0.03)	10^5
	Bacillus cereus (1305)	0.47 (±0.03)	0.52 (±0.04)	10^5
Gram negative bacteria	*Pseudomonas aeruginosa* (2488)	0.49 (±0.02)	0.21 (±0.03)	10^3
	Escherichia coli (DH5α)	0.52 (±0.03)	0.22 (±0.02)	10^3

[a]The detection of bacteria using several FET-based sensors is summarised in Table 4.2.

Table 4.2. Summary of the active materials, receptors, host materials, and LODs for FET-based bacterial sensors.

Bacterium type	Active material	Receptor	Host	Limit of detection (LOD)	Ref.
E. coli	SWCNTs	RNA-based *E. coli* aptamers	*E. coli*	10^7 CFU per mL	13
E. coli O157:H7	SWCNT	Anti-*E-coli O157:H7* antibody	*E. coli O157:H7*	8.2×10^2 CFU per mL	14
Salmonella infantis	CNTs	Anti-Salmonella antibody	Somatic O antigen (O-antigen) and flagellar H antigen (H-antigen)	100 CFU per mL	15
Borrelia burgdorferi (Lyme)	SWCNTs	Anti-p41 antibody	Flagellar antigen	1 ng mL^{-1}	16
Escherichia coli 0157	SiNWs	Polystyrene microbeads	Anti-*E. coli* 0157 antibody	80 cells per mL	17
Salmonella typhimurium			Anti-*S. typhimurium* antibody	20 cells per mL	
Listeria monocytogenes			Anti-*L. monocytogenes* antibody	2 cells per mL	
Bacillus cereus			Anti-*B. cereus* antibody	8 cells per mL	
Staphylococcus aureus			Anti-*S. aureus* antibody	124 cells per mL	
Escherichia coli (*E. coli*)	Graphene	Anti *E. coli*	*Escherichia coli* (*E. coli*)	10 CFU per mL	18
Escherichia coli (*E. coli*)	Reduced graphene oxide (rGO)	AuNPs (gold nanoparticles)	*Escherichia coli (E. coli)*	—	19
E. coli, Salmonella enterica, Staphylococcus aureus, and *Enterococcus faecium*	Graphene	Vancomycin and magainin I	*E. coli* and *S. enterica*	10^0 CFU per mL (1–9 CFU per mL)	20
Streptavidin bacteria	MoS_2	3-Aminopropyl(triethoxy) silane (APTES)	Streptavidin bacterial protein	100 fM	21
E. coli	MoS_2	3-Aminopropyl(triethoxy) silane (APTES)	*E. coli*	10 CFU per mL	22

4.7 Virus Detection by FETs

4.7.1 Carbon Nanotube FETs (CNT-FETs)

A DNA sensor based on CNT-FETs identifies the influenza A virus. A DNA sensor typically consists of two components: a DNA strand that may interact or hybridize with the target to affect the biological characteristics of the sensor surface/electrolyte interface and a transducer that can turn an input stimulus into an electrical signal. In order to create the CNFET sensor, single-stranded DNA probes that are complimentary to the target RNA sequence of the influenza A virus were immobilized on the surface of the CNTs using a straightforward physisorption technique. Figure 4.9(a) illustrates ssDNA probes on the CNTs hybridized with the influenza A virus target RNA sequence causing changes in the electrical conductivity of CNTs that are detected by the CNT-FET sensor. Moreover, this sensor possessed a wide linear detection range from 1 pM to 10 nM with a very low detection limit of 1 pM (Figure 4.9(b)), which is a promising method for viral detection.[24]

Hepatitis C virus (HCV) is detected by developing electrochemical gate-controlled single walled CNT-based FETs (SWNT-FETs) where the SWNT layer is fabricated by the CVD technique. The non-covalent interactions between the peptide nucleic acid (PNA) probes and the SWNTs result in robust functionalization, involving π–π stacking interactions and charge transfer of the amine group in the PNA molecule, resulting in an increase in the transconductance (g_m) with increasing HCV RNA concentration (Figure 4.9(c) and (d)). This biosensor detects label-free HCV RNA sequences within a minute with a LOD of 0.5 pM in real time.[25]

The SWCNT semiconducting material incorporated in a FET device has been found useful in the detection of the COVID-19 (SARS-CoV-2) virus by functionalizing it with an anti-SARS-CoV-2 spike protein antibody (SAb) and an anti-nucleocapsid protein antibody (NAb) to access the presence of two SARS-CoV-2 structural proteins: spike protein (S antigen, SAg) and nucleocapsid protein (N antigen, NAg), illustrated in Figure 4.9(e). Using the dielectrophoresis method, high-purity SWCNTs are deposited on the Si/SiO_2 surface and the antibody is linked covalently by a carbodiimide (EDC)/*N*-hydroxy succinimide (NHS) linker. Upon antibody interaction with the antigens, a positive threshold voltage shift occurs due to the electrostatic gating effect (Figure 4.9(f) and (g)). The LOD is 0.55 $fg\,mL^{-1}$ and 0.016 $fg\,mL^{-1}$ for SAg and NAg in calibration samples, respectively.[26] Another means to recognize the SARS-CoV-2 virus is by detecting the surface spike protein (S1) subunit of the S protein of the virus by immobilizing the SARS-CoV-2 S1 antibody on the CNT surface supported on Si/SiO_2. The antibody immobilization occurs through a non-covalent interaction with the linker molecule 1-pyrenebutanoic acid succinimidyl ester (PBASE), which enables the transfer of a good number of electrons to the biosensor decreasing the drain current (Figure 4.9(i)). Upon the

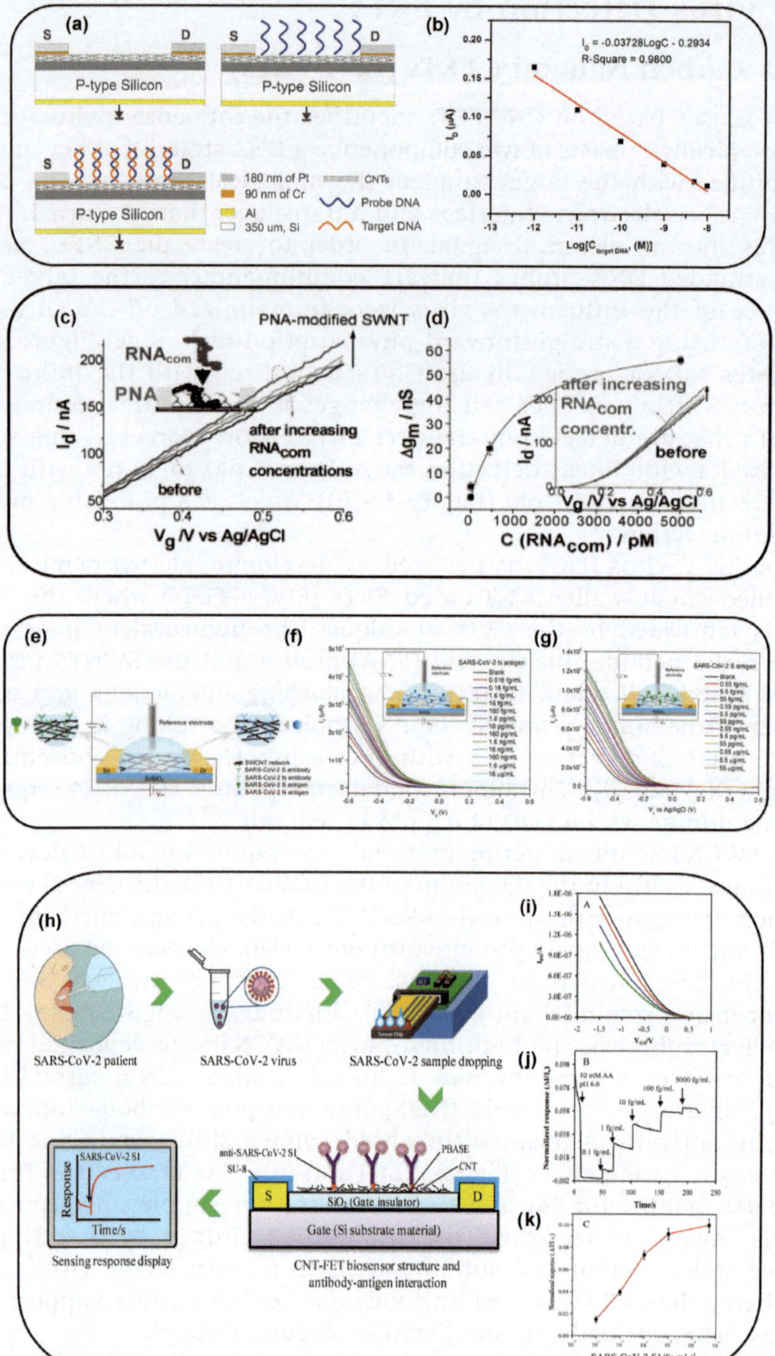
(a)
S
D
P-type Silicon
180 nm of Pt
90 nm of Cr
Al
350 um, Si
CNTs
Probe DNA
Target DNA
(b)
$I_D = -0.03728 \text{LogC} - 0.2934$
R-Square = 0.9800
I_D (μA)
Log[$C_{target\ DNA}$ (M)]
(c)
PNA-modified SWNT
RNA_{com}
PNA
after increasing RNA_{com} concentrations
before
I_d / nA
V_g /V vs Ag/AgCl
(d)
after increasing RNA_{com} concentr.
before
Δg_m / mS
C (RNA_{com}) / pM
(e)
Reference electrode
(f)
SARS-CoV-2 N antigen
(g)
SARS-CoV-2 S antigen
(h)
SARS-CoV-2 patient
SARS-CoV-2 virus
SARS-CoV-2 sample dropping
anti-SARS-CoV-2 S1
PBASE
SU-8
CNT
S
D
SiO_2 (Gate insulator)
Gate (Si substrate material)
CNT-FET biosensor structure and antibody-antigen interaction
SARS-CoV-2 S1
Response
Time/s
Sensing response display
(i)
(j)
(k)
SARS-CoV-2 S1/fg mL^{-1}

antigen–antibody interaction, the drain current further decreases due to electrostatic gating and the Schottky barrier effect. The normalized response curve ($\Delta I/I_o$) measured against each concentration of the SARS-CoV-2 S1 antigen concludes that the detection process is very rapid (within minutes) with a LOD of 4.12 fg mL^{-1} and a sensitivity of 1 pg mL^{-1} (Figure 4.9(j) and (k)). The detection process of the COVID virus is illustrated in Figure 4.9(h).[27]

4.7.2 Nanowire FETs (NW-FETs)

Mo-doped ZnO NW FET-based biosensing can be used for examining the hepatitis virus for medical clinical diagnosis. The device is fabricated using a p-silicon substrate with a thin layer coating of Mo and Zn, followed by heating in a tubular furnace with oxygen and argon gas and channel formation by lithography. The ssDNA is drop-cast onto the nanowire electrode in an argon gas atmosphere for the FET experiment. Target DNA is added to a buffer solution comprising sodium chloride and sodium citrate in order to execute the hybridization process, and this process is carried out on an electrode adapted for the probe. Carrier concentration changes are mostly caused by the lethal thiol group linked to NWs. This PVD-grown Mo-doped FET has the potential for a quick diagnosis of hepatitis B virus in the concentration range of 1 pM to 10 mM (Figure 4.10(a)). The good ability of this sensing element makes it preferable to combining with the PCR test due to the distinction between true positive (TP), true negative (TN), false positive (FP) and false negative (FN) with a detection limit (LOD) of 1 pM.[28]

The SiNW FET combined with RT-PCR technology is promising for quick detection of the most severe dengue virus variety dengue serotype 2 (DEN-2)

Figure 4.9 (a) Schematic of the CNFET-based DNA sensor and (b) the detection range of the DNA sensor. Reproduced from ref. 24 with permission from Elsevier, Copyright 2017. (c) I_d–V_g curves of an SWNT-FET modified with PNA after exposure to RNA_{com} at different concentrations and (d) calibration curves of Δg_m *versus* RNA_{com} concentration. Reproduced from ref. 25 with permission from the Royal Society of Chemistry. (e) Schematic of the SWCNT-FET for SARS-CoV-2 detection, and characteristic curves of (f) the SAb-functionalized SWCNT FET device upon exposure to various concentrations of SARS-CoV-2 SAg and (g) the NAb-functionalized SWCNT-FET device upon exposure to various concentrations of SARS-CoV-2 NAg. Reproduced from ref. 26 with permission from American Chemical Society, Copyright 2021. (h) Schematic diagram of the CNT-FET biosensor and SARS-CoV-2 S1 testing steps. (i) I_{DS} *vs.* V_g characteristics of the bare CNT-FET (black line) and SARS-CoV-2 S1(green line), (j) the characteristic curve of the real-time response after dropping the increasing concentration of the SARS-CoV-2 S1 antigen and (k) the calibration curve of the biosensor response *vs.* applied concentration of the target SARS-CoV-2 S1 antigen. Reproduced from ref. 27, https://doi.org/10.1016/j.bioelechem.2021.107982, under the terms of the CC BY 4.0 license, https://creativecommons.org/licenses/by/4.0/.

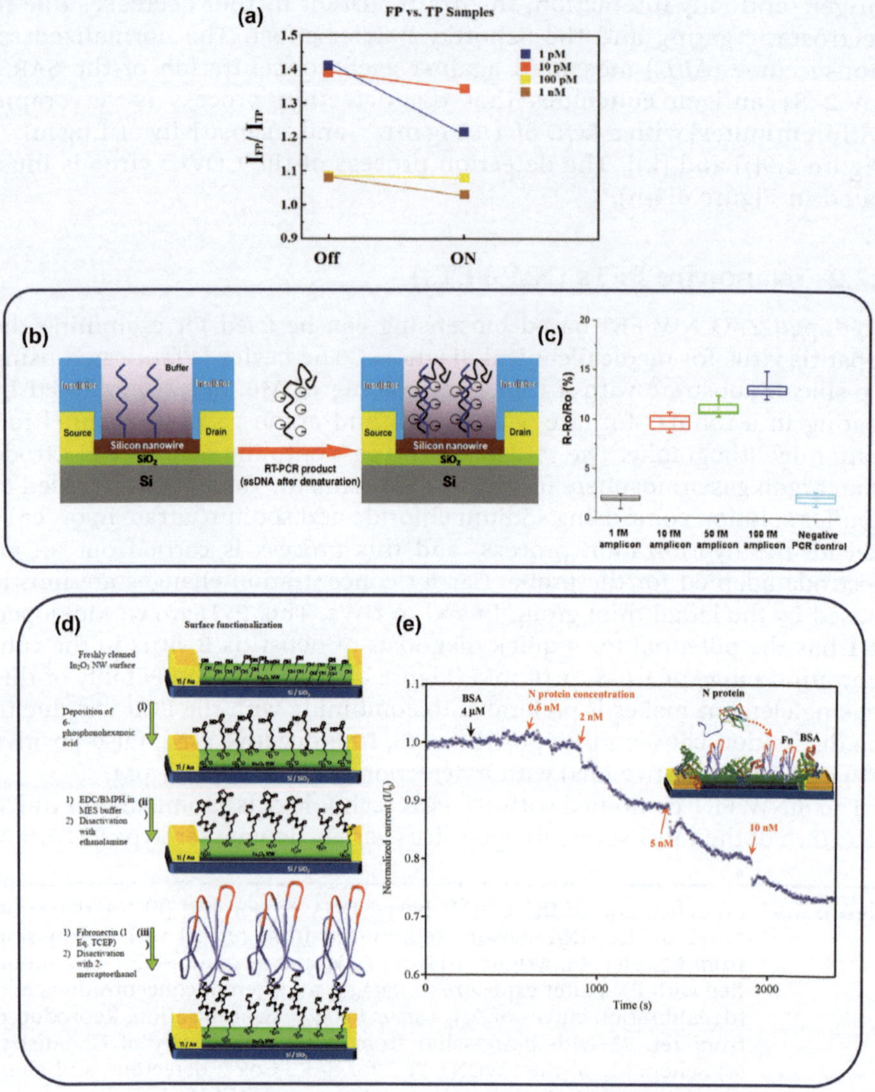

Figure 4.10 (a) The current ratio of FP and TP samples at different concentrations. Reproduced from ref. 28 with permission from Elsevier, Copyright 2022. (b) Schematic showing the hybridization of the DEN-2 RT-PCR product to the PNA-functionalized SiNW sensor and (c) the resistance change *versus* concentration plot of the RT-PCR product of the DEN-2 virus. Reproduced from ref. 29 with permission from Elsevier, Copyright 2010. (d) Overall procedures for surface functionalization of In_2O_3 nanowires and (e) normalized electrical output *versus* time of the single operating device. Reproduced from ref. 30 with permission from American Chemical Society, Copyright 2009.

utilizing electrical detection. By using thermal oxidation of the silicon layer followed by n^+ doping, silicon nanowires are constructed on silicon oxide insulators. Biomolecules are then immobilized on the surface of Si nanowires using the facilitator APTES. Then, amine-terminated aminopropyltriethoxysilane (APTES) and PNA are joined together using the bifunctional linker glutaraldehyde, leaving the other end free. Reverse transcription polymerase chain reaction (RT-PCR) is used to carry out the nucleotide sequence amplification for DEN-2. By applying heat and cooling during the hybridization procedure, the PCR product of DEN-2 transforms into single-stranded DNA (Figure 4.10(b)). A change in resistance before and after the hybridization process serves as the basis for the sensing mechanism. The results of RT-PCR show that while double-stranded DNA (dsDNA) is negatively charged, PNA is electrically neutral. These negatively charged components deplete the FET depletion area, increasing resistance. By using various PCR output concentrations, the detection limit of SiNWs has been studied. The change in resistance with respect to the concentration of the RT-PCR product of DEN-2 is presented in Figure 4.10(c). This rapid detection methodology outperforms the traditional method because of its low detection limit of analyte concentration (less than 10 fM), greater sensitivity, and portable integrated circuit.[29]

Using a capturing element on their surface, NWFET biosensors have been used to identify target biological molecules for over a decade. Antibodies, small ligands, and antibody mimic proteins (AMPs) are included in the capturing components. By using an *in vitro* selection approach, the latter capturing components are evolved. Easy-to-prescribe AMPs can be used to enhance recognition qualities including stability and chemical affinity. Furthermore, they have other benefits that make them an effective capturing agent, including compact dimensions and high production at low cost. Using a metal oxide NW, the SARS virus could be detected by targeting the SARS virus *N*-protein. To achieve this In_2O_3 is used as a base material. A fibronectin-based binding agent that can specifically detect the SARS biomarker N protein is built on In_2O_3 nanowires by several surface functionalization processes (Figure 4.10(d)). The surface of the In_2O_3 NW is positively charged, while the N protein is negatively charged due to the serine group. The conductive current in the FET decreases as a result of the attraction force between two opposite charges. As the N protein concentration rises, the device's conductivity gradually declines, as shown in Figure 4.10(e). This NWFET had an on/off ratio of 4.6×10^3 and a transconductance value of 3.6 s. These beneficial qualities guarantee the device's stability under measurement conditions.[30]

4.7.3 Graphene FETs (G-FETs)

A microfluidic integrated rGO transistor-based DNA sensor can detect the avian influenza virus. It utilizes a flow-through method and can achieve a minimal detection limit of 5 pM more quickly as compared to electrochemical or optical sensing systems because of the extended long probe immobilization strategy. In this method, the stability of DNA probes after hybridization and their effects on the sensitivity of rGO transistors in flowing environments were explored using a variety of DNA probe immobilization techniques, including extended long capture probe *via* π–π stacking, short capture probe *via* stacking and covalent immobilization *via* a linker. For assessing the stability of probes with various immobilization techniques, a PDMS microfluidic chip integrated with a rGO transistor was created (Figure 4.11(a)). The shielding of nucleobases within the negatively charged dsDNA phosphate backbone causes short capture probes with entirely matching sequences to detach and be washed away from the rGO surface in the flowing environment. Extended long capture probes, on the other hand, will continue to adhere to the rGO surface due to the adhesion force from the immobilization section with the rGO surface in the flowing environment, which can guarantee the stability in the flow-through method in microfluidic devices. When the gate of a rGO FET was immobilized with LP, there is a reduction in conductance, indicating that the LP was successfully immobilized on the rGO surface. Figure 4.11(b) shows that the conductance is further lowered by hybridization with the target cDNA of the influenza virus resulting from n-doping due to DNA adsorption on rGO. The relative conductance change of rGO transistors with SP, LP, and pyrene-butyric acid linkers at various cDNA concentrations (10 pM to 1 M) in the same microfluidic flowing environment is depicted in Figure 4.11(c). Compared to absolute conductance change (G), relative conductance change ($\Delta G/G_0$) is a more accurate indicator of the device's sensitivity, as it eliminates the effect of the initial conductance G_0 variation. Clearly, rGO transistors modified with LP have a significantly larger conductance change and greater sensitivity in a microfluidic circulating environment than those modified with SP and pyrene-butyric acid linker.[31]

Ebola virus detection is addressed by developing a FET with 2D graphene utilizing the specific Ebola glycoprotein (EGP) antigen as the capturing agent that is immobilized on the rGO surface by the AuNP anchoring group (Figure 4.11(d)) with a LOD of 1 ng mL^{-1} in real time with the drain current decrease with increasing EGP concentration (Figure 4.11(e) and (f)). Furthermore, the FET biosensor displayed high selectivity towards Ebola virus antigens, with negligible responses from non-specific proteins.[32]

To serve as a platform for the detection of the SARS-CoV-2 virus, a graphene-based biosensing device functionalized with a SARS-CoV-2 spike antibody (COVID-19 FET sensor) stood as a promising strategy. Using 1-pyrenebutyric acid *N*-hydroxysuccinimide ester (PBASE) as the probe linker, the SARS-CoV-2 spike antibody was mounted onto the constructed device. The electrical measurements determine if the SARS-CoV-2 spike antibody was present on the graphene surface. The graphene device current–voltage (I–V) curves are shown in Figure 4.11(g), both before and after antibody attachment, in the range of 0.1 to +0.1 V. The slopes of dI/dV reduce during PBASE functionalization and antibody immobilization on the

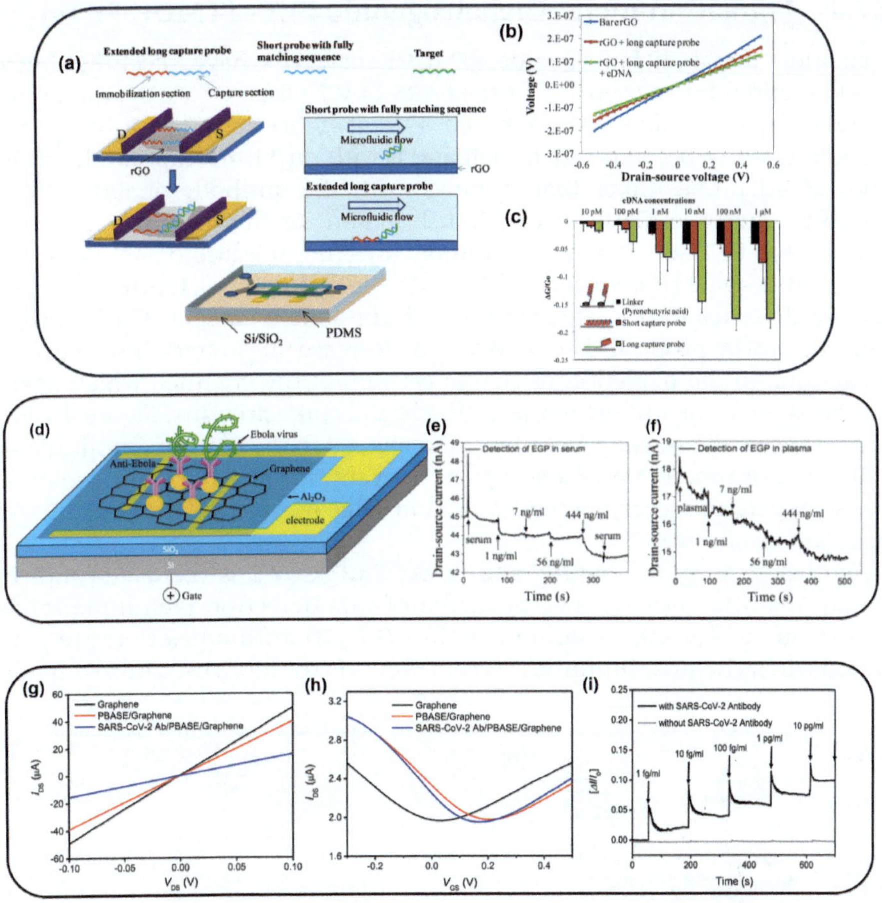

Figure 4.11 (a) Mechanism of the extended long capture probe immobilization method to maintain probe stability in a flowing environment, (b) direct current measurement for cDNA, and (c) after the washing stage the relative conductance change ($\Delta G/G_0$) of rGO transistors. Reproduced from ref. 31 with permission from Elsevier, Copyright 2017. (d) The schematic diagram of the rGO-based FET biosensor, (e) EGP in 0.01 human serum suspension, and (f) EGP in 0.01 human plasma suspension. Reproduced from ref. 32, https://doi.org/10.1038/s41598-017-11387-7, under the terms of the CC BY 4.0 license, http://creativecommons.org/licenses/by/4.0/. (g) Output curve and (h) transfer curve depicting the sensing response by a change in I_{DS} and (i) response of the COVID-19 FET in real time to SARS-CoV-2 antigen protein in PBS. Reproduced from ref. 33 with permission from American Chemical Society, Copyright 2020.

graphene channel. The pyrene group p-doping impact resulted in a noticeable positive shift after PBASE functionalization. However, the transfer curve shifted negatively, indicating that the antibody's positive charge had an n-doping effect on graphene after it was immobilized (Figure 4.11(h). The sensor had a LOD of 1 fg mL^{-1} for SARS-CoV-2 spike protein in PBS (Figure 4.11(i)).[33]

4.7.4 Transition-metal Dichalcogenide FETs (TMDC-FETs)

Transition metal dichalcogenide (TMDC) materials have recently shown good potential for the sensing of the SARS-CoV-2 virus. Of TMDC materials, a tungsten diselenide (WSe_2) based FET is fabricated by utilizing the monolayers of WSe_2 and functionalizing it with an 11-mercaptoundecanoic acid (MUA) probe linker that is intact with the antibody of SARS-CoV-2 providing the base platform for viral detection. In this method, the ionic liquid-gated FET serves as a medium for injecting the liquid sample of the SARS-CoV-2 spike protein for real-time detection (Figure 4.12(a)). The sensing profile relies on the change of the drain–source current. On immobilizing the MUA probe linker to WSe_2, a decrease in source–drain current occurs due to the depletion of charge carriers in the channel, which again increases upon the linking of the antibody and spike proteins. Figure 4.12(b) and (c) shows the sensing technique and change in the current amplitude as a function of time upon increasing the concentration of the spike proteins. The sensor detects a very low concentration value of 1 fg mL^{-1} for SARS-CoV-2 spike protein in PBS.[34]

Another strategy to quantify and detect SARS-CoV-2 is the use of molybdenum disulfide (MoS_2). The principle of this detection technique relies on the quantification of immunoglobin G (IgG) antibodies that are generated when the host immune system responds to the virus. The exfoliated

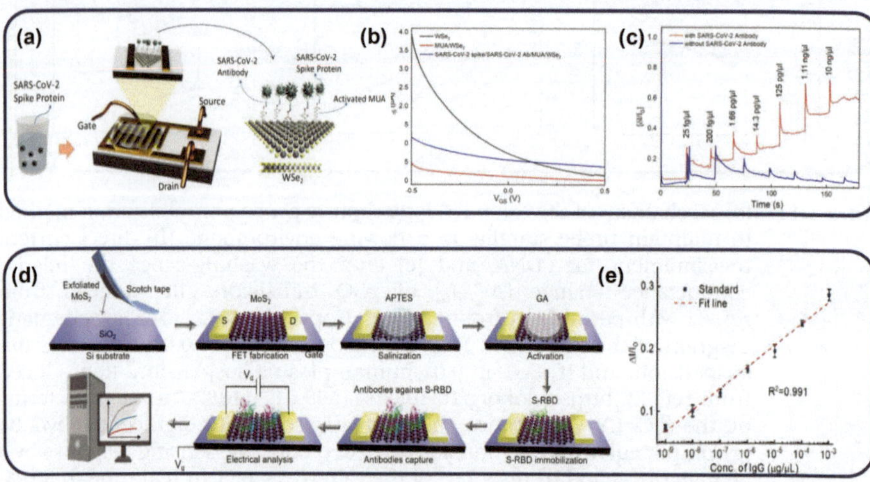

Figure 4.12 (a) Diagrammatic representation of the manufactured monolayer WSe_2-based FET COVID sensor, (b) corresponding I_{DS}–V_{GS} curves of the sensor with $V_{DS} = 1$ V, and (c) real-time response to varying SARS-CoV-2 antigen spike protein concentrations. Reproduced from ref. 34 with permission from American Chemical Society, Copyright 2021. (d) Simplified diagram of MoS_2-FET fabrication, functionalization, and recognition and (e) the concentration measurement device's fit line of I/I_0 *vs.* concentration. Reproduced from ref. 35 with permission from IOP Publishing, Copyright 2014.

Table 4.3 Summary of the active materials, receptors, host materials and LODs for FET based virus sensors.

Virus type	Active material	Receptor	Host	Limit of detection (LOD)	Ref.
SARS-CoV-2	SWCNTs	Anti-SARS-CoV-2 spike protein antibody (SAb) and anti-nucleocapsid protein antibody (NAb)	Spike protein (S antigen, SAg)/nucleocapsid protein (N antigen, NAg)	0.55 fg mL^{-1} and 0.016 fg mL^{-1}	26
SARS-CoV-2	SWCNTs	SARS-CoV-2 S1 antibody	SARS-CoV-2 S1 antigen	4.12 fg mL^{-1}	27
Influenza A	CNTs	ssDNA	Influenza RNA	1 pM	24
Hepatitis C Virus (HCV)	SWCNTs	Peptide nucleic acid (PNA)	HCV RNA	0.5 pM	25
Severe acute respiratory syndrome (SARS) virus	In_2O_3 nanowires	Fibronectin	SARS N protein	—	30
Hepatitis B	Mo doped ZnO	ssDNA probe	dsDNA	1 pM	28
Dengue serotype 2	SiNWs	Peptide nucleic acid (PNA)	ssDNA	Less than 1 fM	29
Avian influenza virus	Reduced graphene oxide (rGO)	Pyrenebutyric acid	cDNA of the influenza virus	5 pM	31
Ebola virus	Graphene oxide	AuNPs (gold nanoparticles)	Ebola glycoprotein (EGP) antigen	1 ng mL^{-1}	32
COVID-19 causative virus (SARS-CoV-2)	Graphene sheet	Pyrenebutyric acid *N*-hydroxysuccinimide ester (PBASE)	SARS-CoV-2 spike protein	1 fg mL^{-1}	33
SARS-CoV-2 virus	WSe_2	11-Mercaptoundecanoic acid (MUA)	SARS-CoV-2 spike proteins	1 fg mL^{-1}	34
SARS-CoV-2 virus	MoS_2	Immunoglobin G (IgG) antibodies	Spike protein receptor binding domain (S-RBD)	10^{-9} µg µL^{-1}	35

MoS_2 on a Si/SiO_2 substrate is deposited as the source and drain electrodes are immersed into a solution of triaminopropyltriethoxy-silane (APTES) involving the cross-linking process including hydrolysis and silanol condensation reactions. The siloxane bond is then formed by the reaction of the surface –OH group of MoS_2 with the silanol of APTES (Figure 4.12(d)). The resulting structures are treated with glutaraldehyde solution and later modified by incubating spike protein receptor binding domain (S-RBD). The S-RBD reacted to the IgG to detect SARS-CoV-2 by showing a change of I_d ($\Delta I = (I - I_0)/I_0$) with varying IgG concentrations as low as 10^{-9} $\mu g\,\mu L^{-1}$ (Figure 4.12(e)).[35]

The detection of several viruses using various FET-based sensors is summarised in Table 4.3.

4.8 Conclusion

In summary, this chapter provides an overview of FET-based pathogen biosensors, their characterization parameters, and the advancement of early label-free detection of a broad range of analytes by employing various inorganic, organic, and 2D structured channel materials on the device. The functionalization of biomolecules on the FET surface is now widely understood, and this has played a role in successful molecule binding *via* antigen–antibody interactions and surface chemical modifications. Integrating nanomaterials and nanostructures as sensing channels, such as NWs, CNTs, graphene, MoS_2, WSe_2, and others, significantly improves the performance of FET-type biosensing devices due to their high surface-to-volume ratio, chemical stiffness, and biocompatibility. The developed biosensors proved to be highly sensitive, with a detection limit ranging from fM to μM. These biosensors are expected to have a significant role in diseases and microorganisms in the future.

Abbreviations

DNA	Deoxyribonucleic acid
RNA	Ribonucleic acid
ssDNA	Single-stranded DNA
dsDNA	Double-stranded DNA
cDNA	Complementary DNA
nM	Nanomolar
pM	Picomolar
fM	Femtomolar
nCoVs	Novel coronaviruses
CFU	Colony forming units
GPB	Gram-positive bacteria
GNB	Gram-negative bacteria
AMP	Antibody mimic protein
APTES	3-Aminopropyl(triethoxy)silane

PMMA	Poly(methylmethacrylate)
RT-PCR	Reverse transcription polymerase chain reaction
PNA	Peptide nucleic acid
EGP	Ebola glycoprotein
PBASE	1-Pyrenebutyric acid *N*-hydroxysuccinimide ester
LOD	Limit of detection
POCT	Point-of-care testing
AuNP	Gold nanoparticle
ABX-GMFET	Antibiotic conjugated graphene micropattern field-effect transistors
PDMS	Polydimethylsiloxane
ELISA	Enzyme-linked immunosorbent assay

References

1. A. Bosch, E. Gkogka, F. S. Le Guyader, F. Loisy-Hamon, A. Lee, L. van Lieshout, B. Marthi, M. Myrmel, A. Sansom, A. C. Schultz, A. Winkler, S. Zuber and T. Phister, *Int. J. Food Microbiol.*, 2018, **285**, 110.
2. F. Pérez-Rodríguez, A. Valero, E. Carrasco, R. M. García and G. Zurera, *Trends Food Sci. Technol.*, 2008, **19**, 131.
3. N. Bhalla, P. Jolly, N. Formisano and P. Estrela, *Essays Biochem.*, 2016, **60**, 1.
4. Y. Chen, R. Ren, H. Pu, X. Guo, J. Chang, G. Zhou, S. Mao, M. Kron and J. Chen, *Sci. Rep.*, 2017, **7**, 10974.
5. Y.-C. Syu, W.-E. Hsu and C.-T. Lin, *ECS J. Solid State Sci. Technol.*, 2018, 7, Q3196.
6. A. Poghossian and M. J. Schöning, *Curr. Opin. Electrochem.*, 2021, **29**, 100811.
7. L. Li, S. Wang, Y. Xiao and Y. Wang, *Trans. Tianjin Univ.*, 2020, **26**, 424–440.
8. N. Masurkar, S. Varma and L. Mohana Reddy Arava, *Electrochem*, 2020, **1**, 260.
9. A. Kalita, N. V. V. Subbarao and P. K. Iyer, *J. Phys. Chem. C*, 2015, **119**, 12772.
10. A. Dey, A. Kalita and P. K. Iyer, *ACS Appl. Mater. Interfaces*, 2014, **6**, 12295.
11. M. N. Khatun, A. Dey, N. Meher and P. K. Iyer, *ACS Appl. Electron. Mater.*, 2021, **3**, 3575.
12. S. Vasimalla, N. V. V. Subbarao, M. Gedda, D. K. Goswami and P. K. Iyer, *ACS Omega*, 2017, **2**, 2552.
13. H.-M. So, D.-W. Park, E.-K. Jeon, Y.-H. Kim, B. S. Kim, C.-K. Lee, S. Y. Choi, S. C. Kim, H. Chang and J.-O. Lee, *Small*, 2008, **4**, 197.
14. X. Zhang, D. Wang, D. Yang, S. Li and Z. Shen, *Engineering*, 2012, **4**, 94.
15. R. A. Villamizar, A. Maroto, F. X. Rius, I. Inza and M. J. Figueras, *Biosens. Bioelectron.*, 2008, **24**, 279.
16. M. B. Lerner, J. Dailey, B. R. Goldsmith, D. Brisson and A. T. C. Johnson, *Biosens. Bioelectron.*, 2013, **45**, 163.

17. H.-R. Kim, J.-H. Bong, M.-J. Kang, W.-B. Shim, J.-S. Lee and J.-C. Pyun, *Talanta*, 2023, **255**, 124203.
18. Y. Huang, X. Dong, Y. Liu, L.-J. Li and P. Chen, *J. Mater. Chem.*, 2011, **21**, 12358.
19. B. Thakur, G. Zhou, J. Chang, H. Pu, B. Jin, X. Sui, X. Yuan, C.-H. Yang, M. Magruder and J. Chen, *Biosens. Bioelectron.*, 2018, **110**, 16.
20. K. H. Kim, S. J. Park, C. S. Park, S. E. Seo, J. Lee, J. Kim, S. H. Lee, S. Lee, J.-S. Kim, C.-M. Ryu, D. Yong, H. Yoon, H. S. Song, S. H. Lee and O. S. Kwon, *Biosens. Bioelectron.*, 2020, **167**, 112514.
21. D. Sarkar, W. Liu, X. Xie, A. C. Anselmo, S. Mitragotri and K. Banerjee, *ACS Nano*, 2014, **8**, 3992.
22. N. Masurkar, N. K. Thangavel, S. Yurgelevic, S. Varma, G. W. Auner and L. M. Reddy Arava, *Biosens. Bioelectron.*, 2021, **172**, 112724.
23. A. Dey, A. Singh, D. Dutta, S. S. Ghosh and P. K. Iyer, *J. Mater. Chem. A*, 2019, 7, 18330.
24. T. L. Tran, T. T. Nguyen, T. T. Huyen Tran, V. T. Chu, Q. Thinh Tran and A. Tuan Mai, *Phys. E*, 2017, **93**, 83.
25. T. Dastagir, E. S. Forzani, R. Zhang, I. Amlani, L. A. Nagahara, R. Tsui and N. Tao, *Analyst*, 2007, **132**, 738.
26. W. Shao, M. R. Shurin, S. E. Wheeler, X. He and A. Star, *ACS Appl. Mater. Interfaces*, 2021, **13**, 10321.
27. M. A. Zamzami, G. Rabbani, A. Ahmad, A. A. Basalah, W. H. Al-Sabban, S. Nate Ahn and H. Choudhry, *Bioelectrochemistry*, 2022, **143**, 107982.
28. M. Shariati, M. Sadeghi and S. H. R. Shojaei, *Anal. Chim. Acta*, 2022, **1195**, 339442.
29. G.-J. Zhang, L. Zhang, M. J. Huang, Z. H. H. Luo, G. K. I. Tay, E.-J. A. Lim, T. G. Kang and Y. Chen, *Sens. Actuators, B*, 2010, **146**, 138.
30. F. N. Ishikawa, H.-K. Chang, M. Curreli, H.-I. Liao, C. A. Olson, P.-C. Chen, R. Zhang, R. W. Roberts, R. Sun, R. J. Cote, M. E. Thompson and C. Zhou, *ACS Nano*, 2009, **3**, 1219.
31. C. Chan, J. Shi, Y. Fan and M. Yang, *Sens. Actuators, B*, 2017, **251**, 927.
32. Y. Chen, R. Ren, H. Pu, X. Guo, J. Chang, G. Zhou, S. Mao, M. Kron and J. Chen, *Sci. Rep.*, 2017, 7, 10974.
33. G. Seo, G. Lee, M. J. Kim, S.-H. Baek, M. Choi, K. B. Ku, C.-S. Lee, S. Jun, D. Park, H. G. Kim, S.-J. Kim, J.-O. Lee, B. T. Kim, E. C. Park and S. I. Kim, *ACS Nano*, 2020, **14**, 5135.
34. P. Fathi-Hafshejani, N. Azam, L. Wang, M. A. Kuroda, M. C. Hamilton, S. Hasim and M. Mahjouri-Samani, *ACS Nano*, 2021, **15**, 11461.
35. J. Wei, Z. Zhao, F. Luo, K. Lan, R. Chen and G. Qin, Sensitive and quantitative detection of SARS-CoV-2 antibodies from vaccinated serum by MoS_2-field effect transistor, *2D Mater.*, 2022, **9**, 015030.

CHAPTER 5

Field-effect Transistor Biosensors Based on Nanomaterials for Zoonotic Pathogen Detection

FAIZA YAHIA,[a,b,c] FARAH NASRI,[d] LYES DOUADJI,[a,b] ATEF THAMRI,[e] DEQIANG WANG*[a,b] AND CHAKER TLILI*[a,b]

[a] Chongqing Key Laboratory of Multi-scale Manufacturing Technology, Institute of Green and Intelligent Technology, Chinese Academy of Sciences, Chongqing, People's Republic of China; [b] Chongqing School, University of Chinese Academy of Sciences, Chongqing 400714, China; [c] University of Chinese Academy of Sciences, Beijing 100049, China; [d] Laboratory of Electronics and Micro-Electronics (EµE) FSM, University of Monastir, Monastir, Tunisia; [e] CRMN, Centre for Research on Microelectronics and Nanotechnology of Sousse, NANOMISENE Lab, LR16CRMN01, Code Postal 4050, Sousse, Tunisia
*Emails: dqwang@cigit.ac.cn; chakertlili@cigit.ac.cn

5.1 Introduction

The etymology of the term "zoonoses" can be traced back to the Greek language, where "zoon" refers to animals and "nosos" means illness.[1] According to the World Health Organization (WHO), any disease or infection that is naturally transmissible from vertebrate animals to humans (anthropozoonosis) or from humans to animals (zooanthroponosis) is classified as

Detection Science Series No. 24
Field-effect Transistor Biosensors for Rapid Pathogen Detection
Edited by Naif H. M. Al-Hardan, Azman Jalar, Muhammad Azmi Abdul Hamid and Mohd Firdaus Raih

Published by the Royal Society of Chemistry, www.rsc.org

zoonosis.[1] Transmission of zoonotic diseases may occur through different routes including direct contact, indirect contact, foodborne/waterborne infection, or vector- or arthropod-borne.[2] Zoonotic diseases can be classified into three categories: endemic zoonoses, which are widespread and affect both humans and animals; epidemic zoonoses, which are sporadic in their distribution over time and space; and emerging and re-emerging zoonoses, which either newly appear in a population or have previously existed but are rapidly increasing in incidence or geographic range. In addition, zoonotic diseases can be categorized based on the etiologic agent, which includes viruses, bacteria, parasites, fungi, or unconventional agents such as prions.[3] There are more than 250 known types of zoonoses according to the WHO Zoonoses Expert Committee and they have revealed that 13% were emerging, of which 73% were zoonotic.[1]

Over the last several decades, zoonotic diseases have continued to pose a major threat to human health, with substantial global economic and social burdens. Zoonotic diseases are generally caused by harmful germs like viruses, bacteria, parasites, and fungi that spread between animals and people. These germs can cause many different types of illnesses in people and animals, ranging from mild to serious illnesses and even death. Each year, zoonotic diseases cause 2.5 billion cases of human illness and 2.7 million human deaths worldwide,[4,5] and most of these deaths occur in low- to middle-income countries.

In the global fight against zoonotic diseases, it is often important to develop accurate, low-cost, and timely diagnosis tools that can be used in resource-limited settings, which would help to make an informed decision about the treatment plan and to prevent the spread of these zoonotic diseases. Traditionally, zoonotic pathogen detection has relied upon culture-based methods, reverse transcription-polymerase chain reaction (RT-PCR), and enzyme-linked immunosorbent assay (ELISA). However, while these diagnostic approaches can achieve high sensitivity and selectivity, they suffer from the drawbacks of being time-consuming, involving sophisticated and bulky instrumentation, requiring special reagent kits and expert handling, and necessitating a high workload, which limit their practical applications in low-resource settings.[6–8] Therefore, there is an urgent need for hand-held and scale-up devices that can eliminate the complexities of traditional methods and provide quantitative measurement in a simple, rapid, accurate, field-deployable, and cost-effective manner to ensure daily diagnosis and effective treatment.

The interface between the biological world and nanomaterials has received tremendous interest in recent years and is considered one of the most dynamic research areas in nanoscience and nanotechnology. Unlike bulk materials, the development of nano/biohybrid systems and biocompatible nanomaterials such as carbon nanotubes (CNTs), organic/inorganic nanowires, graphene, transition metal dichalcogenides (TMDs), and MXenes is rapidly expanding and has excellent prospects for interfacing biological recognition events with electrical/optical signal transduction due to their high electrical and thermal conductivity, exceptional carrier mobility, intrinsically high surface-to-volume ratio and chemical inertness.[9]

Furthermore, the nanoscale size of nanomaterials makes them sensitive to localized entities of similar size, ranging from small molecules to large macromolecules.[10] So far, various methods based on nanomaterials including optical, electrochemical, and mechanical have arisen as an alternative tool for zoonotic pathogen detection.

However, fluorescent or electrochemical tags are required for these detection methods. In contrast, nanomaterial-based field-effect transistor (FET) technology has gained significant attention owing to its outstanding merits of high sensitivity and selectivity, low limit of detection (LoD), fast time response, label-free monitoring capabilities (eliminating the need for fluorescent or electrochemical tags), the ability for large-scale production, and good compatibility with complementary-metal-oxide-semiconductor (CMOS) technology.[11] One of the main factors determining the superior performance characteristics of these low-dimensional nanomaterials is primarily related to their extremely high surface-to-volume ratio and ultimate thinness. As a result, biological events occurring at the surface of the nanomaterial channel can have a significant impact on their physical and chemical properties, thereby providing a mode of signal transduction not necessarily discernible at a bulk structure of the same materials.[12] In general, the concept of sensing is based on detecting small changes in the electrical conductance of the sensor channel caused by an external electric field, which is induced by the interaction of a charged or dipole biological target with an immobilized bioreceptor at the FET surface, resulting in the variation of the drain current.[13]

In this chapter, we aim to present recent advances and developments in the field of major nanomaterial-based FET biosensors for zoonotic pathogen detection. First, we will introduce the different classes of zoonotic diseases. In the next section, we will describe the main nanomaterials used for FET-biosensors for pathogen detection, focusing on several basic phases such as FET device fabrication, the importance of FET surface functionalization, probe (antibodies, nucleic acid, peptides, proteins, aptamers) immobilization, surface blocking to reduce non-specific adsorption, and FET device performance. Also, we will describe the different strategies used in these biosensors including the direct detection of intact pathogens, the detection of pathogen molecular fingerprints (*e.g.*, nucleic acid), and serology through the detection of antibodies, enzymes, and/or cytokines produced by infected patients. Finally, conclusions and outlook for this vibrant field will be presented.

5.2 FET Devices

5.2.1 Brief History of FET Devices

The concept of field-effect transistors (FETs) was initially introduced by Julius Lilienfeld in 1925,[14] but practical devices were not developed until the 1950s. The first FETs were metal-oxide-semiconductor (MOS) devices that used a thin oxide layer to separate the gate electrode from the semiconductor channel. In 1959, at Bell Labs, Dawon Kahng and Martin Atalla independently developed the MOSFET devices, which quickly overtook other FET types due to their notable characteristics of high input impedance and low power requirements.[15] Then, complementary MOS (CMOS) technology was created in the 1970s, enabling the widespread use of low-power digital circuits.[16] More recently, cutting-edge materials such as graphene and other two-dimensional materials have been used in the development of novel field-effect transistors (FETs). The initial demonstration of carbon nanotube FETs (CNT-FETs) was conducted by Odom *et al.* in 1998, followed by the release of graphene FETs (GFETs).[17] These brand-new FET varieties are promising candidates for upcoming electronic devices because they have promising features like high carrier mobility and flexibility.[18]

5.2.2 Working Principle of FET Sensors

Due to their high sensitivity and selectivity, FET transistors have emerged as a promising platform for label-free biosensing. When target biomolecules bind to receptor molecules immobilized on the surface of a semiconductor channel, the electrical conductivity of the channel is modified, and this is the fundamental working principle of a FET sensor. An electric field that controls the flow of charge carriers through the semiconductor channel is created when a voltage is applied to the gate electrode. The electrical characteristics of the channel are altered by the binding of target biomolecules to receptor molecules on its surface, resulting in a change in conductance that can be detected as a signal.[19] The aforementioned changes may be initiated through one or multiple mechanisms, including but not limited to (1) the application of electrostatic gating, (2) the transfer of charge or doping between biomolecules and semiconductor channels, (3) the Schottky barrier effect between the semiconductor channel and metal contacts, (4) charge scattering, and (5) modulation of capacitance. Through careful examination of changes in the transfer curve under such experimental conditions, the fundamental sensing mechanism can be identified or at least narrowed down.[20] Also, various technical parameters, including device geometry, operating circumstances, and receptor molecule density, can be optimized in order to increase the sensitivity and selectivity of FET sensors. For instance, reducing the length and width of the semiconductor channel can increase sensitivity but also raise noise levels.[21] The thickness of the gate dielectric layer is also crucial because a thin layer can both increase sensitivity and raise noise levels.[22] Another crucial factor influencing the performance of FET sensors is receptor molecule density. Although a higher receptor molecule density can increase sensitivity, it can also cause steric hindrance and lower binding efficiency.[23] Furthermore, improving surface functionalization techniques for immobilizing receptor molecules on the graphene surface can enhance sensing performance.[24] To ensure the best sensor performance, operating conditions like temperature, pH, and buffer composition should be carefully managed.[25]

5.2.3 Types of FET Sensor Response Signals

Field-effect transistor (FET) biosensors demonstrate multiple types of response signals, including voltage response, capacitance response, and current response. The voltage response can be further categorized into two distinct types: the Dirac voltage response and the threshold voltage response.[26,27] The concept of capacitance response was initially introduced in 2021 and has the potential to expand the variety of indication signal types,[28] while the current response is based on the charge scattering effect, a phenomenon that involves the scattering of charges by charged biomolecules. This effect has been consistently observed and reported in various studies.[29] The choice of response signal type is determined by the specific properties of the detection material and the molecules being detected. However, researchers usually prefer a specific approach for enhancing sensitivity, rather than adopting a comprehensive approach that encompasses multiple strategies.[30]

5.3 FET Devices for Zoonotic Disease Monitoring

Zoonosis is an infectious illness that has been transmitted from vertebrates to humans. Zoonotic diseases may be bacterial (anthrax, tuberculosis, brucellosis, Lyme disease, and salmonellosis), viral (Ebola, severe acute respiratory syndrome (SARS), rabies, and avian influenza), or parasitic (trichinellosis, toxoplasmosis, malaria, and trematodosis), which have the potential to be transmitted through different ways. These include direct contact with domestic and wild animals, food and water contamination, or bites from ticks and mosquitos.[1] Also, they can lead to a wide range of illnesses in humans, from minor to severe and even death. According to the Centers for Disease Control and Prevention (CDC), bacteria (41.4%) are the most infectious pathogen, followed by viruses (37.7%) and parasites (18.7%), whereas approximately 2% and 0.8% are of fungal and prion origin, respectively.[31]

5.3.1 Bacterial Zoonoses

5.3.1.1 Anthrax

Anthrax is a zoonotic disease that occurs globally and is caused by *Bacillus anthracis* bacteria. It is transmitted to humans either through domestic animals like cattle and goats or through their products like meat, hides, or milk. Annually, a range of 2000–20 000 affected cases are recorded, with a significant concern of a high fatality rate that may rise to 100% in the case of pulmonary anthrax.[32] This bacterium can produce spores, which enables it to survive in the environment for an extended period. As a result, early detection of this bacterium is of the utmost importance.[1]

In this vein, Kim *et al.*[33] have developed an aptamer-immobilized graphene FET for the detection of protective antigens (PA). The sensing tool is based on a higher affinity of immobilized aptamer probes to PA. The FET sensor fabrication started through the self-assembly of graphene oxide (GO) nanosheets onto an aminated glass substrate, followed by chemical reduction in hydrazine vapor. After that, the Ti/Au source–drain electrodes were thermally evaporated using a shadow mask to achieve the desired pattern. In order to prevent leakage current and to define a sensing area, a 200 nm Al_2O_3 layer was deposited on top of the metallic contacts, followed by a deposition of polydimethylsiloxane (PDMS) to further avoid the leakage current. Before the graphene sensing test, a DNA-specific aptamer to the PA antigen was covalently attached to the 1-pyrene-butanoic acid succinimidyl ester (PBSE) linker that was non-covalently adsorbed on the graphene surface through π–π stacking, then the unreacted NHS group was blocked with ethanolamine.

As shown in Figure 5.1, the minimum source–drain current ($I_{ds,min}$) and the minimum conductance at a particular gate voltage ($V_{g,min}$) changed upon exposure to the channel to different concentrations of a protective protein (PA). The authors claimed that the left shift in $V_{g,min}$ and the increase in $I_{ds,min}$ are mainly attributed to the charge doping and the interaction of negatively

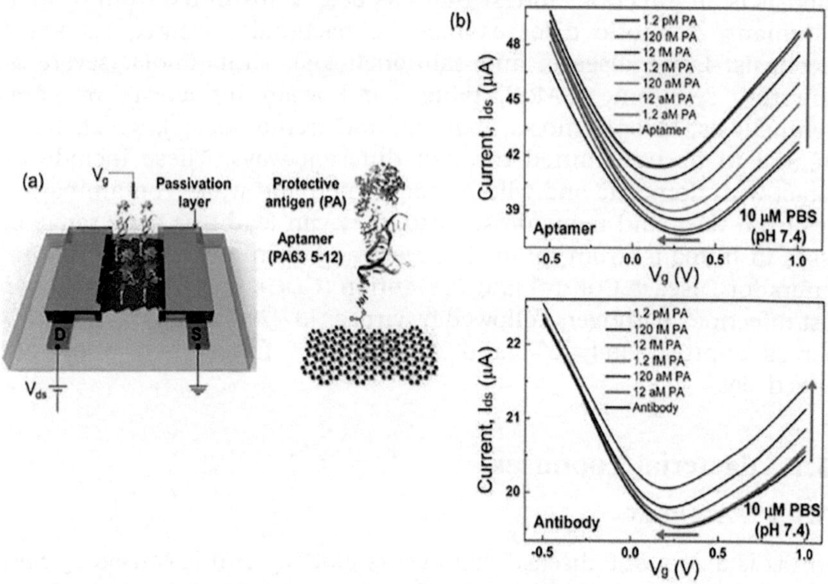

Figure 5.1 (a) Schematic presentation of the aptamer-immobilized graphene FET sensor for PA detection. (b) Transfer characteristics of both aptamer- and antigen-immobilized FETs with PA concentration variation. (a and b) Reproduced from ref. 33 with permission from John Wiley & Sons, Copyright 2013 Wiley-VCH Verlag GmbH & Co. KGaA, Weinheim.

charged PA (pI ~ 5.6) with positively charged impurities in the graphene channel, respectively. The aptamer FET exhibited a dynamic range between 12 aM and 120 fM, with a limit of detection (LOD) of 12 aM, and a high specificity to other proteins. More interestingly, further reduction of the LOD to 1.2 aM was achieved by using secondary aptamer-conjugated gold nanoparticles (AuNPs-PA63 11-2), in a sandwich structure. In addition, the authors reported that DNA aptamers offer certain advantages over sensor channels modified with antibodies, as they are capable of reducing Debye screening by binding and retaining charged particles in closer proximity to the detection surface, as shown in Figure 5.1.

5.3.1.2 Methicillin-resistant Staphylococcus aureus *(MRSA)*

According to the Centers for Disease Control and Prevention, *Staphylococcus aureus* (*S. aureus*) is a Gram-positive bacterium that can be found in various domestic animals and presents a serious menace to public health due to its increasing antibiotic resistance. In 2019, it was ranked seventh among the top ten pathogens responsible for hospitalizations in Europe.[34] Among the Staphylococcus clones, *methicillin-resistant Staphylococcus aureus (MRSA)* remains a significant concern due to its high fatality and morbidity rates, especially for children, athletes, elderly persons, and military personnel.[35] This is attributed to its ability to cause infections in different parts of the human body, ranging from skin and cellulitis infections to bloodstream and invasive bacteremia, as well as its high resistance to β-lactam antibiotics.[36] The prevalence of MRSA exhibited geographical variation, with lower rates found in Scandinavia and high rates in certain regions of Asia and the United States.[37]

Therefore, it is critically important to develop new approaches for early detection of *S. aureus* since conventional detection methods are time-consuming and labor-intensive. Moudgil *et al.* developed a MoS_2/TiO_2 heterostructure FET biosensor based on the binding affinity between vancomycin (Van) and the terminal D-alanyl D-alanine (D-Ala-D-Ala) of the peptidoglycan layer extending from the cell walls of Gram-positive bacteria.[38] As illustrated in Figure 5.2(a), the fabrication process was based on the synthesis of MoS_2 on a Si/SiO_2 wafer by the CVD technique, followed by the deposition of TiO_2 by RF sputtering for forming the MoS_2/TiO_2 heterostructure. Subsequently, optical lithography and reactive ion etching (RIE) have been employed to define the FET channel. This was followed by the deposition of the Cr/Au contact metal pads. Then, the SU-8 photoresist was used to passivate the metallic contact. Finally, the Van molecules were immobilized on the channel region through physical adsorption. As shown in Figure 5.2(b), the increase in bacterial concentration results in a reduction of the I_{ds} current due to the capture of *S. aureus* with the Van-modified FET channel. The authors attributed this phenomenon to the negative charge of *S. aureus* interacting with vancomycin in close proximity to the MoS_2/TiO_2 channel, where the Debye length is less than 2.3 nm (10 mM of PBS as the

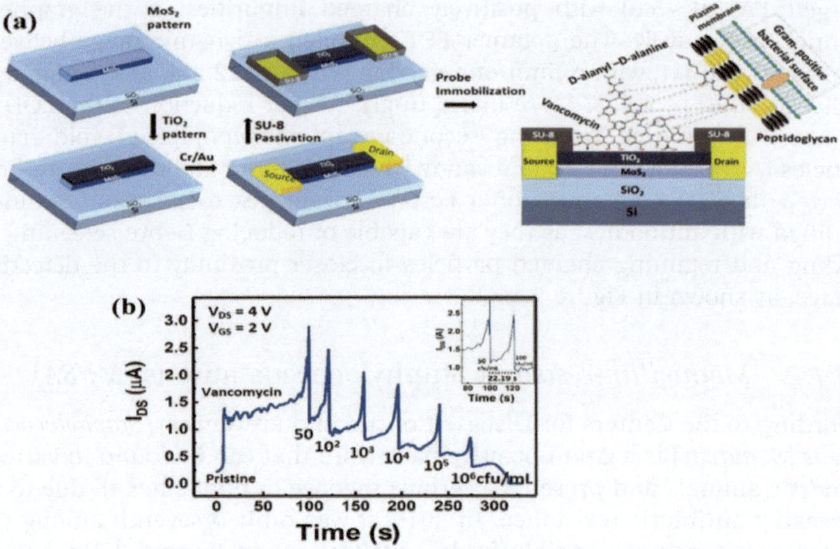

Figure 5.2 (a) Schematic diagram of the fabrication process of the vancomycin-immobilized MOS_2/TiO_2-based FET. (b) The real-time current response to the variation of *S. aureus* concentration. The inset shows a LOD fitting curve. (a and b) Reproduced from ref. 38 with permission from John Wiley & Sons, Copyright 2019 Wiley-VCH Verlag GmbH & Co. KGaA, Weinheim.

supporting electrolyte). This interaction leads to a depletion of charge carriers on the n-type MoS_2/TiO_2 channel in the FET device. According to their findings, the FET biosensor demonstrated the capability to distinguish between Gram-positive and Gram-negative bacteria, as well as exhibit a dynamic range that covered nearly 6 orders of magnitude, as illustrated in Figure 5.2(b). The limit of detection (LoD) was measured at 50 CFU per mL, and the response time was determined to be 20 seconds.

Significantly, the FET biosensor was subjected to assessment under physiological conditions, yielding a remarkable capacity to detect bacteria. The MoS_2/TiO_2 FET biosensor technology exhibits the capability to effectively differentiate between live and dead bacterial cells with precision.

On the other hand, an innovative approach based on a cell-membrane-functionalized FET nanosensor (CM-FET) for sensitive and quantitative detection of MRSA was realized by Gong *et al.*,[39] who coated a semi-conductive carbon nanotube (CNT)-based FET with red blood cell (RBC) membranes through a fusion process, a particular class of cell membranes widely used as alternative bioreceptors due to their specific interaction with numerous pore-forming toxins (PFTs) that have membrane lytic activities, despite their complicated and unknown structures. The binding event between the RCB membranes and the target toxins disturbs consequently the electrical double layer (EDL) within the proximity of the carbon nanotubes, which in turn changes the conductance of the CM-FET. As a proof of concept, they began their study with three distinct PFTs, namely melittin ($M_w = 2.8$ kDa), streptolysin O (SLO, $M_w = 69$ kDa), and alpha-hemolysin (Hlα, $M_w = 33$ kDa). Upon the increase of the toxin concentrations, the conductance of the CM-FET channel decreases and the authors attributed these observations to the interaction between toxin molecules and RCB membranes that decrease the conductance by attracting holes in p-type CM-FETs. The proposed nanosensors exhibited a dynamic range spanning from fM to μM with a limit of detection in the sub-fM range which is substantially lower than those of the currently available whole-cell-based hemolytic assay, ELISA assay, and chemically modified nanowire-based FETs. For the detection of MRSA, three MRSA culture samples obtained from either different incubation times or batches, together with the Hlα reference samples, were subjected to electrical measurement (I_{ds}) with their proposed nanosensors. Consequently, a direct correlation was observed between the bacterial concentration and the increase in current, with a detection limit of $(1.43 \pm 0.83) \times 10^2$ CFU per mL (equivalent to 4 hours). The observed electrical behavior has been ascribed to an increase in hemolytic activity. In addition, the correlation between bacterial concentration and the electrical response of the CM-FET has been experimentally verified by conducting measurements with diverse bacterial strains, including nonhemolytic *Escherichia coli*. The results indicated that the CM-FET did not exhibit any response toward the collected samples of bacterial supernatants. Thus, the aforementioned outcome underscores the distinctiveness of the CM-FET biosensor in detecting hemolytic MRSA bacteria.

5.3.1.3 *Tuberculosis*

Tuberculosis (TB) is the most infectious killer disease globally caused by *Mycobacterium tuberculosis* (M.tb). The Global TB Report 2022 estimated approximately 10.7 million TB cases worldwide. The transmission of tuberculosis in humans can occur through direct contact with infected domestic animals or *via* the consumption of unpasteurized milk or contaminated raw meat.[40] The increase of multidrug-resistant bacteria (MDR) in such regions around the world and the lack of effective techniques to control TB have resulted in significant global economic losses as they affect human lives and livestock health.[41] The most common traditional approaches used for testing tuberculosis include mycobacterial culture, chest radiography, and the Ziehl–Neelsen (ZN) technique. Nevertheless, these methods are known to be time-consuming, less sensitive, and may yield false-negative results.[34] Recent research has demonstrated growing interest in nanobiosensors due to their high accuracy and sensitivity in detecting pathogenic diseases.

In this regard, Jinbiao Ma and his colleagues[42] have developed a highly sensitive silicon nanowire-based field effect transistor (SiNW-FET) immunosensor for the detection of antigen 85 (Ag85), which has been considered as a specific biomarker for the early diagnosis of tuberculosis. In their work, the microfabrication process of SiNW-FET immunosensors involves the manufacturing of a 160-SiNW array using CMOS-compatible microfabrication technology. An oxygen plasma (OP) cleaning procedure was used to enhance the sensors' performance. The silicon nanowire was then functionalized with 3-aminopropyltriethoxysilane (APTES) and activated with glutaraldehyde on the amine-terminated surface. Finally, the antibody against Ag85B was immobilized and the process was completed by blocking with an ethanolamine solution. All the fabrication steps are depicted in Figure 5.3. The sensitivity of the SiNW-FET was evaluated by altering the concentration of Ag85B protein, which was diluted in PBS buffer, ranging from 1 fg mL^{-1} to 100 μg mL^{-1}. A linear relationship between current and concentration was observed, with a limit of detection (LOD) of 0.01 fg mL^{-1} within a response time of 30 s. This observation in the current reduction was assigned to the change in the thickness of the hole accumulation layer upon the binding of the positively charged Ag85B (pI ∼ 5.6) protein to the SiNW surface.[43] In addition, five distinct proteins, two of which bear resemblance to Ag85B, were utilized to evaluate the sensor's specificity. The protein Ag85B demonstrated the highest signal response value of 0.95 when compared to other proteins, thereby confirming high specificity of the sensor. It is noteworthy that the diagnosis of tuberculosis was successfully accomplished through the analysis of sputum samples, which is the most commonly collected sample for clinical diagnostic purposes. Four unprocessed sputum samples, including three positive and one negative, were collected from patients affected by tuberculosis and healthy volunteers, respectively. The proposed immunosensors show a clear differentiation between positive and negative tuberculosis samples, resulting in a significantly reduced detection time of only 2–5 minutes as compared to the existing technologies, as shown in Figure 5.3(c) and (d).

5.3.1.4 Foodborne Infections

Foodborne infections resulting from the consumption of food and water contaminated with pathogenic microorganisms can lead to high morbidity and fatality rates in humans, particularly among children. According to a previous study, an estimated 600 million people are annually affected, with approximately 125 thousand child fatalities reported.[44] Over 200 distinct foodborne diseases have been identified, including but not limited to *Escherichia coli*, Shigella spp., Salmonella spp., and hepatitis A.[45] Moreover, the management of foodborne disease has considerable impact on the global economic deficit, with an estimated loss of approximately 77 billion USD in the United States.[46]

5.3.1.4.1 ***Escherichia coli* (*E. coli*).** *Escherichia coli* (*E. coli*) bacteria are among the most current infectious pathogens that are known to spread

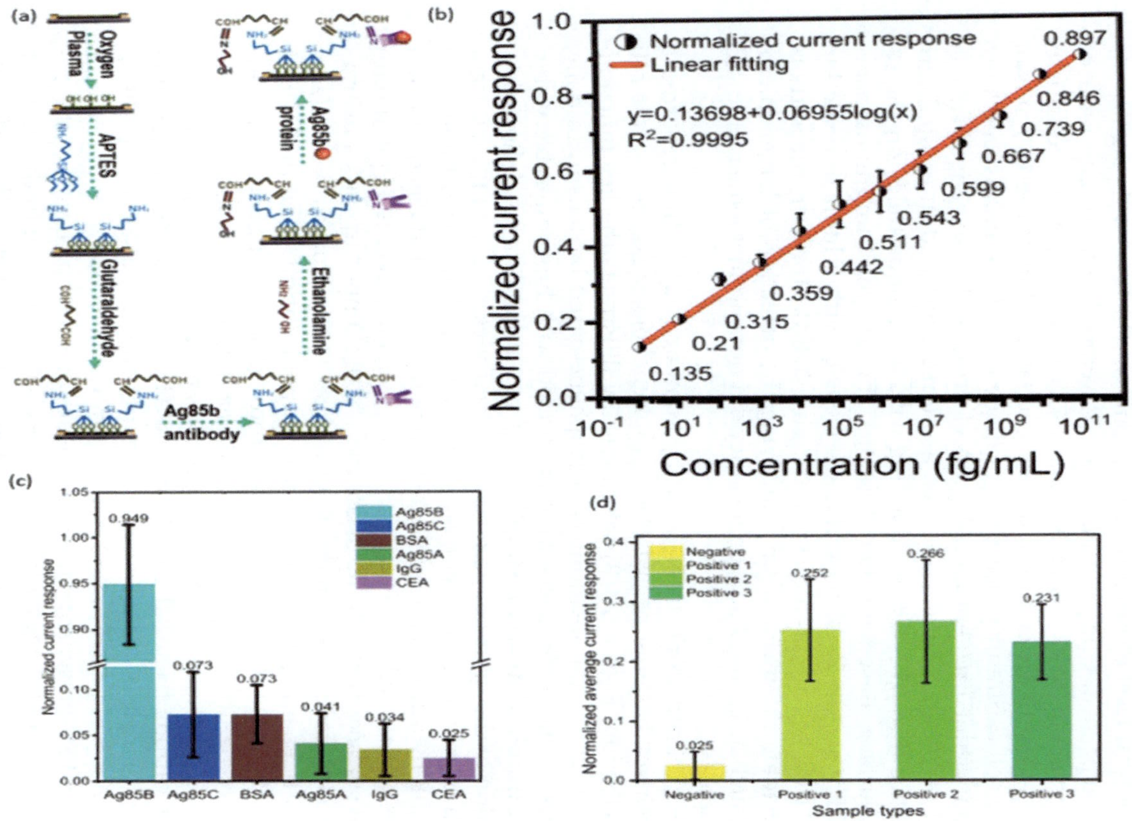

Figure 5.3 (a) Schematic of the SiNW-FET surface modification steps. (b) A linear variation of normalized current with respect to Ag85B protein concentration variation. (c) Current response *versus* protein variation for testing the sensor selectivity. (d) The average current response of negative and positive untreated sputum samples. Reproduced from ref. 42 with permission from American Chemical Society, Copyright 2021.

through food and water, annually affecting approximately 2.8 million cases.[47] They cause 30–50% of hospital-acquired and 80–90% of community-acquired urinary tract infections (UTIs).[48] Among *E. coli* types, *E. coli* O157:H7 is the most vigorous strain of *E. coli* that may lead, in some cases, to death.[49] Therefore, there is a high demand to develop biosensors for real-time monitoring with a high level of discrimination, some of which we will cite.

In this perspective, Chang *et al.*[49] have developed a thermally reduced monolayer graphene oxide (TRMGO) based FET with exceptional sensitivity and selectivity for the detection of *E. coli*. The fabrication of the FET sensor involved the self-assembly process of graphene oxide (GO) nanosheets onto 2-aminoethanethiol (AET)-modified gold electrodes. To selectively detect *E. coli*, anti-*E. coli* antibodies were used as bioreceptors and integrated into the FET device through covalent bonding with Au nanoparticles on the reduced GO surface. Based on their findings, the designed immunosensors displayed high sensitivity and excellent selectivity with a limit of detection (LoD) of 10 CFU per mL, as indicated in Figure 5.4(A). However, the authors did not elucidate the mechanism underlying the detection of larger microorganisms such as *E. coli*. Additionally, they did not conduct a measurement of real samples.

In another related work, carried out by Mannor *et al.*,[50] they reported a highly sensitive, battery-free, wireless graphene biosensor for whole-cell bacterial detection. Figure 5.4(B) schematically illustrates the key functionalities and fundamental operation of a graphene-based biosensor. Initially, the CVD-grown graphene layers were integrated with biocompatible water-soluble silk fibroin films *via* a simple transfer printing process, followed by the deposition of the interdigitated contact electrodes, and the LRC resonant circuit with a gold inductive coil wireless transmission was printed onto graphene–silk hybrid films through a shadow mask-assisted electron beam evaporation system. Subsequently, the sensing element was integrated onto biomaterial surfaces through the dissolution of the supporting silk substrate and intimately attached by van der Waals forces to the biomaterial surfaces such as tooth enamel, followed by the immobilization of antimicrobial peptides (AMP) onto the graphene surface through π-stacking interaction for the efficient recognition of pathogenic bacteria. Finally, the resistance of the AMP-modified graphene-based biosensors was continuously assessed in the presence of various bacterial cell dilutions. The resulting device was able to detect one single *E. coli* bacterium in phosphate buffer saline (PBS). Also, this device was tested in PBS solution for the detection of *S. aureus* and the proposed FET-biosensor shows a linear relationship between the resistance change and the logarithm of *S. aureus* concentration in the range of 10^3–10^8 CFU per mL with a limit of detection of 10^3 CFU per mL in a wireless operation mode. For the physiological condition assessment, the biosensor was evaluated on a bovine tooth and detected *Helicobacter pylori* bacteria in saliva solution with a limit of detection (LoD) of 10 CFU per mL. However, the differentiation of various species of pathogenic bacteria has not been accomplished due to the semi-selective character of the interaction between

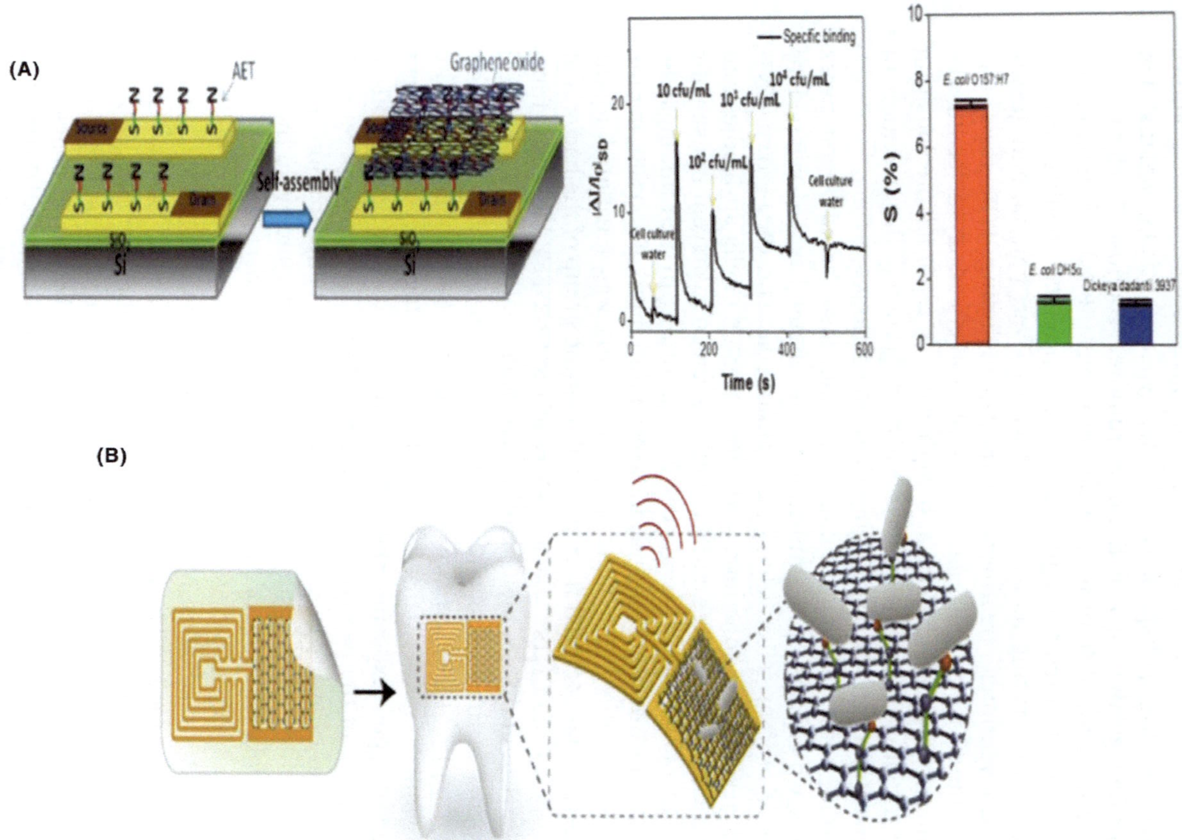

Figure 5.4 (A) Schematic presentation of the TRMGO-based FET fabrication process (left). The sensor response to *E. coli* concentration variation and selectivity performance of the device (right). Reproduced from ref. 49 with permission from the Royal Society of Chemistry. (B) Schematic design of a bio-transferrable AMP-graphene nanosensor. Reproduced from ref. 50 with permission from Springer Nature, Copyright 2012.

antimicrobial peptides (AMPs) and pathogenic bacteria including *E. coli*, *S. aureus*, and *Helicobacter pylori*.

5.3.1.4.2 Salmonella. A recent report released by the European Food Safety Authority (EFSA) announced that *Salmonella enteritidis*, also known as *Salmonella enterica* subsp. or enterica serotype Infantis (*S. infantis*), has been identified as the most commonly rumored zoonosis bacterium in 2021, accounting for 37% of foodborne outbreaks (FBOs).[34] Salmonellosis is frequently transmitted through eggs and egg products, poultry, and unpasteurized milk, leading to symptoms such as fever, nausea, and gastroenteritis. Conventional treatment methods, including PCR tests and enzyme immunoassays, may not be suitable for effectively detecting the bacteria due to their complexity and high cost, as reported previously in ref. 51. Alternative nanobiosensors have then garnered the attention of researchers as a means of overcoming these limitations.

In this vein, a fast, sensitive, and label-free biosensor based on a network of SWCNTs which acts as a semi-conducting channel was developed by Villamizar *et al.*[52] to selectively detect Salmonella. The CVD-grown CNT-based FET devices were functionalized with anti-Salmonella antibodies and subsequently blocked the free surfaces with Tween 20 to prevent nonspecific binding (NSB) of other proteins and bacteria. According to their findings, the proposed CNT-FET devices were able to detect a minimum of 100 CFU per mL of Salmonella within a one-hour timeframe. This was attributed to the charge transfer that occurred from either the amino group of the bacteria or the distortion of the CNTs. The selectivity of the CNT-FET biosensor functionalized with anti-Salmonella antibodies was evaluated *via* electrical measurements in the presence of two additional bacterial strains, namely *S. pyogenes* and *S. sonnei*, which are distinct from Salmonella but share antigenic similarities. A small alteration in the recorded electrical current was observed upon exposing the sensor surface to a concentration of 500 CFU per mL of the two types of bacteria, while a reduction of 50–60% was noted upon exposure to *S. infantis* with the same concentration. These findings are achieved through the utilization of Tween 20 to avoid the NSB and the non-cross-reactivity of the antibody with other bacteria.

On the other hand, Son *et al.* developed a bioelectronic nose based on a carbon nanotube field-effect transistor for the assessment of Salmonella contamination in food.[53] The innovation was founded upon the utilization of a peptide derived from Drosophila odorant binding protein (OBP) to identify the presence of Salmonella-contaminated ham. The confirmation of the biosensor's sensitivity was achieved by recording the current changes in response to variation of the concentration of 3-methyl-1-butanol (3M1BAOH), which was selected as the standard odor indicator for Salmonella-contaminated ham. The limit of detection for this indicator was determined to be 1 fM. The selectivity was further established by the real-time application of six odorant compounds identical to the standard (3M1BAOH) one to the bioelectronic nose sensor. There was just one reaction to 3M1BAOH (Figure 5.5(B)).

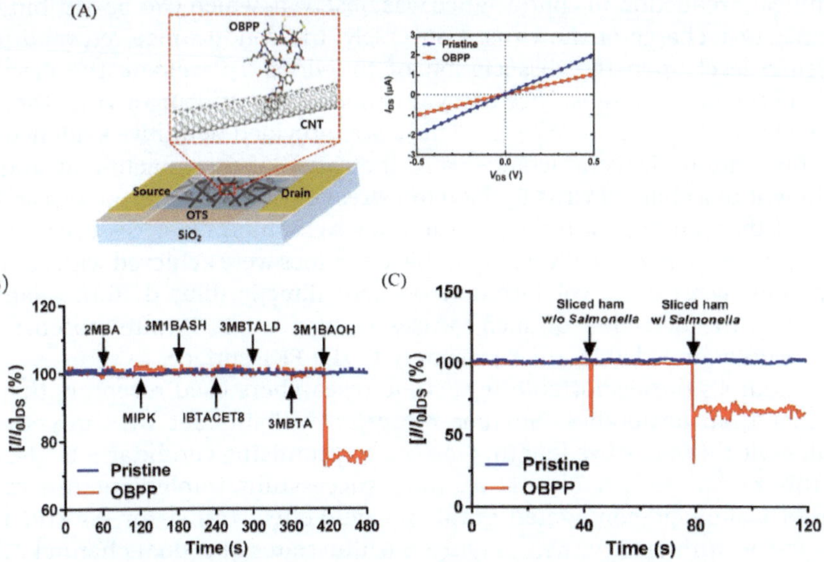

Figure 5.5 (A) Schematic diagram of odorant binding protein-derived peptide functionalized onto the CNT-based FET channel. The inset shows the drain–source current–voltage curves of CNT channels before and after OBPP functionalization. (B and C) Real-time conductance changes toward detection of different odorant compounds and Salmonella in contaminated ham. Reproduced from ref. 53 with permission from American Chemical Society, Copyright 2016.

Finally, genuine sample diagnosis was shown by injecting fresh and infected gammon samples into the biosensor, thereby emphasizing the critical function of OBP in identifying Salmonella bacteria.

5.3.2 Viral Zoonoses

5.3.2.1 Avian Influenza

Influenza is a communicable viral respiratory ailment that is attributed to the influenza virus and is accountable for 1 billion infections, with approximately 4 million severe cases. The primary contributing factor to the elevated morbidity and mortality rates of pandemic diseases is the sporadic modification in surface antigens and the majority of influenza-associated fatalities are attributed to the elderly population, particularly those with pre-existing cardiovascular and respiratory conditions.[11] Patolsky *et al.*[54] developed Si nanowire-based FET devices that were modified with anti-hemagglutinin antibodies for real-time electrical detection of a single influenza virus. The concept of their experiment was based on the conductance change in response to the virus binding/unbinding to the immobilized antibody in diluted PBS buffer. Upon the binding of the virus to the antibody, a significant reduction in conductance was observed, which can be attributed to the negative charge of the virus. Conversely, the conductance returned to its baseline level upon the dissociation of the virus. To validate the electrical measurements, a fluorescence assay was performed utilizing a virus that was labeled with fluorescent markers. This assay provided definitive evidence that the fluctuations in conductance were indicative of the attachment and detachment of a charged virus to the nanowire. For practical use, the researchers noticed that minimal or no purification of virus samples is necessary for these measurements. Specifically, comparable outcomes were achieved with samples that underwent basic gel filtration or were directly diluted from allantoic. However, the detection of such viruses using nano-FET immunosensors is significantly impeded by their proximity to the FET surface.

To address these shortcomings, some researchers used receptors that are smaller than antibodies. Smaller receptors will interact with target molecules within the Debye length. One of the promising candidates is glycans. In this perspective, Sophia *et al.* have successfully implemented a multiplexed sialic acid-conjugated graphene field-effect transistor (SA-GFET) to detect the influenza virus.[55] Figure 5.6 illustrates the dual-channel GFET platform. First, the graphene micropatterns (GMs) were carried out using a conventional microelectromechanical system (MEMS) processing system. Subsequently, the surface of channels 1 and 2 of graphene was functionalized with interfacial chemicals PDI and PDA, respectively, through the mechanism of π–π interaction. Following this, the PDI/PDA-modified

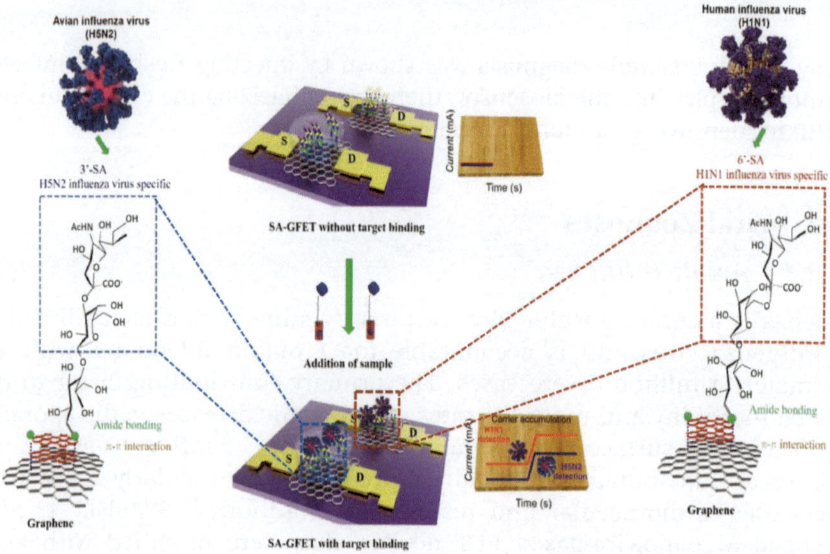

Figure 5.6 Schematic of the dual-channel GFET sensing concept for the detection of H1N1 and H5N2 influenza viruses by using 6′-SA and 3′-SA bioprobes, respectively. Reproduced from ref. 55 with permission from American Chemical Society, Copyright 2023.

graphene surface was used to immobilize the bioprobes 6′-SA and 3′-SA and then block the free surface with BSA to reduce non-specific adsorption. Finally, viral concentrations were monitored in real time by measuring the current (I_{ds}). As a result, the SA-GFET platform demonstrated a noteworthy level of sensitivity in the concentration range of 10^1–10^6 PFU mL^{-1} and attained a limit of detection (LOD) of 10^1 PFU mL^{-1} in buffer solution for H1N1 and H5N2, with a response time of roughly 10 seconds. More importantly, the proposed methodology demonstrated remarkable selectivity towards syncytial viruses (RSVs) and exhibited exceptional discrimination between avian and human influenza.

5.3.2.2 *Hepatitis B Virus (HBV)*

The prevalence of hepatitis B virus (HBV) infection is considered a worldwide health burden, affecting more than one-third of the world's population. This viral infection is known to cause severe liver diseases, including but not limited to cirrhosis, chronic hepatitis, and hepatocellular carcinoma, with approximately 25% of chronic carriers being affected.[11] HBsAg, HBsAb, HBeAg, HBeAb, and HBcAb are significant biomarkers for the diagnosis of HBV from a clinical perspective. HBsAg, which is an envelope glycoprotein, represents the primary serological biomarker in the bloodstream and is considered the most essential biomarker for detecting HBV infection.[56] Thus, the detection of HBsAg is a vital tool for diagnosing HBV infection.

In this vein, Cho and his co-workers[57] have demonstrated the ability of the multidimensional conductive nanofilm (MCNF)-based FET aptasensor in detecting hepatitis B antigen (HBsAg) with high sensitivity and rapid response. This was achieved by vertically growing carboxylic polypyrrole nanowires (CPPy-NWs) on a chemical vapor deposition (CVD)-graphene layer, utilizing an electropolymerization process. Subsequently, the amine-functionalized HBsAg-binding aptamers were covalently immobilized on the carboxyl group-functionalized polypyrrole nanowire surface through acid treatment and DMTMM, an amide group coupling agent. As shown in Figure 5.7, upon introduction of the target analyte (HBsAg) into the chamber, there was an abrupt increase in the I_{sd} current in a negative direction, followed by a slight decrease before the signal eventually reached a saturated value. The rapid decrease in the current suggests that there is a swift reaction occurring between HBsAg and binding aptamers, followed by a prompt transfer of charge to the nanofilm. From the findings, this approach exhibited a broad linear range of detection across a range of HBsAg concentrations spanning from 10 aM to 0.1 μM and has a limit of detection (LOD) lower than 10 aM, with a reaction time of less than 5 s. Moreover, the Apt-MCNF aptasensor displayed remarkable flexibility and durability after 500 binding cycles and exceptional selectivity towards HBsAg even in the presence of interfering molecules, diluted human serum, and artificial saliva, thereby showing potential for the noninvasive diagnosis of HBV infection.

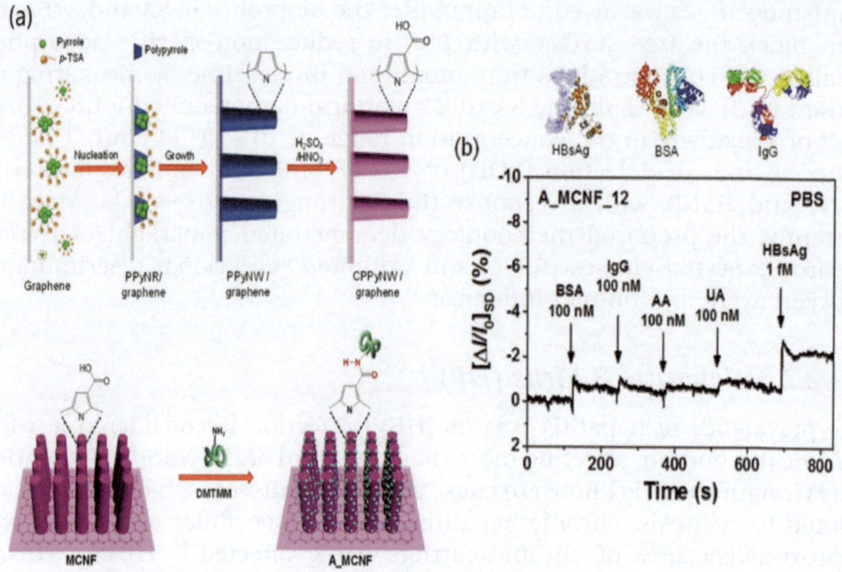

Figure 5.7 (a) Schematic illustration of the growth process of the aptamer-immobilized-multidimensional conductive nanofilm as the sensing channel for the FET sensor (A_MCNF-FET). (b) Real-time selective response of the A_MCNF-based sensor against the target (HBsAg) analyte and non-target (BSA, IgG, ascorbic acid (AA), and uric acid (UA)) analytes. Reproduced from ref. 57 with permission from American Chemical Society, Copyright 2018.

5.3.2.3 *Human Immunodeficiency Virus (HIV)*

Human immunodeficiency virus (HIV) infection is among the most widespread viral infections that affect the immune system of the human body, ultimately leading to the development of acquired immunodeficiency syndrome (AIDS). According to a recent report by the World Health Organization, approximately 1.5 million people were diagnosed with HIV and 650 000 fatalities were attributed to the virus in the year 2021.[58] In fact, the morbidity and mortality rates are currently on the rise due to the lack of accurate and ultrasensitive diagnostic tools, as well as the high genetic variability of the virus within the host cell.[59] The transmission of HIV may occur *via* the exchange of body fluids from HIV-infected people, blood transfusions and tissue transplantation, and sharing contaminated injection equipment. In the absence of treatment, severe diseases such as tuberculosis, and hepatitis B and C can emerge.[58] Antiretroviral therapy (ART), enzyme immunoassays, and ELISA are the most commonly used methods for controlling HIV infection. Nevertheless, these methods are associated with some drawbacks, such as reduced sensitivity and time-consuming procedures. Therefore, the development of a biosensor that is rapid, robust, and highly sensitive is crucial in limiting the spread of the virus.[59]

Typically, the identification of HIV infection is predicated upon the identification of HIV-1 ribonucleic acid (RNA), capsid antigen p24, and anti-HIV antibodies. Nanomaterials-based FET biosensors have been a subject of concern for early detection of the highly virulent HIV-1 virus. A cohort of researchers has successfully developed a label-free graphene FET genosensor for the detection of clinically full genomic length HIV-1 subtype B RNA.[60] After the growth and the wet transfer of graphene nanosheets, the channels of the GFET were delineated through the implementation of photolithography and plasma etching techniques. Subsequently, the surface of the sensor was functionalized with 1-pyrene butyric acid *N*-hydroxysuccinimide ester (PBASE) *via* π–π interaction, followed by the immobilization of 22mer or 80mer HIV-1 amino-modified probe ssDNA. The DNA probes were selected to be complementary to the integrase region of the HIV-1 subtype B pol gene and the HIV-1 subtype B gag gene. In order to evaluate the efficiency of their devices, the researchers conducted measurements of the I_{ds} current upon varying the concentration of HIV-1 RNA (pol gene and gang gene) within the range of 1 aM to 1 pM. The result indicated a linear correlation, with a limit of detection (LoD) of 0.1 aM for 80mer probes and 1 aM for 22mer probes in nuclease-free water, while the limit of detection for the QIAzol reagent was 0.1 aM for both the 22mer and 80mer probes. Furthermore, the selectivity of the GFET genosensor toward the complementary HIV-1 RNA target was proved by testing the biosensor response against the antisense HIV-1 RNA sequence, confirming that the proposed devices were capable of monitoring clinically meaningful low viral loads of HIV RNA in both samples and complex samples. In another work, a novel approach in the production of indium nitride (InN) microtubes, which serve as FET channels for detecting HIV antibodies, has been reported by Song *et al.*[61] through the utilization of rolled-up nanotechnology. The fabrication of rolled-up microtubes involved the use of InN thin films deposited through DC reactive sputtering on SiO_x/SiN_x strained bilayer nanomembranes, which served as the rolling vehicle. Following the growth of the rolled-up InN microtube, the surface of the channel was functionalized using the (3-aminopropyl)trimethoxysilane (APTMS) linker, followed by the immobilization of the HIV gp41 antigen, which is specific to the HIV-1 envelope gp41 biomarker after surface activation with glutaraldehyde (GA), as shown in Figure 5.8. Through experimentation involving the variation of diluted HIV-1 gp41 antibody concentration with PBS solution, the performance of the InN RUM-FET was tested. The results showed a significant current-to-concentration dependency, with a LoD of 2.5 pM for serum samples spiked with HIV gp41 antibodies.

5.3.2.4 Ebola Virus Disease

The Ebola virus is a highly contagious virus that is classified under the Filoviridae family. It is mainly transmitted by human–human contact through infected body fluids.[62] The initial symptoms of the disease are not

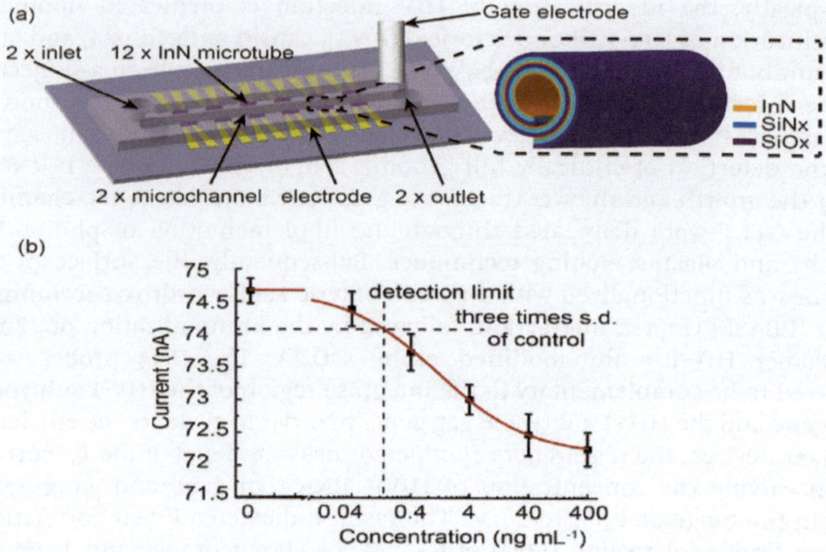

Figure 5.8 (a) Schematic illustration of the fabrication of the rolled-up InN microtube serving as the FET biosensing surface. (b) The sensing performance evaluation of the RUM-InN-FET sensor toward HIV gp1 antibody concentration variation. Reproduced from ref. 61 with permission from Elsevier, Copyright 2021.

clearly defined and may result in hemorrhagic fever and even failure of the human multisystem and it has been found to have mortality rates ranging from 30 to 90%.[62] The Ebola virus surface glycoprotein (GP) GP1,2 is a significant virulence factor responsible for the production of Ebola hemorrhagic fever. The Ebola virus attaches to host cell surfaces through the lectin binding of GP1,2, leading to fusion with host cells and subsequent secretion of significant amounts of a truncated glycoprotein, sGP. The presence of a significant quantity of viral particles in individuals infected with the Ebola virus implies that both antigen and viral RNA detection systems may be useful for Ebola diagnosis. The conventional diagnostic methods employed for the identification of the Ebola virus include reverse transcription polymerase chain reaction (RT-PCR) test, immunofluorescence, and ELISA. However, these techniques have shown their limitations in preventing the spread of Ebola virus due to their time-consuming nature, high-tech equipment requirements, and cost-effectiveness. Therefore, direct detection of the Ebola virus has gained increased interest through the use of reduced graphene oxide (RGO) field-effect transistor (RGO-FET) biosensors.[62] Chen *et al.*[62] have fabricated an rGO-based FET device for the detection of glycoprotein (GP), which constitutes the main virulent factor in Ebola virus disease pathogenesis. The fabrication process of the rGO-FET biosensors involves three steps. First, the adsorbed GO on the cysteamine-modified gold interdigitated electrode was thermally reduced at 400 °C for 10 min, followed by the deposition of an Al_2O_3 thin layer (3 nm) through atomic layer deposition (ALD). Subsequently, gold nanoparticles (NPs) were sputtered onto the passivated layer using a sputter coater, then the NPs were functionalized and activated with cysteamine and glutaraldehyde, respectively. Finally, the anti-Ebola antibodies were immobilized on the functionalized surface for detecting Ebola glycoprotein (EBL). The study revealed a notable degree of sensitivity towards the Ebola glycoprotein (EGP), with a limit of detection (LOD) of less than 1 ng mL^{-1}, achieved within a brief time frame of a few seconds. The study revealed that the sensing performance of the EGP in PBS buffer, human serum, and plasma exhibited a comparable level of similarity, indicating the potential of the sensor for point-of-care (POC) diagnosis. The discriminatory capacity of the reduced graphene oxide (rGO)-based field-effect transistor (FET) sensor was demonstrated through its evaluation against non-specific proteins. The findings of their study indicate that the FET biosensor has the potential to serve as a viable option for promptly detecting EVD patients during the initial phases of the illness. Additionally, the biosensor offers supplementary advantages such as affordability and user-friendliness.

5.3.3 Parasitic Zoonoses

5.3.3.1 Malaria

Malaria remains a serious infectious disease affecting people worldwide, particularly in low- and middle-income countries.[63] The disease is transmitted through the bites of *Anopheles* female mosquitoes and can result in a range of symptoms, from mild to severe, including fever, chills, and headache until jaundice and even death.[64] Annually, approximately 500 million new infections are recorded, with a higher incidence among children under the age of 5 years, resulting in a mortality rate of 1–3 million deaths.[11] Lateral flow-based immunochromatographic tests are widely used as point-of-care diagnostic tools for malaria.[64] Nonetheless, it should be noted that the sensitivity of these rapid diagnostic tests (RDTs) is limited by the elevated temperature in tropical regions, as well as the susceptibility of antibody-based RDTs to changes in heat and humidity.[65] Therefore, there is a pressing need to develop alternative technology that is highly sensitive, simple, and robust in the pursuit of combating malaria.

The utilization of field effect transistor (FET) biosensors has been considered a promising sensing platform for this particular purpose. Singh and co-workers[65] have combined the efficiency and viability of nucleic acid aptamers, as alternative probes to antibodies, with the low-cost and simplicity of FET technology to the *Plasmodium falciparum* glutamate dehydrogenase (*Pf*GDH) malaria biomarker in serum samples. For this reason, the authors have developed their device through the immobilization of a specific aptamer onto interdigitated gold microelectrodes (IDμE) that are connected to the extended gate FET (EgFET) through gold–thiol chemistry,

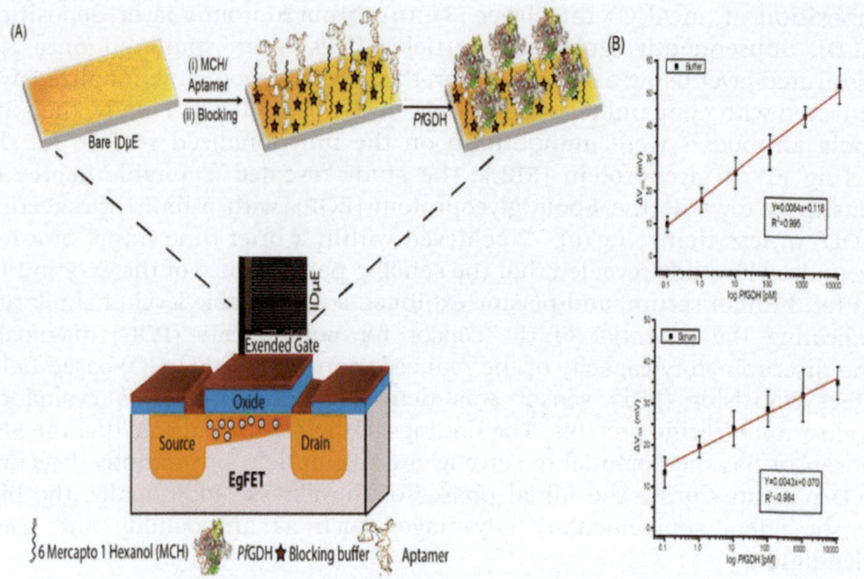

Figure 5.9 (A) Schematic presentation of the aptamer-based EgFET for the detection of the *Pf*GDH biomarker. (B) Calibration curves of the aptaFET biosensor toward the concentration variation of *Pf*GDH in both buffer and serum. Reproduced from ref. 65 with permission from Elsevier, Copyright 2019.

as depicted in Figure 5.9(A). The aptaFET response was examined to determine the capability to detect the *Pf*GDH target molecule at lower concentrations ranging from 100 fM to 10 nM in both PBS buffer and diluted serum, with an established limit of detection (LoD) of 16.7 and 48.6 pM, respectively. Moreover, the selectivity of the aptaFET biosensor has been validated by assessing the sensor response toward different malaria (*Pf*LDH, *Pf*HRP-II) and human (HGDH, HSA) proteins, as displayed in the inset of Figure 5.9(B).

5.3.3.2 Chagas Disease (CD)

Chagas disease, also called American trypanosomiasis, is a deadly tropical infectious disease that results from the transmission of the protozoan parasite *Trypanosoma cruzi* through the bits of triatomine bugs. This disease is also caused by infected blood transfusion as well as the consumption of *T. cruzi*-contaminated food and drink consumption.[66] The CD is a life-threatening disease affecting approximately 6 to 7 million people and causing about 14 000 deaths every year.[67] Indeed, the long-lasting asymptomatic chronic infection with CD and its spread to non-endemic states due to the lack of efficient diagnosis tools pose a worldwide burden.[67] Frequently used techniques for *T. cruzi* parasite detection are based on PCR, ELISA, and immunofluorescence; however, they lead to false-positive tests and

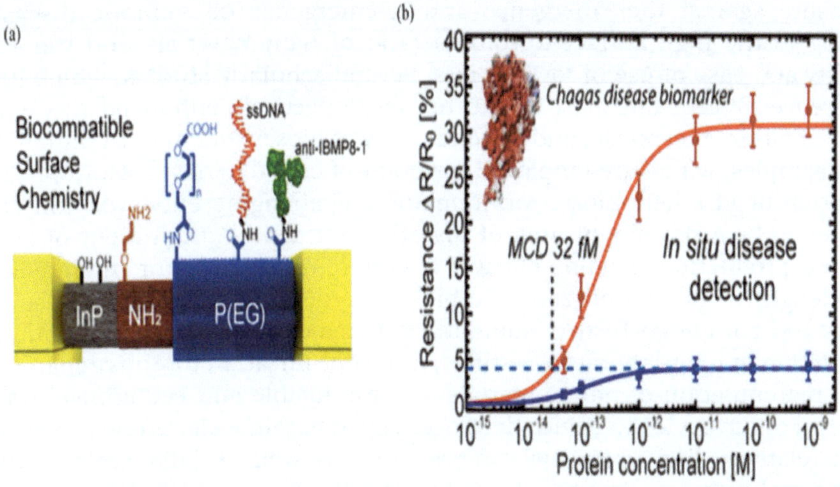

Figure 5.10 (a) Schematic presentation of the surface functionalization of the InP nanowire-based FET biosensor with EA and PEG-linkers for DNA and IBMP 8-1 protein (specific to *T. cruzi*) detection. (b) Biosensor resistance variation *versus* protein concentration of the specific (red color) and non-specific (blue) target molecules. MCD indicates minimal concentration detection. Reproduced from ref. 69 with permission from American Chemical Society, Copyright 2017.

are less sensitive.[68] Developing a sensitive and robust biosensor for *T. cruzi* biomarker detection for asymptomatic patients in the early stage of infection is of the utmost importance, particularly in the absence of a vaccine.[67]

For this concern, a group of researchers[69] have fabricated a InP nanowire-based semiconductor FET to improve selectivity in detecting Chagas biomarkers. The authors have achieved a heightened level of sensitivity by using ethanolamine (EA) and poly(ethylene glycol) (PEG) as alternative cross-linkers for surface functionalization *via* the enhancement of the surface coating and the efficiency of biomolecule detection, as illustrated in Figure 5.10(a). The electric response of the functionalized InP nanowire FET in the presence of specific IBMP 8-1 protein, which is a specific biomarker of positive Chagas disease, and nonspecific glutathione S-transferase (GST) protein. As shown in Figure 5.10(b), the proposed strategy indicated a specificity toward the targeted molecule, with a limit of detection of 5.7 fM and a response time of less than 30 minutes.

5.4 Conclusions and Future Perspectives

The integration of nanotechnology and biosensing applications, specifically concerning our focus on nanomaterial-based FETs, has demonstrated significant promise in the detection, identification, and quantification of zoonotic pathogens such as bacteria, viruses, and parasites. This approach offers high sensitivity and specific selectivity, thereby serving as a preventive measure against the emergence and re-emergence of zoonotic diseases, owing to the high surface-to-volume ratio of nanomaterials and the scalability and ease of use of FET devices. Several scholarly studies, which have referenced recent and past literature, have effectively enhanced the sensitivity, real-time response, and dependability of biosensors in both buffer and real samples, which are employed for point-of-care diagnoses. However, the creation of FET nanobiosensors remains a challenging endeavor and continues to be a prominent area of investigation, as the attainment of widespread production remains elusive. Therefore, there are four fundamental challenges in FET-based zoonotic biosensor development. First, these devices must undergo further standardization measures, such as additional validation in inter-laboratory settings, more optimization for the preparation and manipulation of nanomaterials in reproducible and economical ways, control of the device-to-device heterogeneity of baseline electronic properties and relatively high-cost device fabrication, as well as large-scale studies using real samples. Second, the Debye length screening effect hampers or misleads the signal detection. Possible ways to overcome this issue include the use of antibody fragments or aptamers, increasing the dielectric constant of the ionic solution by modifying the surface channel with polymers, such as polyethylene glycol (PEG), or *ex situ* measurement in low ionic strength buffers. Third, the current understanding of the underlying sensing mechanism for the readout signal is limited; therefore, it is imperative to identify a universal sensing mechanism that can comprehensively elucidate the obtained results. Finally, additional research must be conducted to fabricate an integrated multiplexed FET device into a sensor array with a microfluidic device for an efficient and high-performance assay that can lower the limit of detection, greatly shorten the assay time, and reduce the likelihood of false positive/negative results while ensuring early detection of zoonotic diseases. Hence, it is anticipated that forthcoming research endeavors will facilitate the advancement of FET-biosensor production, thereby augmenting the level of engagement with the industrial sectors. Consequently, this will culminate in the mitigation of the health and economic ramifications stemming from zoonotic diseases.

References

1. M. T. Rahman, M. A. Sobur, M. S. Islam, S. Ievy, M. J. Hossain, M. E. El Zowalaty, A. T. Rahman and H. M. Ashour, *Microorganisms*, 2020, **8**(9), 1405.
2. L. Christou, *Clin. Microbiol. Infect.*, 2011, **17**, 326.
3. B. B. Chomel, *Reference Module in Biomedical Sciences*, 2014.
4. D. Grace, F. Mutua, P. Ochungo, R. Kruska, K. Jones, L. Brierley, M. L. Lapar, M. Said, M. Herrero, P. Pham-Duc, *et al.*, *Zoonoses Project 4 Report to Department for International Development*, UK, 2012, vol. 4, pp. 1–83.
5. A. Carpenter, M. A. Waltenburg, A. Hall, J. Kile, M. Killerby, B. Knust, M. Negron, M. Nichols, R. M. Wallace and C. B. Behravesh, *et al*, *Vaccines*, 2022, **10**, 993.

6. G. G. D. Suminda, S. Bhandari, Y. Won, U. Goutam, K. K. Pulicherla, Y.-O. Son and M. Ghosh, *Comput. Struct. Biotechnol. J.*, 2022, **20**, 5378.
7. L. Yang and R. Bashir, *Biotechnol. Adv.*, 2008, **26**, 135.
8. C. Ménard-Moyon, A. Bianco and K. Kalantar-Zadeh, *ACS Sens.*, 2020, 5(12), 3739.
9. A. Bolotsky, D. Butler, C. Dong, K. Gerace, N. R. Glavin, C. Muratore, J. A. Robinson and A. Ebrahimi, *ACS Nano*, 2019, **13**, 9781.
10. G. Konvalina and H. Haick, *Acc. Chem. Res.*, 2014, **47**, 66.
11. D. Sadighbayan, M. Hasanzadeh and E. Ghafar-Zadeh, *Trends Anal. Chem.*, 2020, **133**, 116067.
12. Y. Ohno, K. Maehashi and K. Matsumoto, *J. Am. Chem. Soc.*, 2010, **132**, 18012.
13. K. Lee, P. R. Nair, A. Scott, M. A. Alam and D. B. Janes, *J. Appl. Phys.*, 2009, **105**, 102046.
14. R. G. Arns, *J. Eng. Educ.*, 1998, 7(5), 233.
15. Y. Zhang, MOSFET amplifier circuits, in *The Tao of Microelectronics*, Morgan & Claypool Publishers, 2014, pp. 4–18.
16. M. Schadt, *Jpn. J. Appl. Phys.*, 2009, **48**(3S2), 03B001.
17. P. Avouris, *et al*, *Proc. IEEE*, 2003, **91**(11), 1772.
18. S. M. Sze and K. N. Kwok, *Physics of Semiconductor Devices*, John Wiley & Sons, 3rd edn, 2006.
19. Y. Cui, Q. Wei, H. Park and C. M. Lieber, *Science*, 2017, **293**(5533), 1289.
20. J. Liu, E. Luong-Van, M. Layani and R. Hadji, *Sensors*, 2018, **18**(4), 1166.
21. N. M. Andoy, M. S. Filipiak, D. Vetter, O. Gutiérrez-Sanz and A. Tarasov, *Adv. Mater. Technol.*, 2018, **3**(12), 1870046.
22. M. Kaisti and R. Kostiainen, *Chem. Soc. Rev.*, 2018, **47**(22), 8740–8761.
23. J. Kim, M. Kim, J. Lee and T. Lee, *Front. Bioeng. Biotechnol.*, 2019, 7, 219.
24. G. Liang, X. Zhang and H. Chen, *Biosens. Bioelectron.*, 2020, **165**, 112361.
25. Z. Li, C. Wang, X. F. Su and L. H. Wang, *Biosens. Bioelectron.*, 2020, **168**, 112554.
26. X. Kong, M. Bahri, K. Djebbi, B. Shi, D. Zhou, C. Tlili and D. Wang, *2021 18th International Multi-Conference on Systems, Signals & Devices (SSD)*, Monastir, Tunisia, 2021, p. 387.
27. H. Ren, K. Liang, D. Li, M. Zhao, F. Li, H. Wang, X. Miao, T. Zhou, L. Wen, Q. Lu and B. Zhu, *Adv. Mater. Interfaces*, 2021, **8**, 2100314.
28. S. Xu, S. Jiang, C. Zhang, W. Yue, Y. Zou, G. Wang, H. Liu, X. Zhang, M. Li and Z. Zhu, *Appl. Surf. Sci.*, 2018, **427**, 1114.
29. M. Bahri, B. Shi, M. A. Elaguech, K. Djebbi, D. Zhou, L. Liang, C. Tlili and D. Wang, *ACS. Appl. Nano. Mater.*, 2022, **5**, 5035.
30. Z. Gao, H. Xia, J. Zauberman, M. Tomaiuolo, J. Ping, Q. Zhang, P. Ducos, H. Ye, S. Wang and X. Yang, *Nano Lett.*, 2018, **18**, 3509.
31. S. J. Salyer, R. Silver, K. Simone and C. B. Behravesh, *Emerg Infect Dis*, 2017, **23**(13), S55.
32. A. K. Goel, Anthrax, *World J. Clin. Cases*, 2015, **3**, 20.
33. D. J. Kim, H. C. Park, I. Y. Sohn, J. H. Jung, O. J. Yoon, J. S. Park, M. Y. Yoon and N. E. Lee, *Small*, 2013, **9**(19), 3352.

34. A. P. M. Tavares, G. V. Martins, M. F. Frasco and M. G. F. Sales, *Sensors*, 2021, **21**, 4547.
35. N. A. Turner, B. K. Sharma-Kuinkel, S. A. Maskarinec, E. M. Eichenberger, P. P. Shah, M. Carugati, T. L. Holland and V. G. Fowler Jr, *Nat. Rev. Microbiol.*, 2019, **17**, 203.
36. I. Choopara, A. Suea-Ngam, Y. Teethaisong, P. D. Howes, M. Schmelcher, A. Leelahavanichkul, S. Thunyaharn, D. Wongsawaeng, A. J. deMello, D. Dean and N. Somboonna, *ACS Sens.*, 2021, **6**(3), 742.
37. A. S. Lee, H. de Lencastre, J. Garau, J. Kluytmans, S. Malhotra-Kumar, A. Peschel and S. Harbarth, *Nat. Rev. Dis. Primers*, 2018, **4**, 18033.
38. A. Moudgil, S. Singh, N. Mishra, P. Mishra and S. Das, *Adv. Mater. Technol.*, 2019, **5**, 1900615.
39. H. Gong, F. Chen, Z. Huang, Y. Gu, Q. Zhang, Y. Chen, Y. Zhang, J. Zhuang, Y. K. Cho, R. H. Fang, W. Gao, S. Xu and L. Zhang, *ACS Nano*, 2019, **13**(3), 3714.
40. D. Kasir, N. Osman, A. Awik, I. El Ratel, R. Rafei, I. Al Kassaa, D. El Safadi, R. Salma, K. El Omari and K. J. Cummings, *et al*, *Diseases*, 2023, **11**, 39.
41. World Health Organization, *Treatment of tuberculosis: guidelines*, World Health Organization, 4th edn, 2010. https://apps.who.int/iris/handle/10665/44165.
42. J. Ma, M. Du, C. Wang, X. Xie, H. Wang, T. Li, S. Chen, L. Zhang, S. Mao, X. Zhou and M. Wu, *ACS Sens.*, 2021, **6**(9), 3367.
43. A. A. Wahid, A. Doekhie, A. Sartbaeva and J. M. H. van den Elsen, *Sci. Rep.*, 2019, **9**, 11409.
44. World Health Organization, *WHO Estimates of the Global Burden of Foodborne Diseases: Foodborne Disease Burden Epidemiology Reference Group 2007–2015*, World Health Organization, Geneva, Switzerland, 2015, p. 265.
45. T. Bintsis, *AIMS Microbiol.*, 2017, **3**(3), 529.
46. R. L. Scharff, *J. Food Prot.*, 2012, **75**, 123.
47. M. A. Ameer, A. Wasey and P. Salen, in *StatPearls*, StatPearls, Treasure Island, 2023.
48. R. Jijie, K. Kahlouche, A. Barras, N. Yamakawa, J. Bouckaert, T. Gharbi, S. Szunerits and R. Boukherroub, *Sens. Actuators, B*, 2018, **260**, 255.
49. J. Chang, S. Mao, Y. Zhang, S. Cui, G. Zhou, X. Wu, C. H. Yangb and J. Chen, *Nanoscale*, 2013, **5**, 3620.
50. M. S. Mannoor, H. Tao, J. D. Clayton, A. Sengupta, D. L. Kaplan, R. R. Naik, N. Verma, F. G. Omenetto and M. C. McAlpine, *Nat. Commun.*, 2012, **27**, 763.
51. A. D. Taylor, J. Ladd, Q. Yu, S. Chen, J. Homola and S. Jiang, *Biosens. Bioelectron.*, 2006, **22**, 752.
52. R. A. Villamizar, A. Maroto, F. X. Rius, I. Inza and M. J. Figueras, *Biosens. Bioelectron.*, 2008, **24**, 279.
53. M. Son, D. Kim, J. Kang, J. H. Lim, S. H. Lee, H. J. Ko, S. Hong and T. H. Park, *Anal. Chem.*, 2016, **88**, 11283.

54. F. Patolsky, G. Zheng, O. Hayden, M. Lakadamyali, X. Zhuang and C. M. Lieber, *Proc. Natl. Acad. Sci. U. S. A.*, 2004, **101**(39), 14017.
55. S. Nazir, K. H. Kim, L. Kim, S. E. Seo, P. K. Bae, J. E. An and O. S. Kwon, *Anal. Chem.*, 2023, **95**, 5532.
56. C. W. Shepard, E. P. Simard, L. Finelli, A. E. Fiore and B. P. Bell, *Epidemiol. Rev.*, 2006, **28**, 112.
57. K. H. Cho, D. H. Shin, J. Oh, J. H. An, J. S. Lee and J. Jang, *ACS Appl. Mater. Interfaces*, 2018, **10**, 28412.
58. World Health Organization. HIV and AIDS. World Health Organization, 2023; https://www.who.int/news-room/fact-sheets/detail/hiv-aids; (accessed May 2023).
59. B. Zhang, H. Mao, H. Zhu, J. Guo, P. Zhou and Z. Ma, *Front. Microbiol.*, 2023, **14**, 1136664.
60. O. O. Dickens, I. Bajwa, K. Garcia-Ramos, Y. Suh, C. Wen, A. Cheng, V. Fethke, Y. Yi, R. G. Collman and A. T. C. Johnson, *Sens. Actuators, B*, 2023, **390**, 134001.
61. P. Song, H. Fu, Y. Wang, C. Chen, P. Ou, R. T. Rashid, S. Duan, J. Song, Z. Mi and X. A. Liu, *Biosens. Bioelectron.*, 2021, **15**, 113264.
62. Y. Chen, R. Ren, H. Pu, X. Guo, J. Chang, G. Zhou, S. Mao, M. Kron and J. Chen, *Sci. Rep.*, 2017, 7, 10974.
63. World Health Organization, *World Malaria Report 2016*, World Health Organization, Geneva, 2016. (http://www.who.int/malaria/publications/world-malaria–report-2016/report/en/) (Last Accessed May 2023).
64. F. D. Krampa, Y. Aniweh, P. Kanyong and G. A. Awandare, *Sensors*, 2020, **20**, 799.
65. N. K. Singh, P. D. Thungoan, P. Estrela and P. Goswami, *Biosens. Bioelectron.*, 2018, **123**, 30.
66. C. Hernández, M. J. Vera, Z. Cucunubá, C. Flórez, O. Cantillo, L. S. Buitrago, M. S. González, S. Ardila, L. Z. Dueñas, R. Tovar, L. F. Forero and J. D. Ramírez, *J. Infect. Dis.*, 2016, **214**(8), 1252.
67. R. Martín-Escolano, M. Etxebeste-Mitxeltorena, J. Martín-Escolano, D. Plano, M. J. Rosales, S. Espuelas, E. Moreno, M. Sánchez-Moreno, C. Sanmartín and C. Marín, *ACS Infect. Dis.*, 2021, 7, 1727.
68. J. A. Pérez-Molina, A. M. Perez, F. F. Norman, B. Monge-Maillo and R. López-Vélez, *Lancet Infect. Dis.*, 2015, **15**(11), 1347.
69. R. Janissen, P. K. Sahoo, C. A. Santos, A. M. da Silva, A. A. G. von Zuben, D. E. P. Souto, A. D. T. Costa, P. Celedon, N. I. T. Zanchin, D. B. Almeida, D. S. Oliveira, L. T. Kubota, C. L. Cesar, A. P. Souza and M. A. Cotta, *Nano Lett.*, 2017, **17**(10), 5938.

CHAPTER 6

Field-effect Transistor Biosensors for Rapid SARS-CoV-2 Detection

A. IBIJBIJEN, F. BESBES AND H. KORRI-YOUSSOUFI*

Paris-Saclay University, ICMMO, UMR CNRS, Site Henri-Moissan, 17 Avenue des sciences, 91405, Orsay France
*Email: Hafsa.korri-youssoufi@universite-paris-saclay.fr

6.1 Introduction

The COVID-19 pandemic has an unprecedented global effect and brought about a significant change in the healthcare industry. With the increasing demand for accurate and fast diagnostic tools, scientists have been exploring various innovative technologies to detect SARS-CoV-2. Biosensors are among the most promising tools in the fight against the virus, providing rapid, accurate, and portable detection methods. BioFETs represent a particularly promising type of biosensor with considerable potential. FET stands for field-effect transistor, and FET-based biosensors use FETs to detect biological molecules such as proteins or DNA. FET-based biosensors have shown great potential for detecting SARS-CoV-2, the virus responsible for COVID-19. These biosensors are designed to detect biological molecules by measuring the changes in the electrical properties of a transistor-based device.

FET biosensors offer several advantages over traditional methods of detection. These advantages include their high speed, sensitivity, and selectivity, as well as their low cost and flexibility. Various BioFETs have been

Detection Science Series No. 24
Field-effect Transistor Biosensors for Rapid Pathogen Detection
Edited by Naif H. M. Al-Hardan, Azman Jalar, Muhammad Azmi Abdul Hamid and Mohd Firdaus Raih

Published by the Royal Society of Chemistry, www.rsc.org

developed for SARS-CoV-2 detection dealing with sensitivity and detection with and without sample treatment.

In this chapter, we will explore the basics of FET-based biosensors and how they are being used in the detection of COVID-19. There are many issues related to integration of SARS-CoV-2 FET biosensors with software applications connected to the Internet interface allowing data transfer to health care centers called Internet of Things (IoT). We will also discuss the advantages and limitations of FET-based biosensors for COVID detection.

6.2 SARS-CoV-2

6.2.1 Origin and Evolution

Coronaviruses have long been known to have a strong potential for pandemic transmission, and they represent a varied group of viruses that can cause mild to severe respiratory infections in humans. This severe acute respiratory syndrome coronavirus 2 (SARS-CoV-2), also referred to as COVID-19 (coronavirus disease 2019) is the ninth coronavirus known to infect humans and the seventh coronavirus to be discovered in the last 20 years.[1] It is extremely contagious and has quickly spread throughout the entire world. In terms of both the number of infected people and the geographic spread of epidemic locations, it has considerably surpassed SARS and MERS[†].[2]

According to epidemiological statistics, the Huanan market in Wuhan was a large and early site of SARS-CoV-2 infection. This market for the sale of wild animals was the source of 28% of all coronavirus disease 2019 (COVID-19) cases and two of the three earliest known cases. SARS-CoV-2 was found in environmental samples taken from the Huanan market after it was shut down, mainly in the western section that dealt in items from domestic and wild animals and in nearby drainage regions.[3]

To date, no intermediate animal host or reservoir for SARS-CoV-2 has been discovered in bats despite the reports.[4,5] This is presumably because the right animal species and/or populations have not been sampled yet, as well as the possibility that any progenitor viruses have a low prevalence. It is also extremely likely that the first cross-species transmission episodes went unnoticed. Just a very small percentage of SARS-CoV-2 case infections will have led to sustained onward transmission and only a very small fraction of spillovers from animals to humans result in major outbreaks.[1]

The COVID-19 outbreak coincided with the Lunar New Year, and virus transmission in China was facilitated by travel between cities prior to the celebration. Soon, other cities in Hubei province as well as other regions of China were affected by this new coronavirus pneumonia. China reached its COVID-19 epidemic peak in February. The overall number of cases increased

[†]Two highly pathogenic coronaviruses with a zoonotic origin, SARS-CoV and MERS-CoV (Middle East respiratory syndrome coronavirus) which emerged in humans in 2002 and 2012, respectively, and caused fatal respiratory illness.

significantly at an average pace of more than 3000 newly confirmed cases every day.[2] This is when the Chinese government implemented unprecedented restrictive public health measures including lockdown and blocked travels from and to Wuhan, and owing to these actions, the daily case number was declining. From the end of February, the spread of COVID-19 across international borders has intensified and an increasing number of countries have reported large clusters of infections. The virus, which spreads very efficiently, has been able to travel across the world quickly due to the widespread nature of international travel. The World Health Organization recognized the outbreak as a pandemic on 11 March 2020.[6,7] According to the COVID-19 dashboard of the Centre for System Science and Engineering at Johns Hopkins University, the number of COVID-19 cases reported globally has exceeded 20 million across all six continents as of August 11, 2020. Over 733 000 individuals have lost their lives due to the pandemic. Areas with limited access to medical services have experienced particularly high mortality rates.[2]

6.2.2 SARS-CoV-2 Symptoms and Mortality

6.2.2.1 Symptoms

Many healthcare facilities in Wuhan, Hubei province, China, reported clusters of patients with pneumonia of unknown origin in late December 2019. These individuals exhibited viral pneumonia symptoms, such as fever, coughing, and chest pain, which were similar to those seen in SARS and MERS patients. In more severe cases, these patients also displayed dyspnoea and bilateral lung infiltration.[2]

From two to fourteen days after exposure to SARS-CoV-2, COVID-19 symptoms, which might include fever, exhaustion, and a dry cough, resemble those of Middle East respiratory syndrome (MERS). Patients can experience aches and pains, nasal congestion, a runny nose, a sore throat, diarrhoea, a new loss of taste or smell, and frequent trembling and chills in certain situations (Figure 6.1). Usually minor, these symptoms appear gradually. Some persons (asymptomatic) contract the infection but do not show any signs of illness or discomfort. Extreme COVID-19 symptoms are linked to immune system tissue damage. One of the severe respiratory symptoms of COVID-19, which is caused by certain immune cells, is chronic lung inflammation. COVID patients exhibiting pneumonia have a high number of T-cells, which indicates that about 70% of the alveolar macrophages, which are common immune cells located in the lungs, harbour SARS-CoV-2.[8]

SARS-CoV-2, a substantial amount of data are required to quantify circulating antibodies. While understanding immune memory is important for assessing long-term protective immunity, it is challenging to make definitive statements based solely on antibody quantification. The respiratory system is highly impacted by COVID-19, with symptoms such as cough and

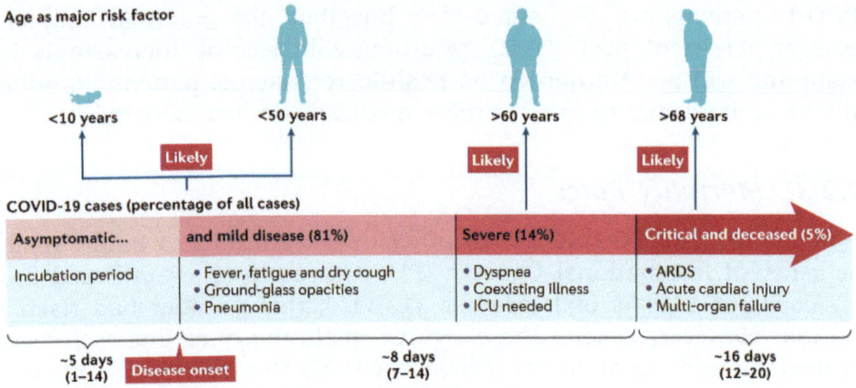

Figure 6.1 Clinical features of COVID-19. Reproduced from ref. 2 with permission from Springer Nature, Copyright 2020.

shortness of breath being common. Medical equipment such as pulse oximeters, chest X-rays, and pulmonary angiograms can be used to monitor lung function and diagnose associated complications like pneumonia and pulmonary embolism. However, there is still much that researchers do not know about the long-term effects of COVID-19 on the body.[8]

6.2.2.2 Post-COVID Effects or "Long COVID"

Most SARS-CoV-2 infected individuals often fully recover within a few weeks. The term "post-COVID-19 syndrome" or "post-COVID-19 conditions" is typically used to describe the phenomenon where certain patients (even those who had minor variants of COVID-19) frequently continue to experience symptoms following their initial recovery. The conditions typically last for longer than four weeks, and older people and those with weaker immune systems are more likely to suffer them. Recent reinfections that have been happening at sporadic intervals are linked to a rapid reduction in antibody levels, which raises severe concerns that immunity against SARS-CoV-2 could fade within weeks of recovery from the infection with the immune system's memory lasting for roughly six to eight months in some people.[8]

Long COVID, also known as persistent post-COVID syndrome, is a pathological condition that is characterized by persistent physical, medical, and cognitive sequelae resulting from COVID-19. This includes persistent immunosuppression, as well as fibrosis of the pulmonary, cardiac, and vascular systems, which can lead to an increased mortality rate and a severely worsened quality of life. The postdrome period, which is referred to as persistent post-COVID syndrome, has received little attention compared to the presymptomatic and prodromal periods of COVID-19; efforts are currently focused on studying the mechanisms and manifestations of COVID-19, and inhibiting the immuno- and fibrosis modulator transforming growth factor beta (TGF-β) may help alleviate these post-COVID sequelae. As more

COVID-19 patients are discharged from hospitals, the potential long-term effects of persistent post-COVID syndrome will become increasingly important and will place a burden on healthcare systems, patients' families, and society in general to care for these medically devastated survivors.[9]

6.2.2.3 *Mortality Rates*

On January 30, 2020, COVID-19 was officially recognized as a Public Health Emergency of International Concern (PHEIC), reaching a confirmed mortality count of 171. As of December 31, 2020, this number had risen to 1 813 188. However, despite these figures, preliminary estimates indicate that the total number of deaths related to COVID-19 worldwide in 2020 was at least 3 million, which represents 1.2 million more deaths than those that were officially reported. The latest data from the WHO put the number of COVID-19 deaths at over 3.3 million, which suggests that there may be a significant undercount of deaths that can be directly or indirectly attributed to COVID-19. COVID-19 deaths serve as an important indicator to track the progress of the pandemic.

Nevertheless, many countries lack the necessary civil registration and vital statistics systems that can provide accurate, complete, and timely data on deaths and causes of death. In order to address these obstacles, numerous countries have resorted to utilizing excess mortality as a more precise indicator of the consequences caused by the pandemic.

Excess mortality is the difference between the expected number of deaths and the actual number of deaths during a crisis. COVID-19 excess mortality accounts for both the direct impact of the virus on mortality as well as the indirect effects, such as disruptions to essential health services and travel.[10]

6.2.2.4 *Virus Variants*

Indeed, viruses such as SARS-CoV-2 undergo genetic mutations or viral recombination during genome replication, resulting in the emergence of different variants from the original virus. These mutations have given rise to new variants that are genetically related, and multiple variants have been observed worldwide during the pandemic. There are about 12 lineage groups with a common ancestor. The well-known variants include Alpha (B.1.1.7 and Q lineages), Beta (B.1.351 and its descendants), Gamma (P.1 and its descendants), Epsilon (B.1.427 and B.1.429), Eta (B.1.525), Iota (B.1.526), Kappa (B.1.617.1), 1.617.3, Mu (B.1.621, B.1.621.1), Zeta (P.2), Delta (B.1.617.2 and AY lineages), and the most recent variant of concern Omicron (B.1.1.529 and BA lineages).[8]

These variants are classified based on their potential to infect the public and evade host immunity by various healthcare associations, including the WHO, as variants of interest (VOIs), variants of concern (VOCs), and variants being monitored (VBMs). Organoid models, such as human airway, alveolar, and intestinal organoids, have been found useful in evaluating the infectivity potential of SARS-CoV-2 variants. Studies have shown that mutations in the S1 domain of the spike protein, in addition to those in the receptor-binding domain (RBD), increase the affinity of the virus for the human ACE2 receptor, thereby increasing its transmissibility. Alpha, Beta, Delta, Gamma, and Omicron are the most common VOCs, with Omicron being the most recent and highly infectious.[11]

6.2.2.5 Therapeutics and Vaccines

6.2.2.5.1 Treatments. Two major categories of treatments can be distinguished based on their objectives: medications that act against viruses and treatments that target the host (see Table 6.1).[12]

6.2.2.5.2 Vaccines. Researchers conducted multiple clinical studies to assess the effectiveness of COVID-19 vaccines in preventing the spread of the virus among asymptomatic individuals for a prolonged duration.[8]

Table 6.1 SARS-CoV-2 treatment categories.

Antiviral agents		Targeting host	
Polymerase inhibitors	Blocking the function of RNA polymerases is an effective way to prevent the replication of several types of viruses, including coronaviruses.	Neutralizing antibody therapy	Convalescent plasma
Protease inhibitors	The processing of coronavirus polyproteins relies heavily on protease, which is a crucial enzyme in this biological process.		Neutralizing monoclonal and polyclonal antibody therapies Can be produced on a large scale and are easily scalable as they accurately target neutralizing sites
Inhibitors of nucleoside and nucleotide reverse transcriptase	Exhibit antiviral activity against a wide range of viruses	Janus kinase inhibitors	COVID-19 is treated with kinase inhibitors because they hinder virus-induced immune activation and the signaling process of inflammation.
Entry and uncoating inhibitors	Prevent the initial phase of viral replication	Steroids	Researchers have shown in the past decade that administering corticosteroids (used by patients with inflammatory diseases such as RA and asthma) can increase the chances of survival for severe cases of COVID-19.

The current vaccines mainly use a nucleic acid-encoded version of the SARS-CoV-2 spike protein to introduce genetic information into host cells.[13]

COVID-19 vaccines are designed according to the whole-microbe approach. This approach uses the virus or bacterium in whole to design vaccines; there are inactivated vaccines, live-attenuated vaccines, and viral vector vaccines.

The inactivated vaccine utilizes either chemical or physical means to render the virus or bacterium that causes the disease inactive or killed. This can be achieved through methods such as psoralens, hydrogen peroxide, gamma irradiation, UV treatment, heat, formaldehyde, or β-propiolactone. Although it can be mass-produced, specialized laboratory facilities are required to cultivate viruses or bacteria, which can be time-consuming. Some potential drawbacks of this vaccine platform include adverse reactions such as fever, a possible link between vaccines and autoimmune disorders, and stress-related responses to immunization.

Live-attenuated vaccines require cultivation of the virus under unfavourable conditions or genetic manipulation through processes such as recombination, deletion mutants, and codon deoptimization. These approaches weaken the living virus or create similar viruses that limit replication, while triggering an immune response similar to natural infections without causing disease. The measles–mumps–rubella (MMR) vaccine, chickenpox vaccine, and shingles vaccine are some examples of successful live-attenuated vaccines.

Vaccines that use viral vectors as a platform utilize modified and safe vector viruses, such as adeno- and poxviruses, to deliver specific sub-parts of the germ of interest into the body, thereby triggering an immune response. These vaccines are classified into two types: non-replication vector vaccines and replication vector vaccines. Non-replication vector vaccines enter cells and produce vaccine antigens but are unable to replicate, while replication vector vaccines can not only produce pathogen-specific antigens but also replicate and produce infectious virus vectors, resulting in a new round of infection. This process allows for the expression of more antigens, replication, and consequently a stronger stimulation of the immune response after the first-time cell infection.[14]

6.2.2.6 Pandemic Current Situation

Presently, there is an ongoing COVID-19 global pandemic caused by SARS-CoV-2, with about 271 million individuals infected and approximately 5.31 million deaths recorded so far. The coronavirus family, to which SARS-CoV-2 belongs, is highly susceptible to mutations. Recently, a new strain of SARS-CoV-2 known as the "Omicron variant" was identified. This variant was first detected in Botswana on November 11, 2021, and South Africa on November 14, 2021. The World Health Organization designated the Omicron variant as a variant of concern on November 26, 2021. This could suggest that the Omicron variant may be more infectious and dangerous than previous variants, potentially leading to reduced expected outcomes. The Omicron variant is the sixth variant of concern on the WHO's current list, along with Delta, Alpha, Beta, Gamma, and Epsilon. It is crucial to determine whether the new variant can evade the antibody neutralization developed by previous variant infections or vaccinations. The SARS-CoV-2 spike protein has 37 amino acid mutations in the Omicron variant, with 25 of them unique to this strain and 12 shared with other variants.[15]

6.2.3 Definition of the Virus and Its Evolution

6.2.3.1 Virus Structural Properties and Biomarkers

SARS-CoV-2 is a β-coronavirus, which is an enveloped non-segmented positive-sense RNA virus (subgenus: Sarbecovirus, subfamily: Orthocoronavirinae).[6] Coronaviruses (CoV) are divided into four genera, including α-/β-/γ-/δ-CoV. α- and β-CoV are able to infect mammals, while γ- and δ-CoV tend to infect birds.[4]

There are four structural proteins that make up the SARS-CoV-2 capsid: the envelope protein (E), the membrane protein (M), the nucleocapsid protein (N), and the spike glycoprotein protein (S) (Figure 6.2). The nucleocapsid protein coats the viral RNA genome and plays a critical role in replication and transcription. The spike protein, which is composed of two subunits (S1 and S2), is responsible for binding to angiotensin converting enzyme-2 (ACE2) receptors present on the surface of various human cells, including those in the lungs, thereby facilitating virus entry. The S1 subunit contains the receptor binding domain (RBD) and is involved in binding to the ACE2 receptor, while the S2 subunit is responsible for viral fusion and entry. These proteins, particularly the spike protein, are the most prominent external features of the virus and thus represent attractive targets for detection systems that do not rely on viral genetic material, such as qRT-PCR.[16]

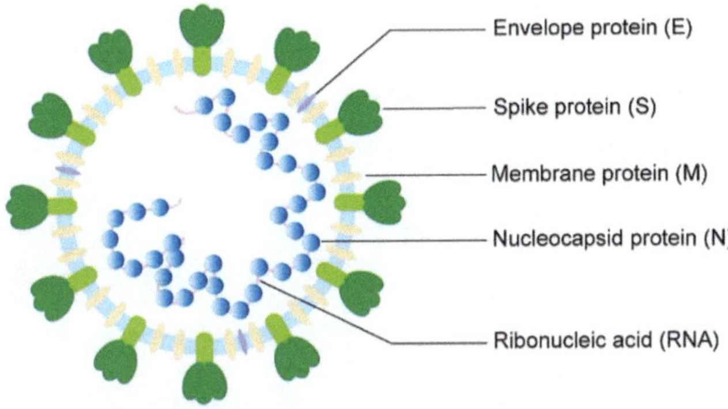

Figure 6.2 Structure of SARS-CoV-2. Reproduced from ref. 17 with permission from American Chemical Society, Copyright 2021.

6.2.3.2 *Transmissibility*

The high transmissibility of SARS-CoV-2 can be attributed to its unique virological characteristics. Unlike SARS-CoV, where transmission mainly occurred after the onset of illness, SARS-CoV-2 viral load in upper respiratory tract samples was highest during the first week of symptoms. This makes the risk of pharyngeal virus shedding very high at the beginning of infection. COVID-19 patients can spread the virus through liquid droplets when speaking, but smaller and more numerous aerosol particles can also linger in the air and be inhaled by others, leading to airborne transmission. Ferret experiments also showed evidence of airborne transmission. The virus can also be transmitted through the ocular surface and persist on surfaces for several days, posing a prolonged risk of infection. These findings explain the rapid spread of COVID-19 and highlight the importance of public health interventions to reduce transmission, as successfully implemented in China and other countries like South Korea.[2]

6.2.3.3 *Current Methods of Detection*

Finding immunological antibodies that can recognize the SARS-CoV-2 virus is a practical and affordable way to diagnose infections and ascertain their present state. After getting a SARS-CoV-2 vaccination, it is crucial to check the number of neutralizing antibodies in order to determine how well it worked. The spike S1 protein is positively charged with an isoelectric point of 8.28, whereas the SARS-CoV-2 spike antibody has a negative charge in foetal bovine serum at a pH level of 6.8–7. This causes a significant electrostatic attraction between the two.[18]

The identification of COVID-19 involves the targeting of biological markers that can indicate the presence of the virus without causing harm to the body (Figure 6.3). In this process, various biomarkers can be targeted

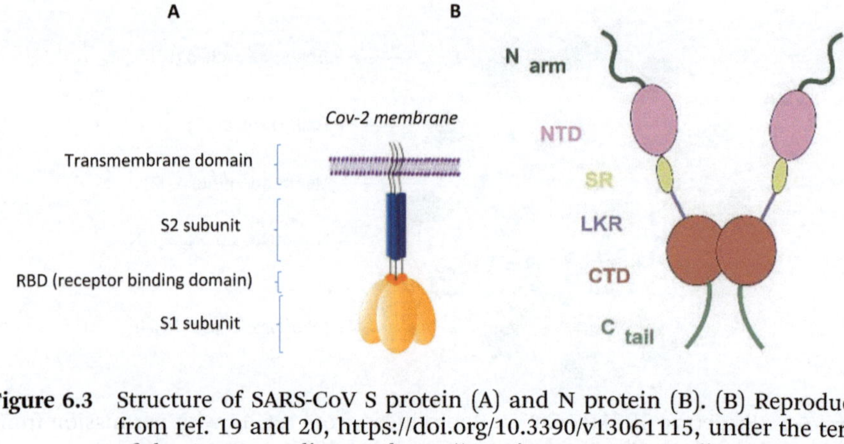

Figure 6.3 Structure of SARS-CoV S protein (A) and N protein (B). (B) Reproduced from ref. 19 and 20, https://doi.org/10.3390/v13061115, under the terms of the CC BY 4.0 license, https://creativecommons.org/licenses/by/4.0/.

structural proteins such as spike (S) and nucleocapsid (N) are commonly used as biomarkers for detecting the virus through using antibodies or the antigen detection method. The detection could be also performed using molecular diagnostics by targeting the genome of SARS-CoV-2.[8]

6.2.3.4 PCR/Molecular Methods

Molecular methods such as the real-time reverse transcription polymerase chain reaction (RT-PCR) technique can also detect SARS-CoV-2. In this method, transcriptase enzymes are used to convert RNA from COVID-19 to cDNA, which is then subjected to real-time PCR for DNA amplification. This method is highly sensitive and considered the gold standard, although it can be expensive and has a detection limit of about 250 genomic copies per mL.

Real-time loop-mediated isothermal amplification (RT-LAMP) is another molecular method that can detect SARS-CoV-2. It does not require a thermocycler and can detect as few as 1.5 genomic copies per μL, making it a more time-efficient option.

Amplicon-based metagenomic sequencing is a PCR-based method that amplifies certain regions in a gene and then uses next-generation sequencing (NGS) for evaluation. The resulting data can be compared against a database, enabling the diagnosis of different coronavirus strains. However, this method is also sensitive and expensive.

Nucleic acid hybridization using microarrays is a molecular method that involves converting RNA to cDNA and adding it to a well containing oligonucleotides specific to the virus. The complex is then purified and the signal is measured. Both the amplicon-based metagenomic sequencing and nucleic acid hybridization techniques are sensitive and expensive.[8]

6.2.3.5 Immunoassays

Measuring antibodies or neutralizing antibodies is crucial and immunoassays play a major role in this process. IgG/IgM antibodies are classified based on their glycoprotein antigen, and the main concern is to ensure that the IgG/IgM antibody test is specific to SARS-CoV-2 and does not cross-react with other coronaviruses. Cryogenic electron microscopy (Cryo-EM) structure analysis has shown that the binding affinity of the SARS-CoV-2 spike protein to ACE2 is much higher than that of the SARS-CoV spike protein, and while some cross-reactive binding may occur, cross-neutralization of live viruses is not as common. The spike protein is unique and ensures the specificity of detecting the neutralizing antibody of SARS-CoV-2, even though the nucleocapsid protein is abundant. Although both IgG and IgM antibodies to the spike protein strongly correlate with virus neutralization, their presence in serum occurs at different phases after infection, and it is important to choose an appropriate time for collecting blood samples for serological tests.[21]

Enzyme-linked immunosorbent assay (ELISA), chemiluminescence enzyme immunoassays (CLIAs), fluorescence immunoassays (FIAs), and lateral flow immunoassays (LFIAs) are the most commonly used immunoassays. All these methods rely on immunoaffinity binding, where the target IgG/IgM antibody binds to an anti-human IgG/IgM antibody and a SARS-CoV-2 antigen, such as the virus spike protein or nucleocapsid protein.[21]

The primary aim of immunoassays is to detect SARS-CoV-2 specific IgG/IgM antibodies using different methods like ELISA, CLIA, FIA, and LFIA. The target IgG/IgM antibody is captured using SARS-CoV-2 antigen, while the anti-human IgG/IgM antibody is conjugated with an enzyme, fluorescent tag, or gold nanoparticles to produce signals. In ELISA and CLIA, the enzyme HRP or ALP is used to produce a colour or light emission that is quantifiable using optical absorbance or detection instruments. Magnetic particles are used in CLIA for automated processes. FIA detects fluorescence signals, requiring a specific fluorescence analyzer. In contrast, LFIA uses immunochromatography, where gold nanoparticles conjugated with the SARS-CoV-2 antigen and the target IgG/IgM antibody produce a colour band available for visual inspection. ELISA and FIA are the gold standard methods for quantitative analysis of antibodies.[21]

Recently, new microfluidic methods have been developed for detecting SARS-CoV-2 antibodies in the fight against COVID-19. One approach developed by Swank *et al.* utilized a microfluidic device with mechanically induced trapping of molecular interaction (MITOMI). In this method, the target anti-SARS-CoV-2 antibodies bind to an immobilized SARS-CoV-2 spike protein and generate fluorescent signals when conjugated with an anti-human IgG antibody labelled with phycoerythrin. This device achieved a limit of detection of approximately 1 nM of IgG and achieved 100% specificity and 98% sensitivity when tested on 289 serum samples, including 155 positive and 134 negative samples. The authors claimed that this device performed as well as, or better than, the standard ELISA. Another device developed by Funari *et al.*[23] used label-free opto-microfluidic sensing with gold nanospikes to detect SARS-CoV-2 spike protein antibodies in human plasma. By measuring the localized surface plasmon resonance wavelength shift caused by the local refractive index change when anti-SARS-CoV-2 antibodies bind to the surface, this device achieved a limit of detection of around 0.08 ng mL^{-1}, which falls within the clinical detection range.

6.2.3.6 Other Detection Methods

Various methods are available for detecting SARS-CoV-2, including point-of-care testing based on either lateral flow or biosensing that mostly relies on nanotechnology and antibodies (IgM and IgG) attached to gold nanoparticles or clustered gold nanoparticles, depending on plasmon band coupling. The most recent technique is the use of clustered regularly interspersed short palindromic repeats (CRISPR) technology coupled with fluorescence, enabling detection within an hour using the specific high sensitivity enzyme reporter unlocking (SHERLOCK) technique. Compared to qRT-PCR, which is also prone to false negative results, the CRISPR technology may be less expensive on an industrial scale, but it is still relatively new and its limitations are not yet fully understood. Technological advancements have made these detection methods possible within 24 hours or less.[8]

Field-effect transistor (FET) biosensors have gained significant attention as label-free sensing devices for biomolecules in various disease screening applications. These biosensors exhibit remarkable sensitivity, can be scaled down to a great extent, and have an exceedingly low limit of detection (LoD) for detecting molecules. They can be seamlessly combined with complementary metal oxide semiconductor (CMOS) technology, and their label-free operation has made them highly desirable in the biosensor community.[24]

6.3 Field-effect Transistors (FETs)

6.3.1 Definition and Basic Structure of FETs

A field-effect transistor, commonly known as FET, is a three-terminal semiconductor device in which the current flowing through one pair of terminals, known as the source (S) and drain (D), is controlled by a voltage applied to the third terminal, known as the gate (G).[25]

To understand how FETs work, it is important to first look at their basic structure. This structure comprises a source, a drain, and a gate. The source and drain are two terminals connected to the ends of a conducting channel, which is typically made of a doped semiconductor material. The gate is a third terminal that is separated from the channel by a thin insulating layer, such as silicon dioxide.

The channel of the FET is the pathway, constructed from a semiconductor material, where electrical charge (electrons, holes) passes through. The channel gives the device its name because it is the feature that is influenced by the presence of an electrical field, as generated by the voltage applied to the gate. The FET channel is usually made of a material such as silicon. When the free electrons within the channel material are subjected to an electric field, they deplete or enhance the number of free electrons available in the channel, thus creating a region of depleted or enhanced conductor.

The gate is a voltage-controlled region surrounding the channel and consists of a material such as polysilicon or other conducting materials. In the presence of an applied voltage, the electrons within the gate region will move and form a conducting layer within the gate. The applied voltage will have a direct effect on the charge stored in the gate and this charge, in turn, will affect the conductivity and resistance of the FET channel.

The source and drain regions located on either side of the gate region provide the inlet and outlet of the FET, allowing current to enter and exit the FET channel. The source and drain regions are usually made from a highly doped semiconductor material and charge is stored within the material when a voltage is applied.

Therefore, knowing the definition and basic structure of FETs is crucial in selecting the appropriate device for a specific application.

6.3.2 Working Principle of FETs

The FET operates on the basic principle of a "voltage-controlled current source (VCCS)".[26] This is achieved by manipulating the threshold voltage of the FET, which is the voltage at which the source–drain current begins to flow. Depending on the type of FET, the threshold voltage can be controlled either by applying a voltage to the gate terminal or by changing the doping concentration of the semiconductor material. This phenomenon, known as the depletion-mode operation, is the basis of how FETs are used to amplify signals.

This means that when no voltage is applied to the gate region, the current flow through the FET is limited. However, when a voltage is applied to the gate region, the current flow is enhanced or decreased depending on the voltage applied. This is due to the electric field created by the applied voltage, creating an enhanced zone of conduction within the channel and allowing current to pass. As a consequence, the FET can be used to amplify or attenuate signals and is used universally in the modulation or switching of electrical signals.

Thus, by understanding the physics behind the operation of FETs, one can gain a deeper appreciation of the device.

6.3.3 Classification of FETs

FETs are classified into two main categories: metal oxide semiconductor field-effect transistor (MOSFET) and ion-sensitive field-effect transistor (ISFET), each having its unique characteristics and applications.

6.3.3.1 Metal Oxide Semiconductor FETs (MOSFETs)

The MOSFET is a key component in the modern technology world. MOSFETs are made up of a metal-oxide layer that acts as a gate to control the current and the electrical charge, a conducting channel between the source and the drain, and a source and a drain connected to the channel to allow current to flow (Figure 6.4).

The operational mechanism of a MOSFET relies on the MOS capacitor, a semiconductor surface positioned beneath the oxide layers connecting the source and drain terminals. By applying a positive or negative gate voltage, the MOS capacitor can undergo an inversion process, transitioning from a P-type to an N-type semiconductor.

In P-channel MOSFETs, the majority of the current flowing is holes moving through the channel, while in N-channel MOSFETs the current carriers are mostly electrons. Figure 6.5 explains how an N-channel MOSFET works. The source and drain are negatively doped and are separated

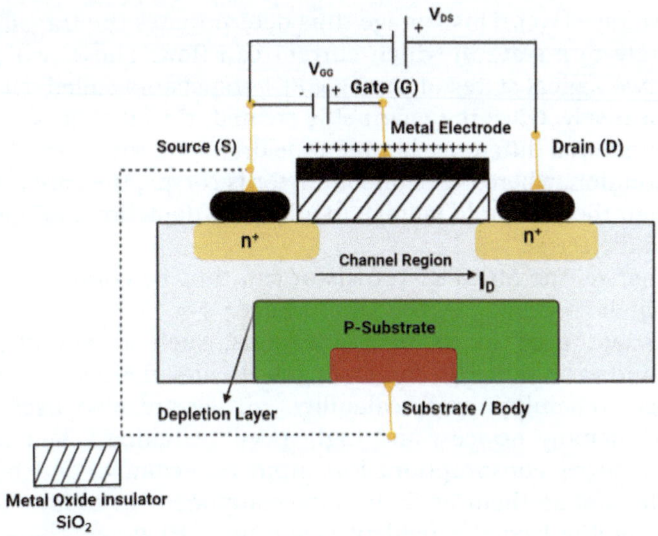

Figure 6.4 Schematic representation of a MOSFET.

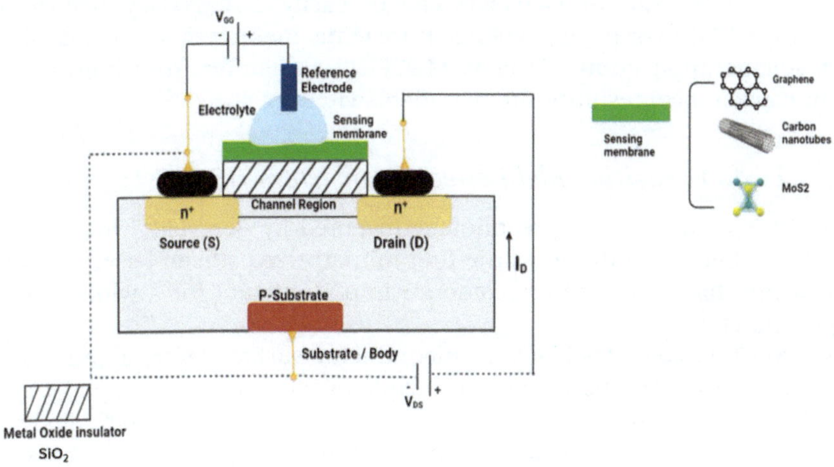

Figure 6.5 Schematic representation of the working principle of an ISFET.

by a P-type semiconductor. When a positive voltage is applied to the gate (V_{GG}), the semiconductor charge carriers which are also positive are repelled to the bottom of the semiconductor, generating a region called the depletion region. As the voltage at the gate increases, the depletion region expands and the minority carriers in the semiconductor, which in this case are electrons, begin to be exposed. If the increase in the gate voltage continues, it reaches a point where electrons are attracted to the top of the semiconductor, creating an N-channel that will allow the current to flow between the source and the drain. The voltage value from which the channel is generated is called the threshold voltage (V_{th}). This voltage thus determines the transition from a blocking state to a state in which current can flow. These two states constitute the two logical states of the MOSFET transistors called 'Off' and 'On' states, respectively. Once the channel is created, the current is produced by applying a potential difference between the drain and the source (V_{DS}). In the opposite situation, where the semiconductor is N-type, the gate voltage must be lower than the threshold voltage to allow the depletion and attract holes close to the dielectric.

To summarize, the MOSFET transistor can thus be compared to an electrostatic switch controlled by the gate voltage.

MOSFETs are used in many applications, such as amplifiers, signal switching, and various digital and analogue circuits, thanks to their superior performance, reliability, and scalability.[27] They are also used in power conversion, memory devices, and high-speed switching. Their advantages include low power consumption, low input capacitance, and high speed. Their use has made them an important component in today's electronics and has revolutionized the field of electronics. However, there are some limitations associated with MOSFETs. The main limitation is that they are temperature sensitive, *i.e.* operate in a limited temperature range. Furthermore, the gate oxide of MOSFETs can be easily damaged by electrostatic discharges (ESDs) or by high voltage transients, making them unsuitable for high-voltage applications. Finally, MOSFETs can suffer from high gate-to-drain capacitance, resulting in degraded switching speed.[28]

6.3.3.2 Ion-sensitive Field-effect Transistors (ISFETs)

The ISFET is an emerging technology invented by Bergveld[29] in the 1970s which has been introduced as the first miniaturized silicon-based chemical sensor and has since then become an important tool for a wide range of applications.

An ISFET is composed of a semiconductor surface layer, a gate, and a reference electrode. The working principle of ISFETs is the same as that of MOSFETs, but in this case the gate is replaced by the solution to be analyzed, the charges within it and a reference electrode (Figure 6.5).[30] The semiconductor surface layer is usually composed of a semiconductor material and a thin insulating layer of a dielectric material, such as silicon dioxide or silicon nitride, which protects the semiconductor from any direct contact with the sample. The gate is connected to the reference electrode, which is usually composed of a metallic material such as gold, silver, or platinum. The gate is then connected to a voltage source, which provides the necessary power to the gate to detect the charges at the surface layer. Both the gate voltage (V_{GG}) and the drain voltage (V_{DS}) will affect the measured drain current (I_D). Indeed, the gate voltage controls the carrier density and thus allows the modulation of the channel resistance by the field effect. In addition to this, the longitudinal polarization between the drain and the source influences the depletion region. The reference electrode is used to keep the voltage of the gate constant, which is necessary to ensure the accuracy of the measurement. When the charge of the sample changes, the charge of the gate is affected and the voltage of the gate is adjusted accordingly. This allows the ISFET to detect changes in the charge of the sample and thus act as a sensor.

ISFETs are increasingly being utilized in modern technology due to their unique sensing capabilities and their ability to detect changes in the concentration of ions in an environment. These transistors are commonly used in the medical industry, such as to test the pH balance of a patient's blood. ISFETs are also used to monitor food safety, water quality, and in agricultural applications. The potential applications of ISFETs are nearly limitless and their use in modern technology will only continue to increase. ISFETs are also applied in chemical and biomedical sensing.[31] By integrating ISFETs in microfluidics and lab-on-a-chip devices, researchers have made a significant progress in advancing the capabilities of drug delivery, medical diagnostics and anatomic studies. This technology has the potential to yield valuable insight into biochemical processes and the overall health of an individual. ISFETs can be easily designed, fabricated, operated and modified, and can provide accurate and reliable measurements, which makes them quite attractive for medical and clinical applications.

6.4 FET-based Biosensors (BioFETs)

6.4.1 Brief History

The BioFET or biosensor field-effect transistor is a technological innovation that has revolutionized the field of biomedical engineering. The technology behind the BioFET originated from the discovery of field-effect transistors (FETs) in 1925 by physicist Julius Lilienfeld. Since then, various improvements have been made to FETs leading to their use in electronic devices such as computers (Figure 6.6). However, it was not until 1952 that researchers began exploring FETs for use in sensing applications. In 1965, Carver Mead and Lynn Conway proposed using FETs for biological sensing

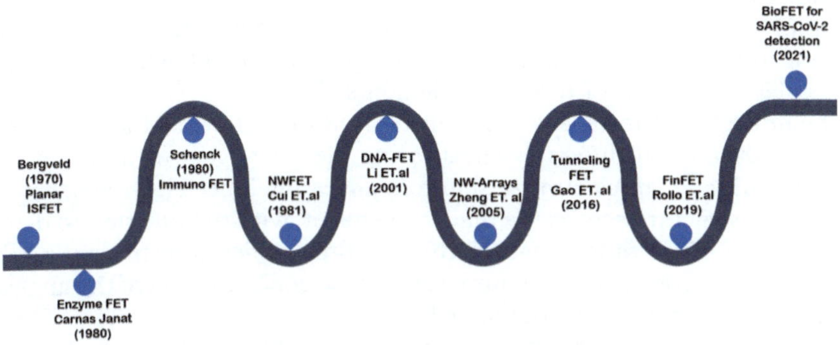

Figure 6.6 Chronological order of the evolution of BioFETs.

applications in their paper "Introduction to VLSI Systems." This proposal paved the way for research into creating a sensor capable of detecting biological molecules.

In the 1970s, two scientists from MIT, Dr. Robert S. Langer and Dr. David H. Koch, further developed the FET biosensor. They created a model that incorporated both a FET and an enzyme to detect glucose in solutions. This was the first example of an enzyme-based FET biosensor, which was then developed further to measure a variety of different biological molecules.

Since then, BioFETs have undergone significant advancements including miniaturization through nanotechnology processes such as silicon micromachining and soft lithography techniques used to create microfluidic channels compatible with nano-biosensors. These devices are capable of detecting and measuring the electrical signals generated by individual cells. BioFETs have a long and turbulent history that has allowed them to become one of the most widely used devices in biomedical sciences.

In recent years, BioFETs have been used in medical diagnostics and drug discovery research. They are becoming increasingly popular due to their high sensitivity, accuracy, and portability. The use of BioFET biosensors has enabled researchers to develop rapid, accurate and low-cost detection methods for a variety of diseases, viruses, toxins, and allergens. In addition, BioFETs are also being used in biosecurity applications to detect hazardous materials quickly and efficiently.

6.4.2 Definition and Basic Structure of BioFETs

A field effect transistor biosensor ("FET biosensor") also known as a biosensor field-effect transistor (Bio-FET or BioFET) or a field-effect biosensor (FEB) is a bio-sensing device that is able to detect and measure the presence of a specific molecule or biomolecule in a biological sample. This type of sensor is based on the principle of "conduction through a field effect transistor". A FET biosensor is defined as a type of transducer that converts a biological signal into a measurable electrical signal, which can be used to detect, measure, and monitor biological molecules and compounds[32] (Figure 6.7).

A BioFET typically consists of three main components: the FET structure, the transduction element, and the biorecognition element, which allows the molecular interaction of the target elements.[33]

The BioFET architecture is primarily derived from the ISFET and MOSFET designs, although it differs by substituting the metal gate with an ion-sensitive membrane, an electrolyte solution, and a reference electrode.

The FET structure acts as the main transducer, with the function to detect the presence of the target analyte. This could be achieved by measuring the electrical properties of the medium, such as the concentration of the analyte, in order to generate an output signal.

The source electrode (S) (typically made from gold): The source electrode in a FET biosensor is the active electrode that is responsible for providing

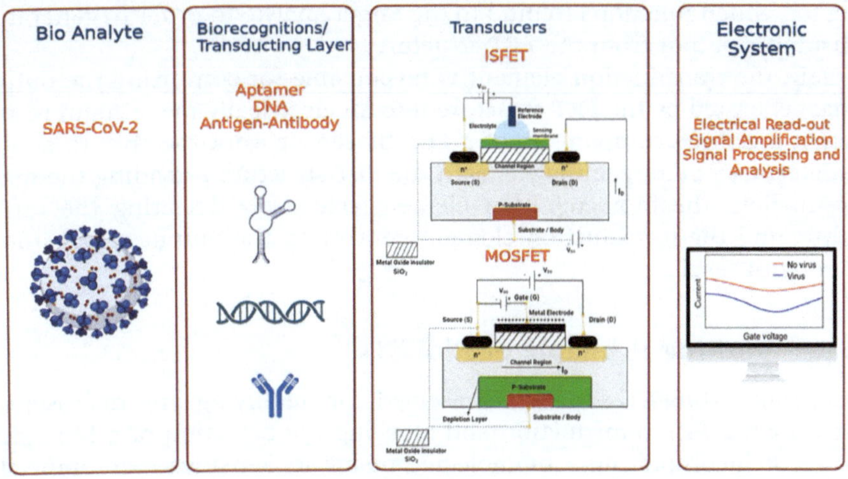

Figure 6.7 Illustration of a FET biosensor.

the electrical current to the device. It is also responsible for providing the necessary voltage to the gate of the FET, which is used to control the current flow through the device. The source electrode is also responsible for providing the necessary electrical connection between the FET and the biological material being measured.

The drain electrode (D): The drain electrode in a BioFET is typically made of a conductive material such as gold, platinum, or carbon nanotubes. The drain electrode in a FET biosensor acts as a current collector, allowing current to flow from the source electrode to the drain electrode. This current is used to measure the electrical properties of the biological material that is being tested. It is also used to control the voltage applied to the device, allowing for precise measurements.

The gate electrode (G): The gate electrode acts as a control element, allowing the current to flow through the channel between the source and drain electrodes when a voltage is applied. This component could be also modified to contain an immobilized DNA or antibody, allowing it to recognize and react to the analyte. When a voltage is applied, the gate enables the surface of the source and drain electrodes to bind to the analyte. The electrons then move from the source to the drain, creating a current that can be detected and measured. It is used to control the amount of current flowing through the channel, allowing for precise control of the biosensor's response.

These three electrodes act in a similar way to regular transistors, but with an optimized design for molecular recognition. In a regular FET, the source and drain electrodes are typically electrically conductive, allowing current to flow between them when a voltage is applied.

The biorecognition element is responsible for selectively detecting the target analyte from the sample. This is typically an antibody, DNA, or a receptor, which functions to bind to the target analyte in order to generate a measurable output from the FET structure.

Lastly, the transduction element is responsible for converting the output signal generated by the FET structure into an electrically meaningful result. Overall, the three components of a FET biosensor work together to enable the detection of a target analyte, with the FET structure providing the main transduction, the biorecognition element selectively detecting the target analyte, and the transduction element converting the output signal into a meaningful result.

6.4.3 Working Principle of BioFETs

The acronym BioFET should be reserved for qualifying any microsensor based upon a FET transduction and allowing the detection of a biological species or incorporating a biomolecule inside its sensitive membrane. By studying the basic structure of a BioFET and its components, we are able to gain insight into its design and performance.

Two distinct categories of BioFETs exist, distinguished by their respective transducers: ISFETs and MOSFETs.

ISFET-based BioFETs are a type of FETs in which the gate is replaced by an ion-sensitive membrane (Figure 6.8). When the target biomolecule binds to the membrane, the surface charge on the membrane changes, leading to a change in the electrical potential of the gate. This change is then amplified and sensed by the source and drain electrodes, allowing for the detection of the target molecule. The principle behind ISFET-based BioFETs is based on the fact that the charge on the ion-sensitive membrane changes as a result of the interaction between the target biomolecule and the membrane.

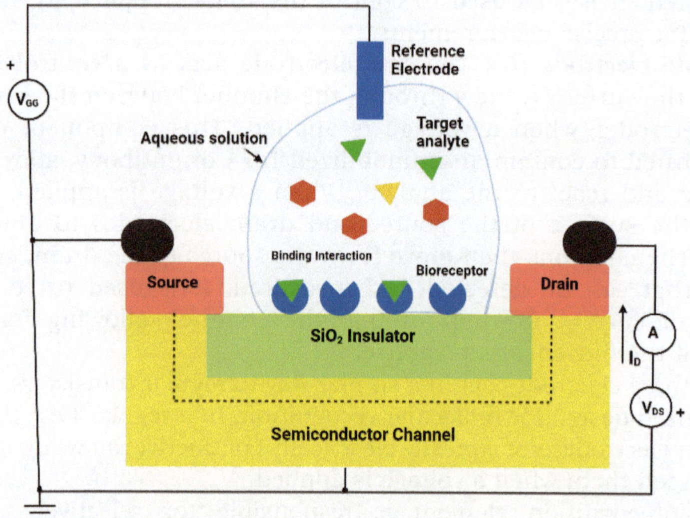

Figure 6.8 Schematic representation of an ISFET-based BioFET sensor.

On the other hand, MOSFET-based BioFETs are a type of FETs in which the gate is replaced by a biomolecular recognition element (such as an antibody or a DNA probe) (Figure 6.9). When the target biomolecule binds to the recognition element, it causes a charge transfer between the recognition element and the gate, which results in a change in the electrical potential of the gate. This change in potential is then sensed by the source and drain electrodes, allowing for the detection of the target molecule. The principle behind MOSFET-based BioFETs is based on the fact that the binding of the target biomolecule to the recognition element changes the electrical properties of the gate, which can be detected by the FET.

In summary, both types of BioFETs work based on the principle of detecting changes in the electrical potential of the gate as a result of the interaction between the target biomolecule and the transducer. However, the specific mechanism of interaction between the target biomolecule and the transducer is different for each type of BioFET.

Biosensing in FETs is performed through three processes: biorecognition – the identification of an analyte containing biomolecular species (such as proteins, antibodies, proteins, *etc.*) through modifications of the electronic properties of the sensor; a transductor operation, which converts the change in electronic properties into an electrical flag; and amplification of the signal for extended reading by the customer.

A numerical representation of the initial process or biorecognition event (BE) is given by

$$\mathrm{BE} = \mathrm{d}c\mathrm{d}\sigma/\mathrm{d}c \tag{6.1}$$

where dc is the change in the concentration of the analyte, considering the change in the conductivity dσ of the sensor due to the charge transfer process.

Thus, eqn (6.1) maps the sensor's response to the presence of the analyte through modifications to its electronic characteristics, *i.e.* its conductivity.

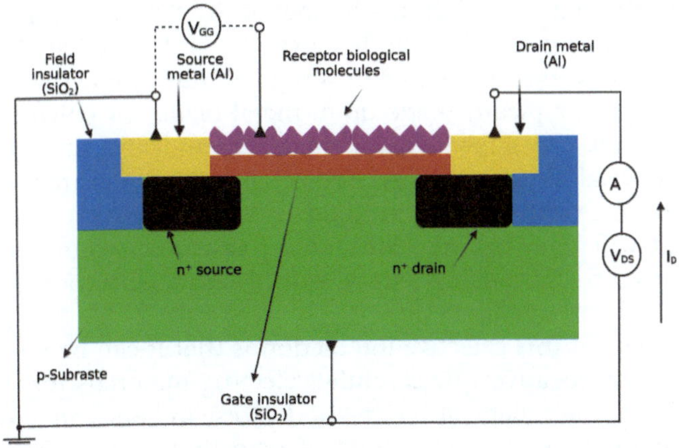

Figure 6.9 Schematic representation of a MOSFET-based BioFET sensor.

The viability of the sensor typically determines the first step in biosensing and the sensitivity of the sensor.

The following mathematical formula expresses the converter's next operation (TR):

$$\mathrm{TR} = \mathrm{d}V_{\mathrm{EG}}/\mathrm{d}\sigma(\mathrm{d}I_{\mathrm{D}}\ \mathrm{d}V_{\mathrm{EG}})/I_{\mathrm{D}} \quad (6.2)$$

Here the change in conductivity changes the threshold voltage and consequently the effective gate voltage V_{EG}, resulting in a change in the drain current I_D. Sensitivity is correlated to the magnitude of the change in this measurable parameter, the drain current, which indicates the presence of the biomolecule. Various types of BioFETs are made and proposed, relying more on BE or on the TR parameter, or on both.

6.5 BioFET Biosensing Devices for SARS-CoV-2

Presently, the detection of infectious diseases like SARS-CoV-2 is of the utmost importance. BioFETs have emerged as a promising tool for the detection of SARS-CoV-2 due to their high sensitivity, low cost, and real-time monitoring capabilities. In this chapter, we will discuss and explore the use of BioFETs for the rapid detection of SARS-CoV-2, highlighting their potential as a valuable tool in the fight against the COVID-19 pandemic.

6.5.1 ISFET Biosensors

In the midst of the COVID-19 pandemic, accurate and rapid detection of SARS-CoV-2 has become paramount. Luckily, advancements in biosensor technology have paved the way for innovative solutions to this global crisis. One such solution is ISFET biosensors – highly sensitive devices that can detect even trace amounts of the virus.

To use an ISFET biosensor for detecting SARS-CoV-2, a sensitive layer is deposited on the source and drain channel of the device (Figure 6.10). This layer is crucial as it determines both sensitivity and selectivity by immobilizing biomolecules like antibodies or antigens that interact with viral proteins. These layers, typically made up of metal oxides or polymers, act as a medium between the analyte being detected and the underlying semiconductor material. When samples containing SARS-CoV-2 are introduced to an ISFET biosensor with specific antibodies immobilized on its surface, any virus present will bind to these antibodies, causing a change in ion concentration around the sensing area which generates electrical signals that can be detected and analyzed.

One advantage of this selective interaction is that it can help to minimize false positives or negatives. By carefully selecting materials for these layers and optimizing their deposition methods, researchers can create ISFET biosensors with incredibly low limits of detection – some even down to single-digit femtomolar concentrations! The importance of these sensitive

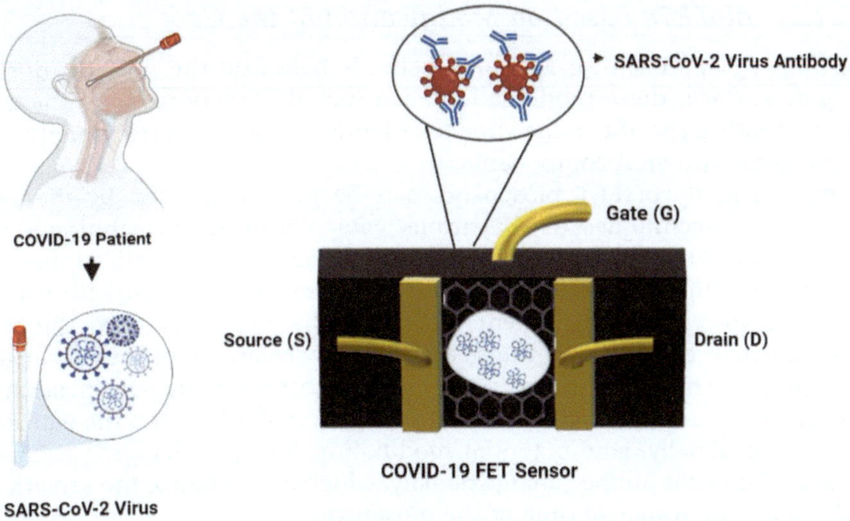

Figure 6.10 Scheme of a graphene-based BioFET for detection of COVID-19 (ISFET).

layers cannot be overstated when it comes to achieving accurate and reliable sensing devices for detecting SARS-CoV-2.

ISFET biosensors hold great promise for future disease diagnosis and monitoring technologies, thanks to their unique features and ability to be integrated into portable point-of-care testing systems.

6.5.1.1 BioFETs Based on Modification of the Channel (Source–Drain)

Channel (source/drain) modification plays a key role that allows more accurate detection of viruses, SARS-CoV-2.

A recent study has developed a BioFET that could accurately detect the virus in only 5 minutes, thanks to a modification of the source–drain. In a study conducted by Wasfi *et al.* (2022), a layer of gold was introduced onto the device along with the bimetallic nanoparticles of platinum and palladium. The synthesis of these nanoparticles was achieved using a sputtering and inert-gas condensation technique within an ultra-high vacuum (UHV) compatible system.[34] The findings indicate that the GO-FET biosensor was enhanced by incorporating bimetallic nanoparticles consisting of platinum and palladium. This modification aimed to explore the enhancement in sensor sensitivity, particularly in detecting the COVID-19 spike antigen at a low detection limit (LoD) of 1 $fg\,mL^{-1}$ in PBS solution. Additionally, the biosensor's performance was investigated by studying the effects of magnetic labelled SARS-CoV-2 spike antibodies, with the goal of further enhancing sensor capabilities. The results demonstrate the successful development of a highly promising field effect transistor biosensor for COVID-19 diagnosis.

6.5.1.2 *BioFETs Based on Modification of the Gate*

A promising approach for FET biosensing is based on the modification of the gate surface; these modifications can include functionalizing the gate with antibodies specific to the virus to enhance the surface area available for interactions with viral components.

The sensitivity of FET biosensors can be greatly affected by the gate modification techniques used. Various gate modification techniques are available for FET biosensors,[35] including surface functionalization, gate dielectric modification, and gate potential modulation. Surface functionalization involves the attachment of specific biomolecules onto the gate surface, which can enhance the selectivity and sensitivity of the biosensor. Gate dielectric modification, on the other hand, involves the modification of the gate dielectric layer to improve the charge transfer between the gate and the analyte. Finally, gate potential modulation involves changing the gate voltage to alter the surface charge density, which can enhance the sensitivity and reduce the response time of the biosensor.

Recently, several articles have proposed different examples of gate modification for SARS-CoV-2 detection. One of the approaches proposed by Toral-Lopez *et al.* (2022)[36] used a multiscale numerical approach to investigate the performance of graphene-based BioFETs. The authors utilized a combination of molecular dynamics simulations and quantum mechanics calculations to model the interaction between graphene and biomolecules such as *angiotensin converting enzyme 2* (*ACE2*). The results of their simulations showed that the presence of biomolecules can significantly affect the electronic properties of graphene-based BioFETs. The authors also found that the sensitivity of the BioFETs to the presence of biomolecules is highly dependent on the size and charge of the biomolecules.

6.5.2 MOSFET Biosensors

In the perplexing fight with coronavirus, technology and science are assuming an essential job. MOSFET-based biosensors are significant advancements which are being used as a weapon to battle SARS-CoV-2.

The development of MOSFET biosensors for SARS-CoV-2 detection started from isolation of viral RNA, which serves as the target molecule for the detection process. A MOSFET biosensor comprises a transistor that is coated with a specific molecular probe that is complementary to the target sequence. When the viral RNA molecule binds with probe molecules, it creates an electrical signal on the transistor (Figure 6.11). The MOSFET biosensor measures and records the changes in electrical proprieties induced by the viral RNA, thus yielding highly specific reading.

For example, in their recent articles, Biswas *et al.* (2022)[37,38] described a novel biosensor known as Bio-RFET which incorporates both hetero-material (HM) and hetero-dielectric (HD) layers. This biosensor is designed to detect biomolecules such as proteins and nucleic acids with high sensitivity and

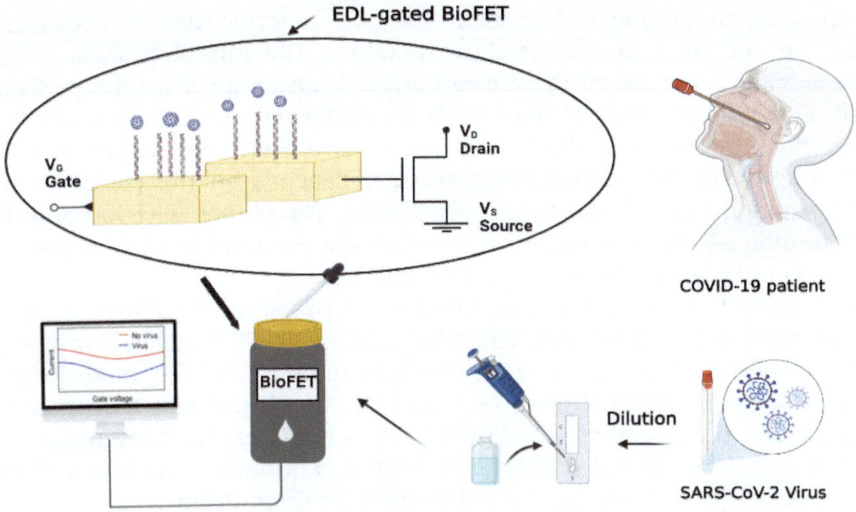

Figure 6.11 Scheme of an EDL-gated BioFET for detection of COVID-19 (MOSFET).

selectivity. The Bio-RFET biosensor consists of a high-κ dielectric layer, which enhances the sensitivity of the device, and a low-κ dielectric layer, which improves the selectivity. The hetero-material layer is composed of a combination of two different materials, such as silicon and germanium, which creates a built-in potential due to the difference in their work functions. This built-in potential allows for the detection of charged biomolecules. The hetero-dielectric layer is made up of two different dielectric materials, such as silicon oxide and silicon nitride, which creates a selective barrier for the detection of specific biomolecules.

The HD-HM-RFET biosensor operates on the principle of modulating the electrical conductivity of the transistor channel *via* hybridization of target biomolecules with the probe DNA immobilized on the gate electrode.

The proposed Bio-RFET biosensor with hetero-material and hetero-dielectric layers presents an innovative solution to this challenge with its ability to detect biomolecules through the label-free dielectric modulation technique. The research article further demonstrates the effectiveness of (SiO_2 + TiO_2) and an AlGaAs/Ge-based HD-HM-RFET in biosensing. As the world continues to grapple with the effects of COVID-19 and other potential health crises, the development of efficient and cost-effective biosensors will remain a crucial area of research and innovation.

6.5.3 Biorecognition Elements in SARS-CoV-2 BioFETs

A BioFET is a biosensor that uses a field-effect transistor (FET) to detect biomolecules. The biorecognition element is a crucial component of the BioFET sensor. It is responsible for the selective binding of target biomolecules to the sensor surface, which leads to changes in the electrical properties of the FET. The biorecognition element can be an antibody, an aptamer, an enzyme, or any other biological molecule that can specifically recognize and bind to the target biomolecule. The immobilization of the biorecognition element on the sensor surface is also critical for the performance of the BioFET sensor. Various methods, such as covalent binding, physical adsorption, and self-assembled monolayers, have been used for immobilization. The choice of immobilization method depends on the nature of the biorecognition element and the sensor surface. The proper selection and immobilization of the biorecognition element are essential for the successful detection of biomolecules using BioFET sensors.

Various FET biosensors can be categorized based on the biorecognition element utilized for detection. These include the En-FET, which involves an enzyme modification of the FET, the Immuno-FET, which incorporates immunological modifications, the DNA-FET, which involves DNA modifications, and the aptamer-FET, which utilizes aptamers for recognition.

Moreover, biorecognition elements, such as antibodies, aptamers, DNA, and enzymes, have been extensively studied for their ability to bind specifically to SARS-CoV-2 target molecules and facilitate virus detection.

6.5.3.1 Antibody-based FET Biosensors

Antibody-based FET biosensors have emerged as a promising diagnostic tool for the detection of SARS-CoV-2. An antibody-based BioFET has been developed based on its ability to recognize specific SARS-CoV-2 spikes. The antibody could target the N-protein or S-protein and where both proteins are present. However, various studies demonstrated that the N-protein is present before symptom apparition and could be a better biomarker (21).

According to Krishna *et al.* (2022),[39] utilizing a silicon-based label-free electrical device, like the reduced graphene oxide field-effect transistor (RGO FET or GFET), that has been specifically modified with monoclonal antibodies (mAbs) targeting SARS-CoV-2, has demonstrated encouraging outcomes when detecting the antibody–antigen interaction associated with SARS-CoV-2 across various molar ranges. This characteristic showcases its potential applicability for point-of-care (PoC) diagnostic tests, emphasizing its suitability for rapid and on-site detection. This study shows that the device characteristics have been observed to different extents and have been found to be sensitive to charged biomolecules, including those absorbed by the receptor. The authors also found that GFETs can detect the SARS-CoV-2 virus at concentrations as low as 2×10^{-18} M, indicating their suitability for PoC diagnostic tests. These findings suggest that RGO FETs have the potential to be used as a rapid and sensitive diagnostic tool for the detection of the SARS-CoV-2 virus.

A recent study by Seo *et al.* (2020)[40] presents a novel biosensor based on FETs for COVID-19 diagnosis. The biosensor is composed of a graphene FET (GFET) functionalized with anti-SARS-CoV-2 antibodies. When the GFET comes into contact with a sample containing the virus, the binding of the antibodies to the virus causes changes in the electrical properties of the GFET, which can be measured and used to determine the presence of the virus. The FET-based biosensor was evaluated using both antigen protein and clinical samples. The study found that the FET-based biosensor had a high sensitivity and specificity in detecting the antigen protein. The FET device exhibited the ability to identify the presence of the SARS-CoV-2 spike protein at an extremely low concentration of 1 $fg\,mL^{-1}$ in phosphate-buffered saline and 100 $fg\,mL^{-1}$ in clinical transport medium. Furthermore, the FET sensor effectively detected SARS-CoV-2 in culture medium with a limit of detection (LOD) of 1.6×10^1 PFU per mL and in clinical samples with an LOD of 2.42×10^2 copies per mL. This FET device shows great potential as a valuable tool for the early diagnosis of COVID-19. These findings suggest that the FET-based biosensor may be a reliable tool in detecting clinical samples and could potentially be used in clinical settings.

In another study by Wei *et al.* (2022),[41] the fabrication of a functionalized MoS_2-field effect transistor (FET) for sensitive and quantitative detection of antibodies against SARS-CoV-2 spike protein receptor binding domain (S-RBD) in vaccinated serum specimens was shown. The device was modified by SARS-CoV-2 *S*-RBD fusion protein on the surface and enabled rapid detection of SARS-CoV-2 antibodies. In addition, an on-chip calibration method was established for quantitative measurement. Furthermore, this method was applied to measure the levels of *S*-RBD antibodies in serum specimens from vaccinated donors. The devices showed no response to negative control samples from individuals who did not receive vaccination, suggesting the high specificity of this method.

Additionally, Payandehpeyman *et al.* (2021)[42] introduced a graphene-based nanoresonator sensor for SARS-CoV-2 detection functionalized with a specific antibody against the SARS-CoV-2 spike S1 antigen.

6.5.3.2 *Aptamer-based FET Biosensors*

Aptamers, which are synthetic nucleic acids, can be designed to recognize specific target molecules with high affinity and selectivity. Aptamers were also designed for targeting spike proteins and BioFETs for SARS-CoV-2 detection were developed. Thus, a silicon field-effect transistor (FET) functionalized with aptamers for detecting the spike protein of SARS-CoV-2 was developed. This FET-based sensor using aptamers as bioreceptors offers better sensitivity compared to traditional antibody-based sensors and provides results comparable to other FET-based sensors using graphene, but requires further optimization to reduce the testing time.[43]

In a recent study by Ban *et al.*,[44] the authors developed a DNA aptamer-based FET biosensor for detecting the SARS-CoV-2 virus. The biosensor utilized a graphene FET platform, which was functionalized with specific aptamers that bind to the various proteins on the virus *S*-protein and N-protein.

The biosensor was able to detect the virus with LODs and LOQs of 1.28 and 3.89 plaque-forming units (PFU) per mL for the S protein and 1.45 and 4.39 PFU per mL for the N protein, respectively, which is more sensitive than traditional PCR-based methods.

Farrow *et al.*[43] proposed a low-cost and rapid solution for COVID-19 detection using an aptamer-functionalized Schottky-FET. A Schottky-FET is a semiconductor device that can detect changes in the electrical properties of a material. The study demonstrated that the aptamer-functionalized Schottky-FET could detect the SARS-CoV-2 virus within a few minutes with high sensitivity and specificity. In fact, the SFET has been designed to attach to the spike protein of SARS-CoV-2, and it has been shown to provide a linear response in the concentration range of 100 fM to 10 pM. By using a larger range of source–drain potentials, it is possible to optimise the response of the SFET.

6.5.3.3 DNA-based FET Biosensors

DNA-based FET biosensors are based on the interaction between the DNA probe and the target molecule, which causes a change in the charge density on the FET gate surface. This change in charge density results in a shift in the threshold voltage of the FET, which can be measured and correlated to the concentration of the target molecule.[45] The use of FET biosensors in the detection of DNA is particularly useful in the diagnosis of genetic diseases and genetic fingerprinting. In conclusion, DNA-based FET biosensors have the potential to revolutionize the field of biosensing due to their high sensitivity, real-time detection, and ability to detect multiple targets simultaneously.

A direct nucleic acid assay using a graphene field-effect transistor (g-FET) with Y-shaped DNA dual probes (Y-dual probes) has shown promising results.[46]

The development of the Y-shaped DNA dual-probe transistor assay offers a promising solution for the rapid screening of infected individuals from a large population in the midst of the COVID-19 pandemic (Figure 6.12). With its direct nucleic acid detection capabilities, this assay eliminates the need for time-consuming extraction and amplification procedures that have hindered the efficiency of previous diagnostic techniques. The use of Y-dual probes modified on g-FETs ensures a high recognition ratio and an extremely low limit of detection, making it one of the most sensitive nucleic acid assays available.

Additionally, the dual-probe design of this assay provides increased specificity and sensitivity compared to single-probe assays. This is due to the fact that dual probes bind to adjacent regions on the target DNA molecule, increasing the likelihood of detection. With its ability to provide results in less than a minute, DNA has the potential to significantly improve epidemiological efforts aimed at containing outbreaks such as COVID-19.

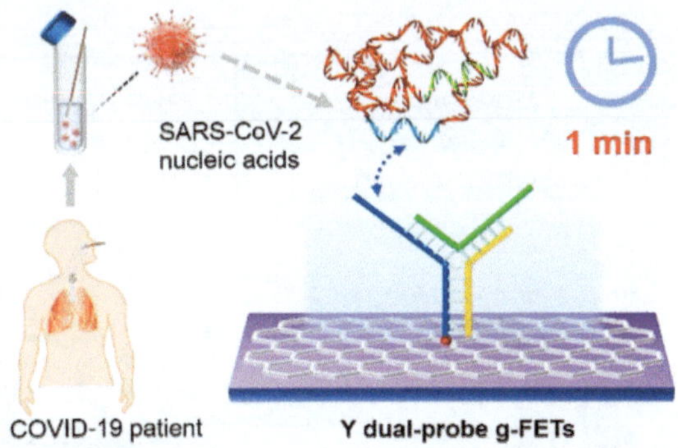

Figure 6.12 Scheme of the BioFET based on a DNA-dual probe. Reproduced from ref. 46 with permission from American Chemical Society, Copyright 2021.

6.5.4 Nanomaterial-based BioFETs for SARS-CoV-2 Detection

6.5.4.1 FET-based Carbon Nanotubes (FET-CNTs)

FET-based carbon nanotubes (CN-FET) have recently been reported by M. A. Zamzami *et al.* in 2022,[47] which enabled highly sensitive and specific detection of the SARS-CoV-2 virus in fortified saliva samples. The CN-FETs are based on Si/SiO_2 technology where carbon nanotubes are printed between the source and drain channel and the antibody anti-SARS-CoV-2 spike protein S1 is immobilized through non-covalent interaction using 1-pyrenebutanoic acid succinimidyl ester. The SARS-CoV-2 spike protein S1 was detected with a limit of detection of 4.12 fg mL^{-1} and with a wide range of concentrations from 0.1 fg mL^{-1} to 5.0 pg mL^{-1} (Figure 6.13). In addition, the CN-FET biosensors offer the advantage of being able to discriminate between different types of viral proteins such as SARS-CoV-1 S1 and MERS-CoV S1, which is important for determining the effectiveness of treatments. It is asserted that these biosensors are comparatively affordable and simple to use, rendering them attainable to researchers worldwide. They also have the ability to assess SARS-CoV-2 in clinical and environmental samples.

Another work involved the use of a high-purity semiconducting single-walled carbon nanotube-based field-effect transistor to detect SARS-CoV-2 antigens in clinical nasopharyngeal samples. The functionalized SWCNT FET sensors demonstrated high sensitivity and specificity and could potentially be used as a rapid and low-cost COVID-19 antigen diagnostic tool.[17]

6.5.4.2 Graphene-based FETs (G-FETs)

Graphene-based FET (G-FET) biosensors are one of the most developed BioFETs, and various forms and nanostructures of graphene have been used in this way from graphene oxide to graphene nanosheets, doped graphene, crumpled graphene, laser induced graphene, and MXene–graphene.

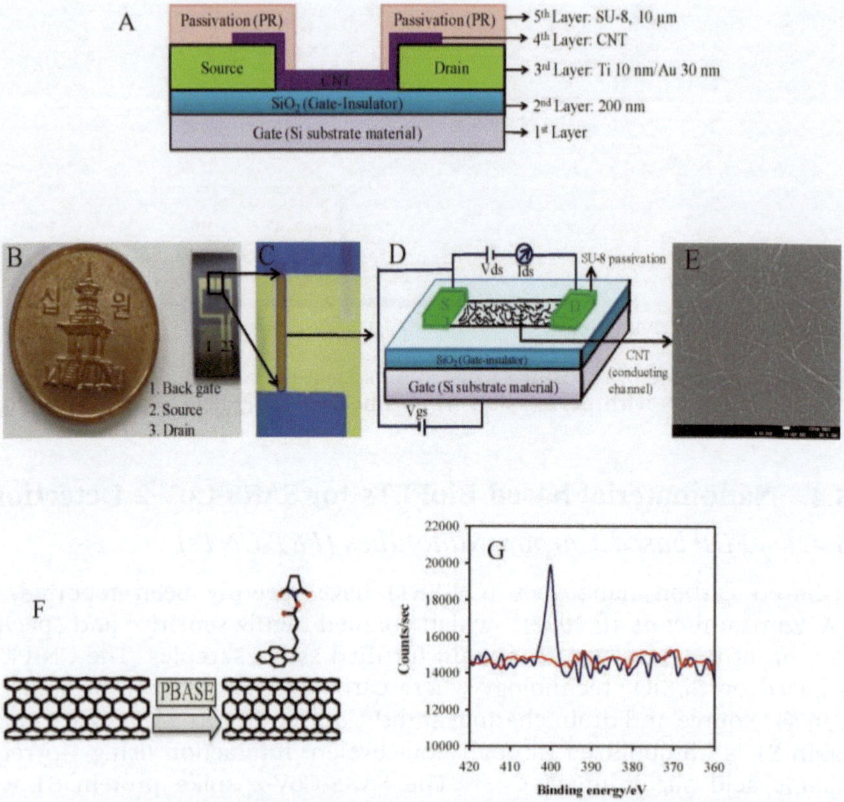

Figure 6.13 Scheme presenting the BioFET based on CNTs. Reproduced from ref. 47, https://doi.org/10.1016/j.bioelechem.2021.107982, under the terms of the CC BY 4.0 license, https://creativecommons.org/licenses/by/4.0/.

Krsihna *et al.* developed a label-free electrical device, reduced graphene oxide field-effect transistor (rGO FET), for the detection of SARS-CoV-2 antigens using monoclonal antibodies (mAbs). This biosensor detects coronavirus antigens, and the presence of the SARS-CoV-2 virus affects the electrical properties and sensor characteristics of the rGO-FET. The device exhibited good performance, high sensitivity and low limit-of-detection (LoD). Although the developed device was developed for SARS-CoV-2, it can be used for rapid screening of virus proteins, small biomolecules, and different biomarkers, making it useful for future biomedical applications such as virus infection diagnosis and lab-on-chip platforms (50).

Sadighbayan *et al.*[22] state that the use of laser-induced GFET biosensors in COVID-19 detection has several advantages over traditional methods. Firstly, laser-induced GFET biosensors can detect the virus at low concentrations, making them highly sensitive. Secondly, they can detect specific biomolecules, providing high selectivity. Thirdly, they have a rapid response time, allowing for quick and accurate detection of the virus. Additionally, LI-GFET biosensors can be easily integrated into portable devices, making them ideal for use in point-of-care testing. Furthermore, LI-GFET biosensors require minimal sample preparation, reducing the risk of contamination and increasing the accuracy of detection.

TR Cui *et al.*[48] have recently reported on the development of a COVID-19 LIG-FET sensor that can detect the virus with high sensitivity and specificity. The sensor was designed to detect the SARS-CoV-2 nucleocapsid protein, which is a key component of the virus. The LIG-FET sensor was fabricated using a simple and low-cost process that involves laser irradiation of a polymer film to create graphene. The graphene was then functionalized with antibodies that can specifically bind to the nucleocapsid protein. The sensor was able to detect the virus in PBS solvent and the real serum of people at an LOD of 1 $pg\,mL^{-1}$. Importantly, the sensor showed no cross-reactivity with other common respiratory viruses, indicating its high specificity for SARS-CoV-2. The LIG sensor platform is also cost-effective and easy to fabricate, making it a practical option for widespread use. Furthermore, the platform can be easily integrated with portable electronic devices, allowing for on-site detection and timely diagnosis. These results demonstrate the potential of LIG-FET sensors as rapid and accurate diagnostic tools for COVID-19.

Novodchuk *et al.*[49] demonstrated the successful detection of the SARS-CoV-2 nucleocapsid (N) protein using a FET biosensor based on a boron nitride–graphene oxide (BN–GO) gel in buffer as a proof-of-concept. This innovative technology combines the benefits of a graphene oxide (GO) gel with the specificity of nucleoprotein antibodies to detect and quantify specific biomolecules. The use of a boron and nitrogen co-doped graphene oxide gel transducer functionalized with nucleoprotein antibodies has resulted in a highly sensitive response, capable of detecting the N-protein in buffer with a limit of detection of 10 $ag\,mL^{-1}$ and a detection range of 10 $ag\,mL^{-1}$–1 $\mu g\,mL^{-1}$ within minutes. They modified the graphene surface with antibodies specific to the N protein, allowing for the selective binding of the viral protein to the FET surface. They found that the graphene-based FET was capable of detecting the N protein at concentrations as low as 1 $ng\,mL^{-1}$, with a response time of approximately 30 seconds. This sensitivity and rapid response make graphene-based FETs an attractive option for the development of point-of-care diagnostic tools for SARS-CoV-2.

Furthermore, one promising approach for biosensing SARS-CoV-2 is the use of graphene oxide–graphene van der Waals heterostructures. This type of heterostructure combines the unique properties of graphene oxide and graphene to create a highly sensitive biosensor. Graphene oxide has excellent biocompatibility and high surface area, which allows it to bind to a wide range of biomolecules, while graphene provides high electrical conductivity and mechanical strength. The combination of these two materials creates a biosensor with high sensitivity, selectivity, and stability. Gao *et al.*[50] developed an ultrasensitive biosensor composed of graphene oxide (GO) and graphene (Gr) nanosheets to detect SARS-CoV-2 proteins (Figure 6.14). The GO/Gr biosensor is integrated on microfluidic devices and exhibited a limit

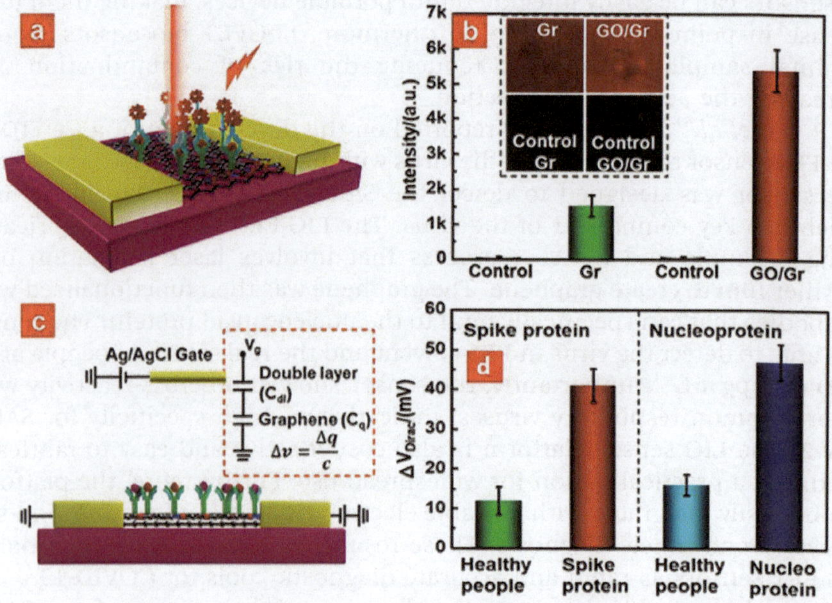

Figure 6.14 (a) Schematic representation of the BioFET based on GO/Gr. (b) Fluorescence signal after adding 100 pg mL^{-1} SARS-CoV-2 antigen on Gr and GO/Gr. Inset: The scanned fluorescence mapping patterns on Gr and GO/Gr. (c) Schematic diagram of the electric double layer model of the biosensor. (d) Detection results of spiked SARS-CoV-2 proteins in healthy human throat swab solution by GO/Gr FET biosensors. Reproduced from ref. 50 with permission from Elsevier, Copyright 2022.

of detection of 0.01 fM, which is significantly lower compared to other reported methods. The high sensitivity of this biosensor can be attributed to the large surface area of GO/Gr nanosheets, which can effectively capture SARS-CoV-2 proteins through antibody–antigen interactions. The GO/Gr biosensor is also easy to operate and requires minimal sample preparation, making it a promising tool for rapid and accurate detection of SARS-CoV-2 proteins.

Researchers have explored the use of a crumpled graphene field-effect transistor (CGFET) for COVID-19 detection by immobilizing N- and S-protein antibodies on them. According to MT Hwang *et al.* (2021),[51] CGFET devices were fabricated using a simple and scalable method, and their performance was evaluated for biosensing applications such as SARS-CoV-2 detection. The CGFET biosensor is based on a crumpled graphene sheet that is deposited on a silicon substrate having a unique hierarchical structure that provides a large surface area for specific biomolecules and is then functionalized for target detection, thus increasing the sensitivity of the sensor. Additionally, the crumpled structure provides a conductive pathway between the graphene sheets, allowing for efficient electron transfer and signal amplification. Additionally, the CGFET devices showed stable and reproducible performance over multiple cycles of sensing and regeneration. The crumpled graphene FET biosensor can detect the SARS-CoV-2 spike protein with a limit of detection of 1 fg mL^{-1}, which is more sensitive than the conventional PCR method. The biosensor can also distinguish between the SARS-CoV-2 spike protein and other proteins with similar structures, thus reducing false positives.

MXene–graphene field-effect transistor sensing of SARS-CoV-2 is a recent development in the field of virus detection and sensing.[52] This innovative sensor combines a high chemical sensitivity MXene and large-area high-quality graphene to form an ultra-sensitive virus-sensing transduction material. MXenes, a type of two-dimensional (2D) materials, have been used as the channel material in FET sensors due to their unique properties, such as high conductivity and large surface area. Graphene, another 2D material, has been used as the sensing material due to its high sensitivity to changes in the local environment. The combination of the MXene and graphene in the FET structure allows for high sensitivity and selectivity in the detection of analytes. Moreover, these sensors have a large surface area and high electron mobility, which enhance their sensitivity and selectivity towards virus detection. The use of MXene–graphene FET sensors can also reduce the cost of virus detection as they require minimal sample and reagent volumes. The researchers functionalized the FET surface with antibodies specific to the SARS-CoV-2 spike protein, allowing for the detection of the virus at concentrations as low as 10 pg mL^{-1}. This demonstrates the potential of MXene–graphene FETs for use in medical diagnostics and disease detection.

In a recent study, Piccinini *et al.*[53] demonstrated the successful biofunctionalization of graphene-based FET sensors using heterobifunctional nanoscaffolds. The nanoscaffolds were designed to bind to both the graphene surface and the target antigen, allowing for the efficient capture and detection of COVID-19 antigens. The biosensors demonstrated high specificity, a fast response time, and a wide dynamic range of detection with limits of detection of 0.74 and 0.23 nm for the SARS-CoV-2 spike protein and human ferritin, respectively, and can detect a significant fraction of clinical samples from COVID-19 patients.[53]

A nanosensing platform for the detection of SARS-CoV-2 RNA has been developed using a gold nanoparticle-decorated graphene field-effect transistor (G-FET) sensor.[54] This technology is highly sensitive and can detect the virus in human throat swab specimens and serum samples within just 2 minutes. This biosensing platform offers several advantages over other methods, including a low detection limit of 0.37 fM, an excellent anti-interference capability, and precision in undiluted biological samples. Additionally, this nanosensor has demonstrated outstanding reliability and accuracy in testing RNA extracts from real samples.

6.5.4.3 *Silicon Nanowire-based BioFETs*

A novel SiNW-FET sensor is designed in this study to detect the COVID-19 spike protein using a semiempirical approach. The electronic transport properties such as the transmission spectrum, conductance, and electronic current are investigated by a combination of semiempirical modelling and nonequilibrium Green's function. Furthermore, the sensor's selectivity is tested for other viruses such as influenza, rotavirus, and HIV. The results show that the SiNW-FET sensor can accurately identify COVID-19 with high sensitivity and selectivity.[34]

6.5.4.4 *TMDC-based BioFETs*

In recent years, 2D transition metal dichalcogenides (TMDCs) have gained significant attention in the field of biosensors due to their unique electrical and optical properties.

Fathi-Hafshejani *et al.*[55] conducted a study on the effectiveness of 2D TMDC-based field-effect transistors (FETs) as biosensors. Their results showed that 2D TMDC-based FETs have high sensitivity and selectivity towards biomolecules such as DNA, RNA, and proteins. The high surface-to-volume ratio of 2D TMDCs and their unique electronic properties are the key factors that contribute to their effectiveness as biosensors. Theoretical studies have also supported the experimental findings, suggesting that the strong interaction between 2D TMDCs and biomolecules can be utilized for sensitive detection of biomolecules. The authors utilized MoS_2-based FETs functionalized with antibodies specific to the SARS-CoV-2 spike protein. The biosensor showed a significant change in conductance upon exposure to the target protein, indicating the successful detection of SARS-CoV-2.

In another study, researchers have developed a FET biosensor using WSe_2 monolayers that can detect SARS-CoV-2 with a limit of detection of 25 fg μL^{-1} using a monoclonal antibody against the spike protein of the virus. The device shows potential for detecting other SARS-CoV-2 biomarkers, such as cytokines.[55]

A scalable and cost-efficient lithography-free process was used to fabricate FET biosensors with an In_2O_3 channel and a stable enzyme reporter. The phosphatase enzymatic reaction generated a pH change, which was then detected and converted to an electrical signal by the In_2O_3 FETs, which showed a much better stability than urease. The biosensors were able to detect the SARS-CoV-2 spike protein and IgG antibodies in both buffer and whole blood, exhibiting selectivity for biomarkers in physiological fluids and potential for real clinical diagnosis applications.[56]

6.6 Other Detection Methods

6.6.1 VOCs Related to SARS-CoV-2

Deploying animals to detect COVID-19 is a challenging solution due to the scale of the pandemic. Nevertheless, electronic noses (eNoses) have been engineered to imitate the olfactory capabilities of animals and have the potential for large-scale deployment. In an effort to examine whether

SARS-CoV-2 infection can be identified through body odour, Snitz *et al.* positioned a standard eNose at a drive-through testing station. The eNose measurements were then subjected to analysis using a deep learning classifier, resulting in real-time detection of SARS-CoV-2 infection in both symptomatic and asymptomatic individuals, surpassing random chance significantly. This proof of concept involving a standard eNose indicates that an optimized eNose may offer effective real-time diagnostics, providing substantial relief during the COVID-19 pandemic.[57]

6.6.2 Modified Structure of BioFETs

The design and working of a vertical tunnel field-effect transistor (VTFET)-based dielectric modulated biosensor have been presented in the article by VD Wangkheirakpam *et al.*[58] The device consists of a vertical TFET structure that is modified with a dielectric layer. The dielectric layer changes the effective thickness of the channel region, which modulates the electric field in the channel. The modulation of the electric field affects the carrier transport in the channel, leading to the detection of biological molecules.

The VTFET is employed for the detection of SARS-CoV-2 in clinical samples through the examination of spike, envelope, and DNA proteins. This particular sensor exhibits a notable level of sensitivity, reaching an order of magnitude of 10^6. Moreover, the analysis of sensitivity is carried out with respect to the charge density of DNA.[58]

Kumar *et al.* also proposed the use of a dual gate dielectric modulated field-effect transistor (DGDMFET) biosensor to detect SARS-CoV-2 by sensing the dielectric constant of the converted-DNA and spike protein. This developed biosensor is highly sensitive and has the potential to detect various protein molecules, offering a promising approach to overcome the limited sensitivity of biosensors operated using surface interaction with virus biomolecules.[59]

6.6.3 Combination with Other Technologies

Various other technologies can be combined with BioFETs for SARS-CoV-2 detection for improving the sensitivity of detection or detection in real samples. For example, a portable detection cell using loop-mediated isothermal amplification at room temperature (RT-LAMP) was combined with crumpled graphene field-effect transistors (cgFETs) to detect the SARS-CoV-2 virus in clinical samples. The device was found to be more sensitive, making it a promising diagnostic tool for rapid and accurate COVID-19 detection. The biosensor does not need kits for RNA purification, gold nanoparticles, or functionalization of molecules on the graphene surface, which considerably reduces the cost.[60] This work demonstrates the benefit of combining biosensors with other methods for detection in real samples without any sample treatment.

6.6.4 IA and IoT BioFETs for SARS-CoV-2

Machine learning and Internet of Things are two new technologies that could add value for following, predicting and tracking COVID 19 diseases during the pandemic situation. Many current efforts are using these technologies, which have been previously employed during the pandemic.

Various approaches have been developed using machine learning in combination with biosensors for diagnostics. For example, Wang *et al.* developed a, termed Edge and Structure information for Medical Image Inpainting (ESMII), which addresses the challenge of inpainting medical images. The proposed method extracts features from images using multiscale residual blocks at three different scales. The ESMII model outperforms existing methods, generating high-quality structures with improved texture. Additionally, this model effectively addresses the issue of image distortion, which is common in medical image.[61]

Kumar and colleagues introduced an approach to detect latent features of pulmonary disease and track their progression during the early stages to mitigate further growth and enhance patient outcomes. Their method employs P-SVM computation to sequence high-dimensional pulmonary datasets. The proposed model demonstrated 83% accuracy in predicting the characterization of lung sickness.[62]

Another approach for detecting pneumonia from chest X-rays has been developed through the use of convolution neural networks (CNNs) and Nesterov accelerated Adaptive moment estimation (Nadam). The proposed model is able to differentiate between bacterial and viral phases of pneumonia by using Nadam for training. The model achieves a high validation accuracy of 94% and 93% for training data, and an accuracy of 87% for test data.[63]

Casaccia *et al.* proposed a real-time location system (RTLS) that integrates inertial measurement unit (IMU) sensors for social distancing and physical activity measurement. The system monitors and tracks social distancing using this instrumentation, continuously assessing the interpersonal distance to ensure that it does not exceed a maximum of 1.54 m. The system collects data from the accelerometers, which are filtered using discrete wavelet transform, and measures the activity levels of elderly people by setting a threshold value.[64]

Concerning the detection of SRA-Cov 2, few examples are described, we can cite the work of Karthickraja, R. *et al.* describing a novel approach for predicting COVID-19 cases by utilizing wearable sensor devices. Their model integrates clinical therapy with IoT-enabled wearable devices to monitor symptoms and test suspected cases. The researchers also conducted risk factor analysis to identify the association between oxygen saturation (SpO2) and COVID-19 using chi-square distribution. The wearable sensors used in the system provide real-time data, which are processed and analyzed using machine learning algorithms to predict COVID-19 cases. The proposed approach could potentially aid in early detection and treatment of the disease, thus mitigating its spread and impact.[65]

Kumar and colleagues introduced an approach using the adaptive neuro-fuzzy inference system (ANFIS), a type of machine learning technique, to predict the outbreak and monitor the spread of COVID-19. The ANFIS system employs prior datasets gathered from cloud computing to track the growth of the epidemic. The authors reported a high degree of accuracy of 86% for this approach.[66]

Hemamalini and colleagues proposed a biosensor system consisting of four different sensors: a blood pressure sensor, a graphene-FET biosensor, an electrochemical sensor, and a potentiometric sensor for detecting COVID-19 biomarkers in patient samples. The authors used a multi-layer perceptron (MLP) AI model to classify the virus into three categories: COVID-19 positive, COVID-19 negative, and viral pneumonia influenced by the COVID-19 viruses. The flow rate of the biosensors was measured based on bio-perception compounds integrated with the target analyte, and the calibration parameters of biomedical sensors were compared in terms of the detection limit. The study demonstrates the potential of AI in predicting the signature of COVID-19. However, the paper could benefit from more detailed information on the methodology of sensor integration and the accuracy of the biosensors in detecting COVID-19 biomarkers (Figure 6.15). Additionally, the authors could provide more clarity on the criteria used for selecting the specific sensors and the rationale behind their selection.[67]

In the case of IOT technology, there is a real weakness of healthcare systems across the world and a need for new innovative IoT solutions in diagnosis.

A study by Maniam *et al.* introduced a BioFET sensor interface module coupled with IoT technology for detecting and monitoring SARS-CoV-2. This module utilized a microcontroller to read signals and identify the presence of the virus. Additionally, it transmitted data to the cloud through Wi-Fi, allowing storage and display on a dashboard. The prototype module underwent simulations in PSpice for signal amplification and utilized affordable components for hardware implementation. Notably, they achieved consistent results in both the simulations and hardware experiments. By employing the AD620 instrumentation amplifier, they successfully amplified negative and positive BioFET signal outputs from 15.9 mV and 45.8 mV to 1.59 V and 4.58 V, respectively. Moreover, they effectively collected and recorded location, time, date, output voltage, and SARS-CoV-2 presence data, which were then displayed on the cloud dashboard.[68]

6.6.5 Pros and Cons of BioFET Sensors

BioFET sensors present many advantages; the label-free approach of detecting biomolecules is intrinsically based on the fact that most of them are charged. Additionally, the electronic signal generated in this process is easily measurable and reportable through conventional electronics, without the need for optical alignment or moving parts, which makes the devices quite robust. The advanced manufacturing technology for electronics enables easy

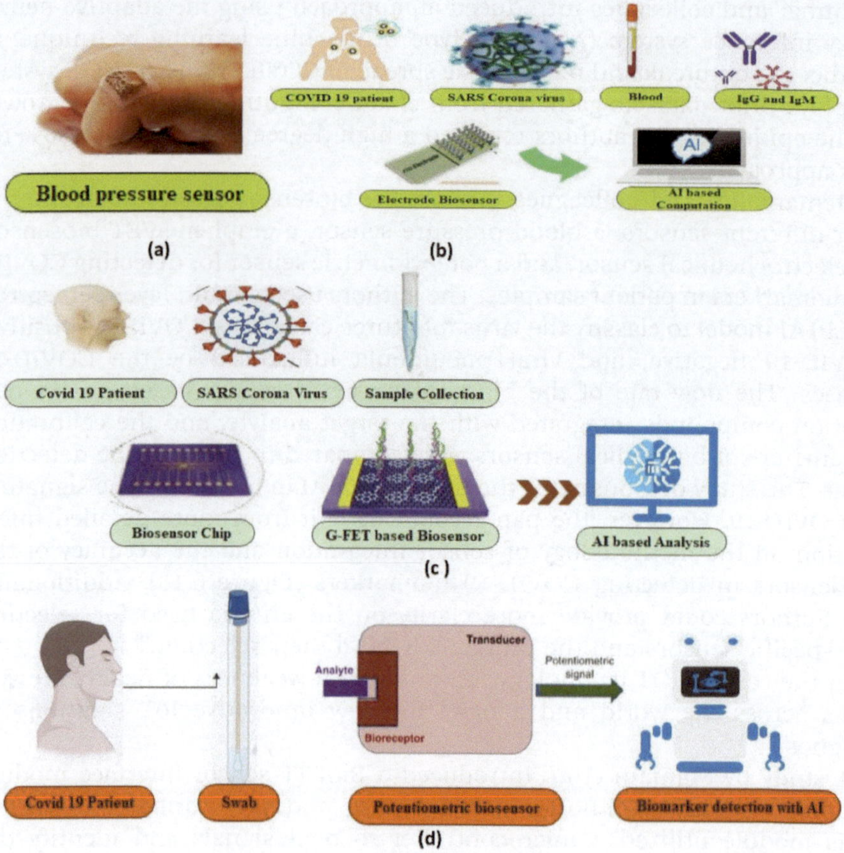

Figure 6.15 Scheme presenting the AI model for predicting the categories of COVID-19 diseases. Reproduced from ref. 67 with permission from Elsevier, Copyright 2022.

miniaturization and scaling up for fabrication, which further enhances the potential of this approach.[69,70]

The limitation that is likely to be most constraining for BioFET sensors is the Debye screening, which is the phenomenon of shielding of the charged analyte by the dissolved ions in the solution. Typically, the solution is to employ buffers with lower ionic strength during measurement, following various washing stages. However, this method may impact the stability and functionality of biological species, especially regarding modifications in the conformation of the receptor/target molecules and the strength of the interaction.[71]

6.7 Conclusion

This chapter provides an overview of COVID pandemic spread and detection methods with a focus on recent developments in the application of BioFETs for diagnosis of COVID-19. The chapter reports the most important information for understanding the SARS-CoV-2 pandemic situation. It explains the basic methods of FET and BioFET constituents and working principles. It also presents a recent achievement in BioFETs for SARS-CoV-2 detection in diagnosing COVID-19. We believe that many efforts have been made in the development of BioFETs for COVID-19 with high sensitivity and low sample treatment. However, further efforts should be made in the development of BioFETs with low cost, ease of use, multiplexing, and integration into devices for real implementation in the market replacing the classical methods such as PCR.

Acknowledgements

The authors thank the Embassy of France for « France-Malaysia Collaboration Programme Joint Research » MyTIGER 2023 Grant N MT29-11 and Ministère de l'Europe et des Affaires étrangères (MEAE), Ministère de l'enseignement supérieur, de la recherche et de l'innovation (MESRI) France for PHC Grant No. 22MAG20 and CNRS for interdisciplinary MITI funding.

References

1. E. C. Holmes, S. A. Goldstein, A. L. Rasmussen, D. L. Robertson, A. Crits-Christoph, J. O. Wertheim, S. J. Anthony, W. S. Barclay, M. F. Boni, P. C. Doherty, J. Farrar, J. L. Geoghegan, X. Jiang, J. L. Leibowitz, S. J. D. Neil, T. Skern, S. R. Weiss, M. Worobey, K. G. Andersen, R. F. Garry and A. Rambaut, *Cell*, 2021, **184**, 4848–4856.
2. B. Hu, H. Guo, P. Zhou and Z.-L. Shi, *Nat. Rev. Microbiol.*, 2021, **19**, 141–154.
3. World Health Organization, *WHO-convened global study of origins of SARS-CoV-2: China Part*, 2021.
4. Y.-R. Guo, Q.-D. Cao, Z.-S. Hong, Y.-Y. Tan, S.-D. Chen, H.-J. Jin, K.-S. Tan, D.-Y. Wang and Y. Yan, *Mil. Med. Res.*, 2020, **7**, 11.
5. S. Temmam, K. Vongphayloth, E. Baquero, S. Munier, M. Bonomi, B. Regnault, B. Douangboubpha, Y. Karami, D. Chrétien, D. Sanamxay, V. Xayaphet, P. Paphaphanh, V. Lacoste, S. Somlor, K. Lakeomany, N. Phommavanh, P. Pérot, O. Dehan, F. Amara, F. Donati, T. Bigot, M. Nilges, F. A. Rey, S. van der Werf, P. T. Brey and M. Eloit, *Nature*, 2022, **604**, 330–336.
6. D. Fisher and D. Heymann, *BMC Med.*, 2020, **18**, 57.
7. World Health Organization, *Coronavirus disease 2019 (COVID-19) Situation report – 51*, 2020.
8. N. Al-Awwal, F. Dweik, S. Mahdi, M. El-Dweik and S. H. Anderson, *Pathogens*, 2022, **11**, 368.
9. B. Oronsky, C. Larson, T. C. Hammond, A. Oronsky, S. Kesari, M. Lybeck and T. R. Reid, *Clin. Rev. Allergy Immunol.*, 2021, **64**, 66–74.
10. World Health Organization, The true death toll of COVID-19 Estimating global excess mortality, 2020.

11. M. H. Boroojerdi, T. Al Jabry, S. M. J. Mirarefin and H. Albalushi, *Virol. J.*, 2023, **20**, 37.
12. Y. Yuan, B. Jiao, L. Qu, D. Yang and R. Liu, *Front. Immunol.*, 2023, **14**, 1125246.
13. A. Souiri, S. Lemriss, B. El Maliki, H. Falahi, E. El Fahime and S. El Kabbaj, *Vaccines*, 2023, **11**, 590.
14. Z. Zhou, Y. Zhu and M. Chu, *Front. Immunol.*, 2022, **13**, 898192.
15. A. Vitiello, R. La Porta, U. Trama, F. Ferrara, A. Zovi, A. M. Auti, M. Di Domenico and M. Boccellino, *Naunyn-Schmiedeberg's Arch. Pharmacol.*, 2022, **395**, 1159–1165.
16. N. H. Al-Hardan, M. Firdaus-Raih, M. A. A. Hamid and A. Jalar, *Crit. Rev. Solid State Mater. Sci.*, 2024, **49**, 1–27.
17. W. Shao, M. R. Shurin, S. E. Wheeler, X. He and A. Star, *ACS Appl. Mater. Interfaces*, 2021, **13**, 10321–10327.
18. H. Kang, X. Wang, M. Guo, C. Dai, R. Chen, L. Yang, Y. Wu, T. Ying, Z. Zhu, D. Wei, Y. Liu and D. Wei, *Nano Lett.*, 2021, **21**, 7897–7904.
19. Z. Bai, Y. Cao, W. Liu and J. Li, *Viruses*, 2021, **13**, 1115.
20. Cube Biotech, SARS-CoV-2 Spike Protein Set (LMNG, DIBMA Glycerol & Nanodisc Complex).html, https://cube-biotech.com/products/active-proteins/membrane-proteins/corona-virus-proteins/sars-cov-2-spike-protein-set-lmng-dibma-glycerol-nanodisc-complex/28704 (accessed 18 May 2023).
21. H. Hartanto, M. Wu, M. L. Lam and T.-H. Chen, *Biomicrofluidics*, 2020, **14**, 061507.
22. Z. Swank, G. Michielin, H. M. Yip, P. Cohen, D. O. Andrey, N. Vuilleumier, L. Kaiser, I. Eckerle, B. Meyer and S. J. Maerkl, A high-throughput microfluidic nano-immunoassay for detecting anti-SARS-CoV-2 antibodies in serum or ultra-low volume dried blood samples, Infectious Diseases (except HIV/AIDS), 2020.
23. R. Funari, K.-Y. Chu and A. Q. Shen, *Biosens. Bioelectron.*, 2020, **169**, 112578.
24. A. Panahi, D. Sadighbayan, S. Forouhi and E. Ghafar-Zadeh, *Biosensors*, 2021, **11**, 103.
25. L. Bürgi, H. Sirringhaus and R. H. Friend, *Appl. Phys. Lett.*, 2002, **80**, 2913–2915.
26. K. ElMahgoub and A. Z. Elsherbeni, in *2012 IEEE/MTT-S International Microwave Symposium Digest*, IEEE, Montreal, QC, Canada, 2012, pp. 1–3.
27. M. Shur, T. A. Fjeldly, T. Ytterdal and K. Lee, *Solid-State Electron.*, 1992, **35**, 1795–1802.
28. D. J. Frank, R. H. Dennard, E. Nowak, P. M. Solomon, Y. Taur and H.-S. Philip Wong, *Proc. IEEE*, 2001, **89**, 259–288.
29. P. Bergveld, *IEEE Trans. Biomed. Eng.*, 1970, **17**, 70–71.
30. C.-S. Lee, S. Kim and M. Kim, *Sensors*, 2009, **9**, 7111–7131.
31. M. J. Schöning and A. Poghossian, *Analyst*, 2002, **127**, 1137–1151.
32. J. J. Xu, X. L. Luo and H. Y. Chen, *Front. Biosci.*, 2005, **10**, 420430.
33. M. Suryaganesh, T. S. Arun Samuel, T. Ananth Kumar and M. Navaneetha Velammal, in *Contemporary Issues in Communication, Cloud and Big Data*

Analytics, ed. H. K. D. Sarma, V. E. Balas, B. Bhuyan and N. Dutta, Springer, Singapore, Singapore, 2022, vol. 281, pp. 273–284.
34. A. Wasfi, F. Awwad, J. G. Gelovani, N. Qamhieh and A. I. Ayesh, *Nanomaterials*, 2022, **12**, 2638.
35. S. Cheng, S. Hideshima, S. Kuroiwa, T. Nakanishi and T. Osaka, *Sens. Actuators, B*, 2015, **212**, 329–334.
36. A. Toral-Lopez, D. B. Kokh, E. G. Marin, R. C. Wade and A. Godoy, *Nanoscale Adv.*, 2022, **4**, 3065–3072.
37. A. Biswas, C. Rajan and D. P. Samajdar, *Silicon*, 2022, **14**, 9533–9541.
38. A. Biswas, C. Rajan and D. P. Samajdar, *J. Electron. Mater.*, 2022, **51**, 6388–6396.
39. B. V. Krsihna, S. Ahmadsaidulu, S. S. T. Teja, D. Jayanthi, A. Navaneetha, P. R. Reddy and M. D. Prakash, *Silicon*, 2022, **14**, 5913–5921.
40. G. Seo, G. Lee, M. J. Kim, S.-H. Baek, M. Choi, K. B. Ku, C.-S. Lee, S. Jun, D. Park, H. G. Kim, S.-J. Kim, J.-O. Lee, B. T. Kim, E. C. Park and S. I. Kim, *ACS Nano*, 2020, **14**, 5135–5142.
41. J. Wei, Z. Zhao, F. Luo, K. Lan, R. Chen and G. Qin, *2D Mater.*, 2022, **9**, 015030.
42. J. Payandehpeyman, N. Parvini, K. Moradi and N. Hashemian, *ACS Appl. Nano Mater.*, 2021, **4**, 6189–6200.
43. T. Farrow, S. Laumier, I. Sandall and H. van Zalinge, *Biosensors*, 2022, **12**, 347.
44. D. K. Ban, T. Bodily, A. G. Karkisaval, Y. Dong, S. Natani, A. Ramanathan, A. Ramil, S. Srivastava, P. Bandaru, G. Glinsky and R. Lal, *Proc. Natl. Acad. Sci. U. S. A.*, 2022, **119**, e2206521119.
45. Y. Zhang, D. Feng, Y. Xu, Z. Yin, W. Dou, U. E. Habiba, C. Pan, Z. Zhang, H. Mou, H. Deng, X. Mi and N. Dai, *Appl. Surf. Sci.*, 2021, **548**, 149169.
46. D. Kong, X. Wang, C. Gu, M. Guo, Y. Wang, Z. Ai, S. Zhang, Y. Chen, W. Liu, Y. Wu, C. Dai, Q. Guo, D. Qu, Z. Zhu, Y. Xie, Y. Liu and D. Wei, *J. Am. Chem. Soc.*, 2021, **143**, 17004–17014.
47. M. A. Zamzami, G. Rabbani, A. Ahmad, A. A. Basalah, W. H. Al-Sabban, S. Nate Ahn and H. Choudhry, *Bioelectrochemistry*, 2022, **143**, 107982.
48. T.-R. Cui, Y.-C. Qiao, J.-W. Gao, C.-H. Wang, Y. Zhang, L. Han, Y. Yang and T.-L. Ren, *Molecules*, 2021, **26**, 6947.
49. I. Novodchuk, M. Kayaharman, I. Prassas, A. Soosaipillai, R. Karimi, I. A. Goldthorpe, E. Abdel-Rahman, J. Sanderson, E. P. Diamandis, M. Bajcsy and M. Yavuz, *Biosens. Bioelectron.*, 2022, **210**, 114331.
50. J. Gao, C. Wang, Y. Chu, Y. Han, Y. Gao, Y. Wang, C. Wang, H. Liu, L. Han and Y. Zhang, *Talanta*, 2022, **240**, 123197.
51. M. T. Hwang, I. Park, M. Heiranian, A. Taqieddin, S. You, V. Faramarzi, A. A. Pak, A. M. Zande, N. R. Aluru and R. Bashir, *Adv. Mater. Technol.*, 2021, **6**, 2100712.
52. Y. Li, Z. Peng, N. J. Holl, Md. R. Hassan, J. M. Pappas, C. Wei, O. H. Izadi, Y. Wang, X. Dong, C. Wang, Y.-W. Huang, D. Kim and C. Wu, *ACS Omega*, 2021, **6**, 6643–6653.

53. E. Piccinini, G. E. Fenoy, A. L. Cantillo, J. A. Allegretto, J. Scotto, J. M. Piccinini, W. A. Marmisollé and O. Azzaroni, *Adv. Mater. Interfaces*, 2022, **9**, 2102526.
54. J. Li, D. Wu, Y. Yu, T. Li, K. Li, M.-M. Xiao, Y. Li, Z.-Y. Zhang and G.-J. Zhang, *Biosens. Bioelectron.*, 2021, **183**, 113206.
55. P. Fathi-Hafshejani, N. Azam, L. Wang, M. A. Kuroda, M. C. Hamilton, S. Hasim and M. Mahjouri-Samani, *ACS Nano*, 2021, **15**, 11461–11469.
56. M. Chen, D. Cui, Z. Zhao, D. Kang, Z. Li, S. Albawardi, S. Alsageer, F. Alamri, A. Alhazmi, Moh. R. Amer and C. Zhou, *Nano Res.*, 2022, **15**, 5510–5516.
57. K. Snitz, M. Andelman-Gur, L. Pinchover, R. Weissgross, A. Weissbrod, E. Mishor, R. Zoller, V. Linetsky, A. Medhanie, S. Shushan, E. Jaffe and N. Sobel, *PLoS One*, 2021, **16**, e0252121.
58. V. D. Wangkheirakpam, B. Bhowmick and P. D. Pukhrambam, *J. Mater. Sci.: Mater. Electron.*, 2022, **33**, 10323–10334.
59. S. Kumar, R. K. Chauhan and M. Kumar, *Silicon*, 2022, **14**, 11453–11462.
60. I. Park, J. Lim, S. You, M. T. Hwang, J. Kwon, K. Koprowski, S. Kim, J. Heredia, S. A. Stewart de Ramirez, E. Valera and R. Bashir, *ACS Sens.*, 2021, **6**, 4461–4470.
61. Q. Wang, Y. Chen, N. Zhang and Y. Gu, *Measurement*, 2021, **185**, 110027.
62. R. Kumar, F. Al-Turjman, L. Anand, A. Kumar, S. Magesh, K. Vengatesan, R. Sitharthan and M. Rajesh, *Interdiscip. Sci.: Comput. Life Sci.*, 2021, **13**, 192–200.
63. S. S. Sai Koushik and K. G. Srinivasa, *Measurement*, 2021, **176**, 109153.
64. S. Casaccia, R. Naccarelli, S. Moccia, L. Migliorelli, E. Frontoni and G. M. Revel, *Measurement*, 2021, **184**, 109946.
65. R. Karthickraja, R. Kumar, S. Kirubakaran and R. Manikandan, *Int. J. Pervasive Comput. Commun.*, 2022, **18**, 499–507.
66. R. Kumar, F. Al-Turjman, L. N. B. Srinivas, M. Braveen and J. Ramakrishnan, *Neural Comput. Appl.*, 2023, **35**, 7207–7220.
67. V. Hemamalini, L. Anand, S. Nachiyappan, S. Geeitha, V. Ramana Motupalli, R. Kumar, A. Ahilan and M. Rajesh, *Measurement*, 2022, **194**, 111054.
68. G. Maniam, J. Sampe, R. Jaafar, A. A. Hamzah and N. Mohamad Zin, *Int. J. Online Biomed. Eng.*, 2022, **18**, 70–88.
69. R. Stine, S. P. Mulvaney, J. T. Robinson, C. R. Tamanaha and P. E. Sheehan, *Anal. Chem.*, 2013, **85**, 509–521.
70. B. J. Polk, in *Proceedings of IEEE Sensors*, IEEE, Orlando, FL, USA, 2002, vol. 1, pp. 732–735.
71. S. Rollo, D. Rani, W. Olthuis and C. Pascual García, *Biophys. Rev.*, 2019, **11**, 757–763.

CHAPTER 7

The Future of Commercializing FET-based Biosensors

H. BEN HALIMA,[a] A. ERRACHID[b] AND N. JAFFREZIC-RENAULT*[b]

[a] University Bourgogne-Franche-Comte, UTINAM Institute, 25030 Besancon Cedex, France; [b] University of Lyon, Institute of Analytical Sciences, 69100 Villeurbanne, France
*Email: nicole.jaffrezic@univ-lyon1.fr

7.1 Introduction

Biosensors are analytical devices that combine a biological component with a transducer for the detection of a chemical substance. Immunosensors are based on the immobilization of a specific antibody, a glycoprotein produced by the immune system to neutralize foreign substances (antigens) in the body. The antibody can then identify and bind to antigens with extremely high specificity. The immunosensor can detect the formation of the antigen–antibody complex through the modification of the transducer signal: electrochemical, optical, gravimetric, thermal, *etc.*, without any labeling, leading to a simplified immunosensing procedure and a fast response time.

Aptamers consist of artificial oligonucleotide sequences (peptides or nucleic acids) which can recognize and bind to a wide range of targets; they have drawn much attention as promising alternatives to conventional antibodies. Aptamers are produced through the *in vitro* selection and amplification of populations of random sequence oligonucleotide libraries, known as the SELEX process (selection evolution of ligands by

Detection Science Series No. 24
Field-effect Transistor Biosensors for Rapid Pathogen Detection
Edited by Naif H. M. Al-Hardan, Azman Jalar, Muhammad Azmi Abdul Hamid and Mohd Firdaus Raih

Published by the Royal Society of Chemistry, www.rsc.org

exponential enrichment). When specific sequences are identified, aptasensors can then replace immunosensors. Affinity sensors include immunosensors, aptasensors for detecting chemical substances such as biomarkers, toxins, *etc.* and nucleic acid sensors for detecting DNA or RNA sequences.

Among various electrochemical transducers, those based on field-effect transistors (FETs) have attracted considerable attention because of their potential for miniaturization, multi-sensing, fast response time, and integration with electronic manufacturing processes, such as complementary metal–oxide semiconductors (CMOS). The concept of an ion-sensitive FET (ISFET) was introduced by Bergveld[1] in the early 1970s; it was derived from a metal-oxide-semiconductor FET (MOSFET) by replacing the metal gate with a reference electrode. The detection of different target analytes results in a myriad of varying sensor system combinations. The affinity sensors based on an ISFET can be presented along the generic structure illustrated in Figure 7.1. The information obtained from the sample corresponds to the concentration/activity of the target analyte, which is transduced to an electrical signal *via* the field effect. Then, the signal can be amplified, processed, and displayed depending on the application.

A simplified potential diagram at the different interfaces of an affinity sensor based on a semiconductor field-effect transistor is shown in Figure 7.2. The observed responses originate from the charge σ_0 resting at the sensing surface (the insulator surface grafted with the receptor molecule). This charge sees capacitances on both sides: the capacitance is presented as a series combination of the double-layer and sensor (*i.e.*, FET) capacitances, C_{DL} and C_{FET}, respectively, where C_{FET} comprises the insulator, C_{IN}, and

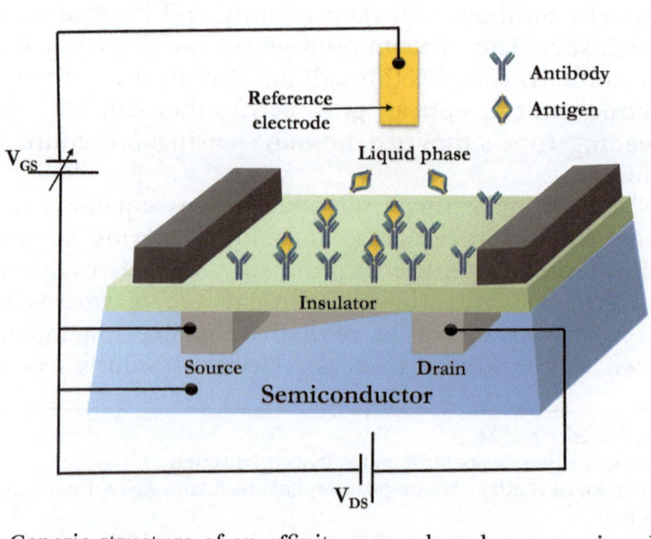

Figure 7.1 Generic structure of an affinity sensor based on a semiconductor field-effect transistor.

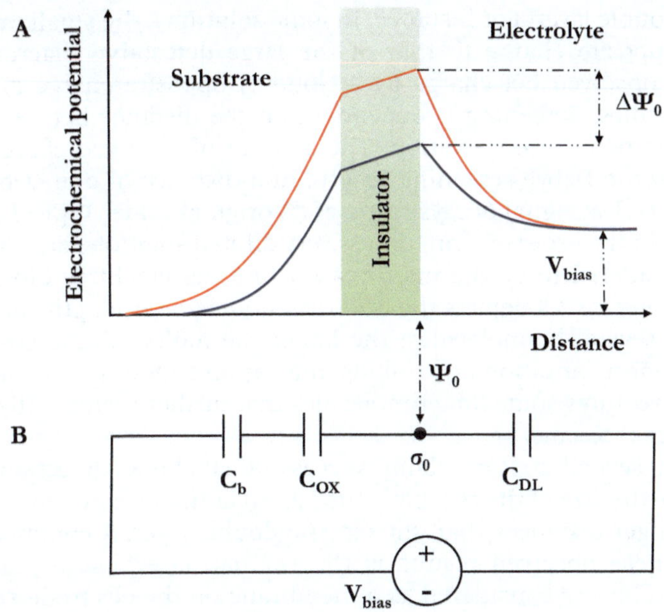

Figure 7.2 (A) Potential over a simplified model of an electrochemical cell with an insulator as the interfacial material. The charge binding on the surface creates a potential shift, denoted by $\Delta\Psi_0$, at the interface. (B) Equivalent circuit model of the sensor where C_{DL}, C_{IN}, and C_C are the double-layer, gate insulator, and channel capacitances, respectively. The interfacial charge, σ_0, is surrounded by capacitors on both sides.

channel, C_C, capacitances. The potential change at the sensing surface can be approximated by

$$\Psi_o = \Delta\sigma_o/(C_{DL} + C_{FET}) \tag{7.1}$$

Depending on the transistor biasing, one or another capacitance can dominate. In weak inversion (low density of carriers in the channel), C_C will clearly be smaller and determine the overall sensor capacitance, in contrast to strong inversion (high density of carriers in the channel), where C_C can be omitted and C_{IN} is relevant. In all cases, these capacitances are usually clearly smaller than the double-layer capacitance. Hence, we can conclude that the transistor capacitance has a negligible effect on the sensitivity at the sensing interface.

First, the charge at the sensing surface alters the insulator electric field, changing the potential at the outer surface of the electrode. Second, this potential shift subsequently alters the semiconductor drain current, which is the transconductance effect of a FET.

A widely debated issue is the label-free detection of biomolecules using FET-based biosensors. Originally, it was believed that the fact of biomolecules carrying intrinsic charge allowed their detection using field-effect devices. Despite significant efforts, the results were unsatisfactory due to the electrical double layer. For instance, in ionic solutions, the small ions, which carry an opposite charge to that of the large detectable macromolecule, screen the observed net charge by a cloud of opposite charge around the macromolecules. Screening is dependent on the distance between the surface and the point of observation. The amount of observed charge is characterized by the Debye screening length. At a distance of one Debye length (λ_D), the electrical signal decays to $1/e$ of its original value. Typical screening lengths are in the order of 1 nm unless very diluted solutions are considered. This is considered to be the main reason limiting label-free biosensor development. Figure 7.3 depicts the electrical double-layer length compared to the size of several biomolecules; the larger the molecule, the stronger the screening effect. Additionally, linking the capture molecule to the surface commonly requires some linker molecules that further increase the distance to the target molecule.[2]

However, several reports claim successful label-free detection even in cases where the size of the complex, such as an antibody–antigen complex, is at vastly longer distances than the electric double layer. A common explanation for these observed results is the Donnan effect. According to this theory, proteins are considered as a membrane on the electrode surface. In addition, small ions can shuffle between the solution and this protein membrane. Then, when a fixed charge is present due to the target, a difference in ion concentration appears on the interface between the membrane and the solution. This redistribution of ions creates a detectable change in the interfacial potential. Moreover, the change in this Donnan

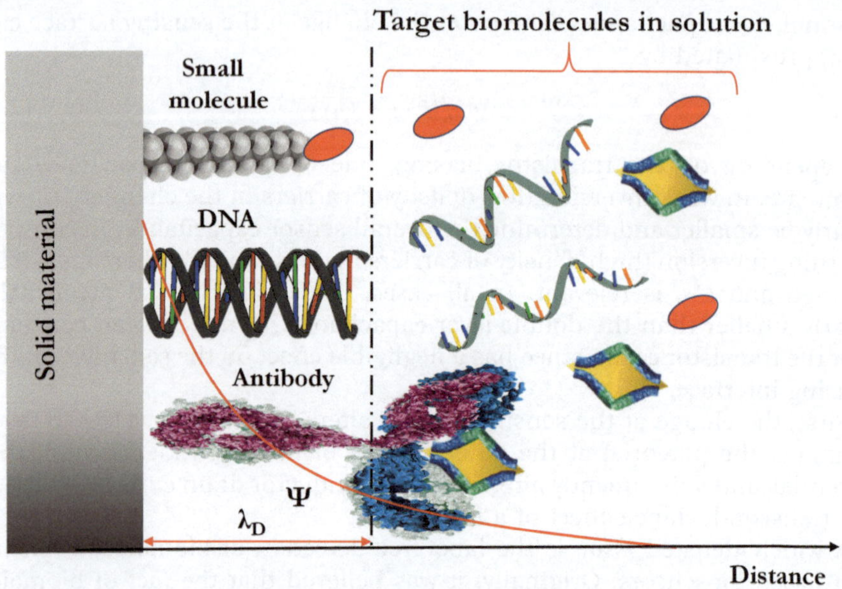

Figure 7.3 Electrical double-layer length in the presence of different targets (dimensions are not scaled).

potential also causes a shift in pH. Thus, the total response combines the surface pH response and the Donnan potential.

During the years 2020 and 2022, many review papers about biosensors for virus detection were published due to the COVID-19 epidemic period.[3–19] Among the cited affinity biosensors, FET-based biosensors were cited, the most cited one being from Seo *et al.*[3] Other recent reviews were devoted to biosensors for the diagnostics of other health deficiencies, among which were FET based affinity biosensors, such as chronic obstructive pulmonary disease,[20] immunodeficiency,[21] neurodegenerative diseases,[22] diabetes mellitus,[23,24] cardiac failure,[25] pathogenic bacteria,[26] acute ischemic stroke,[27] cancer[28–30] and different pathologies.[31] Two recent reviews were devoted to biosensors for the detection of contaminants in environmental matrices, such as pesticides/herbicides[32] and pharmaceuticals and endocrine-disrupting compounds.[33]

Reviews about micro- and nanoscale biosensors reported some FET devices. Some are based on graphene,[34,35] self-assembled monolayers,[36,37] AlGaN/GaN heterostructures,[38] nanowires,[39] nanotechnologies[3] or on molecularly imprinted polymers.[40]

Since 2015, 14 reviews have been devoted to field-effect transistors for biodetection. They were all oriented towards one type of field effect transistors: CMOS-type FETs,[2,41] conductive polymer-based FETs,[42–45] carbon nanotube-based FETs,[46] graphene-based FETs,[47–49] nanowire-based FETs,[50] and nanomaterial-based FETs.[51–53]

This chapter provides a detailed description of the works reported since 2015 on affinity sensors based on field effect transistors, whatever the nature of the nanomaterial-based FETs (silicon, graphene, SWCNTs, Si nanowires, organic FETs, nanomaterials, GaN). The objective of this chapter is to study the relationship between the analytical performance and the characteristics of the FETs (materials, structure and working parameters).

7.2 Silicon-based Field-effect Transistors for Affinity Sensing

Since 2015, only twelve papers have been published in this field. Seven of them were conventional ISFETs; they were devoted to the detection of biomarkers of diseases such as lung cancer,[54] asthma,[55] mental stress,[56] oxidative and nitrosative stress,[57] and heart failure,[58–64] and to the detection of viruses.[65]

They were fabricated through CMOS technology, the biofunctionalization being performed afterwards. The receptor molecule (antibody, amino-modified aptamer) was directly covalently grafted on the insulator surface (SiO_2 or Si_3N_4) through its silanization. In ref. 54 and 65, several FETs were fabricated on the same chip, allowing multi-detection. Variation of the gate–source voltage was the measured signal, depending on the concentration of the biomarker. In ref. 6, the impedance of the channel was measured, leading to a lower detection limit of TNF-α and a larger dynamic range compared to the results obtained by measuring the variation of the gate–source voltage.[58] When the frequency is high enough, the ions in the solution are unable to form the double layer after the AC perturbations. This allows probing further into the solution, and it also decreases the sensitivity towards small absorbates within the double layer.[2] ALP-labelled secondary antibody was used in a low-cost electrical ELISA test based on an ISFET which then becomes a low-cost portable system.[55]

Extended gate (EG) field-effect transistors were presented in six papers. This configuration improves passivation and packaging compared to the conventional ISFET. As shown in Figure 7.4, the EG immunoFET consists of two parts: the gate of a MOSFET is connected to the extended gate. This one is an electrode modified with the receptor molecule (antibody) in contact with the solution. The MOSFET can be commercially available once well packaged, avoiding any damage from humidification and no influence of the light. For the five published papers, the electrode used was gold.

The presented EG-FET allows the detection of different biomarkers of diseases, such as thyroid carcinoma,[66] breast cancer,[67] bacteria (*E. coli*),[68] Alzheimer's disease,[69] and bacterial infection.[70] For the detection of DNA,[71] PNA probes were used for efficient hybridization in low salt conditions, which is required to limit the ion screening effect.

In ref. 70, an enhanced electric-double-layer (EnEDL) field-effect-transistor-based immunosensor was built (Figure 7.5) to successfully detect proteins in high ionic strength solutions. The results show that as the EDL is enhanced at higher gate voltage or higher salt concentration, the sensitivity of detection increases. A detection limit of 0.5 mg of C-RP was obtained in whole blood and in PBS without any dilution step.

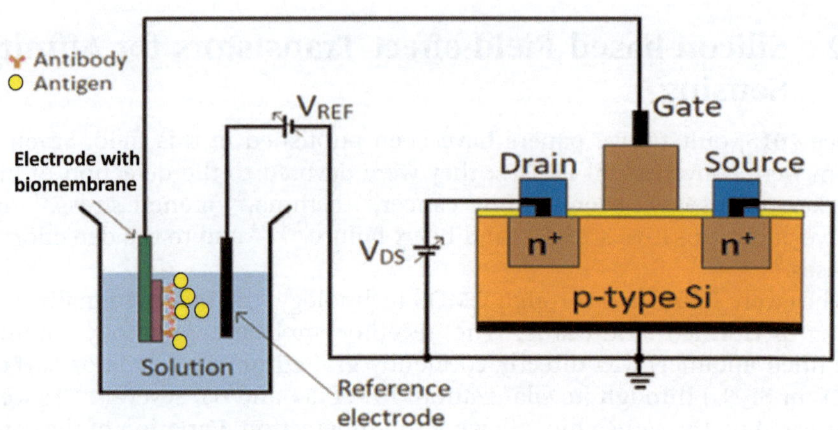

Figure 7.4 General scheme of an extended gate immunoFET.

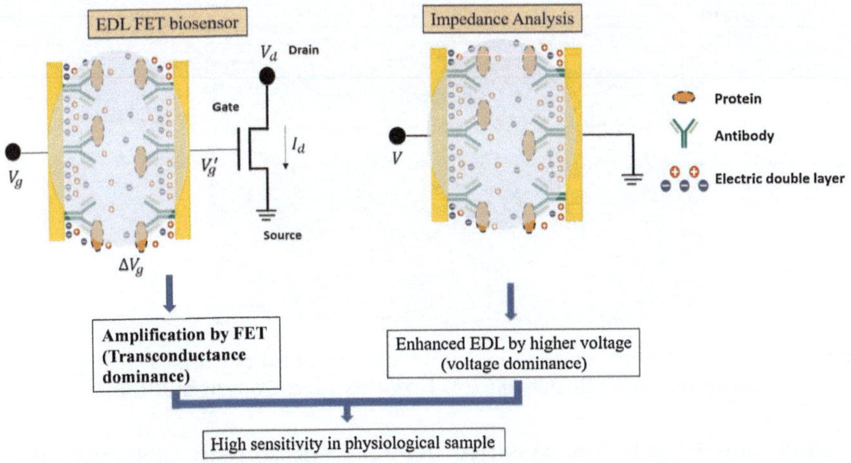

Figure 7.5 Scheme of an enhanced electric-double-layer (EnEDL) field-effect transistor and impedance analysis. Reproduced from ref. 70 with permission from Elsevier, Copyright 2021.

7.3 AlGaN/GaN High Electron Mobility Transistors (HEMT) for Affinity Sensing

Compared to silicon-based field effect transistors, GaN-based high electron mobility transistors (HEMT) present several advantages for affinity sensing:

- an exceptional chemical and physical stability of GaN, allowing its use in liquid media without passivation;
- a very high two-dimensional electron gas (2DEG), due to high polarization in AlGaN/GaN heterostructures (Figure 7.6), results in high sensitivity;
- being environment-friendly and bio-friendly with non-toxic surface content, GaN performs well for interaction with biological species;
- GaN-based sensors do not need surface passivation and respond to polar molecules instantaneously.

Since 2015, nine papers have been published on AlGaN/GaN HEMT for the detection of biomarkers of different diseases: bacterial infection,[72] Zika virus,[73] prostate cancer,[74,75] heart failure,[76] mental stress[77] and DNA.[78]

In ref. 73 and 76, electric-double-layer HEMT were fabricated to defeat the severe charge-screening effect caused by high ionic strength in solution. The antibodies or the aptamers were grafted on the gold electrodes in ref. 74, where the authors used a disposable, extended gate. In order to enlarge the sensing capacity beyond the Debye-screening limit, Wang *et al.*[75] used magnetic beads to immobilize the antibody. An electronic ELISA test was then performed for PSA detection. A detection limit of 1 fg mL^{-1} was obtained, which is one hundred times lower than that obtained with the

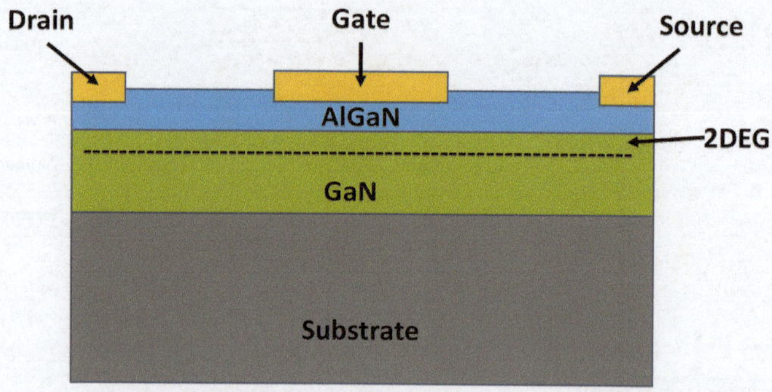

Figure 7.6 Scheme of an AlGaN/GaN high electron mobility transistor. Reproduced from ref. 38 with permission from Elsevier, Copyright 2021.

extended gate HEMT.[74] A sensitive detection of cortisol (detection limit: 1 pM) was achieved by the generation of a photocurrent through laser illumination of the HEMT sensor[77]).

7.4 Organic Field Effect Transistors (OFETs)

The organic field effect transistor (OFET) structure resembles that of the basic silicon-based field-effect transistors, where silicon is replaced by an organic semiconductor (OSC) and the gate is biased through the substrate for back-gated OFETs (Figure 7.7(a)). Another configuration is the electrolyte gated OFET (EGOFET) in which the gate is applied to the electrolyte (Figure 7.7(b)). The recognition molecule (mainly antibody or ss-DNA) is immobilized on top of the OSC layer. Some OFETs with an extended gate (EG) allow the immobilization of the receptor molecule on the extended gate, as presented in Figure 7.7(c). Several works published since 2015 were based on EG-OFETs.[79–83] Different types of organic semiconductors were recently used: P3HT (poly-3-hexylthiophenepentacene),[84–86] PBTTT (poly(2,5-bis(3-hexadecylthiophene-2-yl)thieno[3,2-*b*]thiophene)),[79] pentacene,[80,81] TIPS-pentacene (6,13-bis(triisopropylsilylethynyl)pentacene),[87,88] poly(DPP-DTT) (poly(*N*-alkyldiketopyrrole dithienylthienol[3,2-*b*]thiophene)),[89] PDVT-8,[90] diF-TES-ADT (2,8-difluoro-5,11-bis(triethylsilylethynyl)anthradithiophene),[91] and PANI.[92]

Since 2015 there have been twelve papers published on OFET-based affinity sensors for the detection of biomarkers of different diseases: bacterial infection,[84,85] endocrine tumors,[79] sepsis,[86] brain injuries,[80] mental stress,[81,87] the immune response against drugs,[88] liver cancer,[90] rheumatoid arthritis,[91] and rectal colon cancer.[92] One EGOFET based on immobilized algae in a hydrogel allows the detection of herbicides.[89]

Several OFET-based affinity sensors could detect the biomarkers with higher sensitivity, even in high ionic strength media such as serum. To increase the Debye screening length beyond bioanalyte/bioreceptor complex dimensions, a phospholipid layer,[70] a polyethylene glycol (MW 5000) layer[80] or a poly(styrene-*co*-methacrylic acid) (PSMA) layer[81] were used to change

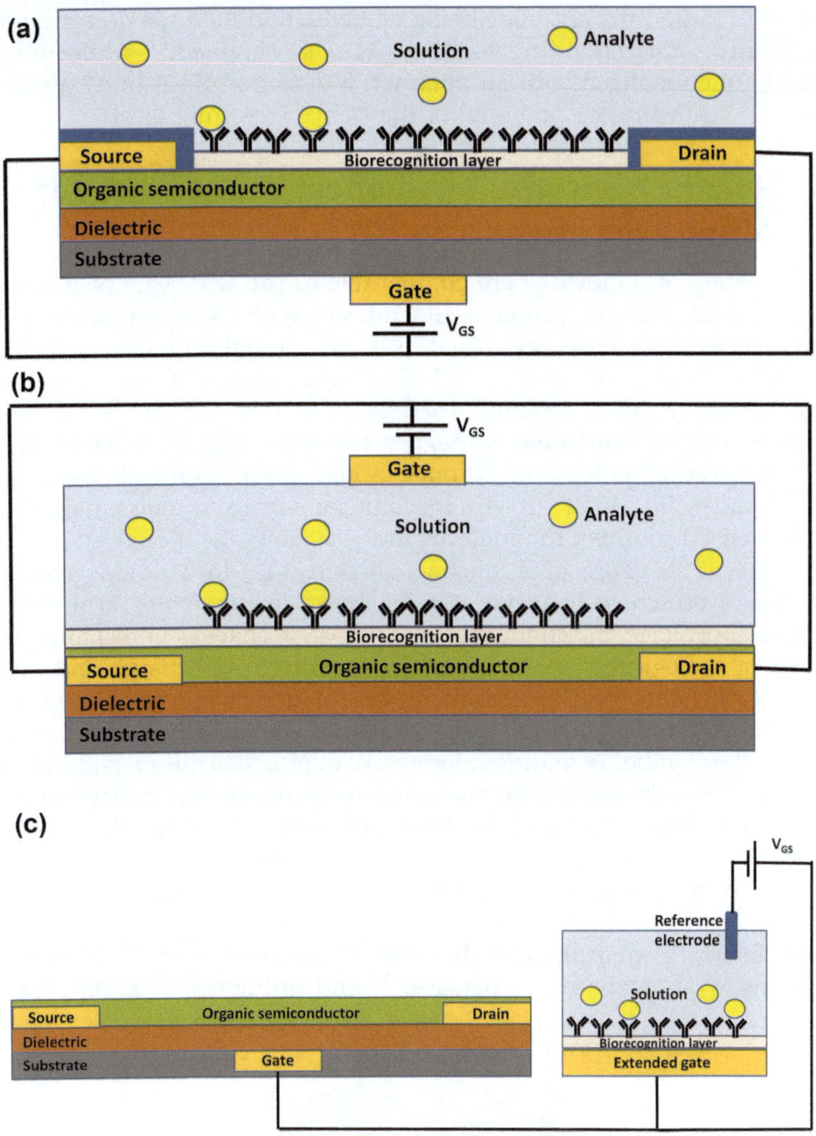

Figure 7.7 (a) Schematic representation of an organic FET (back-gated). (b) Schematic representation of an electrolyte-gated organic FET structure. (c) Schematic structure of an extended-gate type OFET. Reproduced from ref. 41 with permission from Elsevier, Copyright 2019; reproduced from ref. 79, https://doi.org/10.3390/s16122033, under the terms of the CC BY 4.0 license, http://creativecommons.org/licenses/by/4.0/.

the properties in aqueous solution and then to increase the "effective Debye screening length" by keeping a low ion density environment for the binding protein. A larger drain current could then be obtained. When cortisol was detected in buffer, the antibody being embedded in PSMA, a detection limit of 1 pg mL^{-1} corresponding to 2.75 pM was obtained,[81] while with an EGOFET functionalized with an aptamer, a detection limit of 27.3 pM was obtained,[87] showing the potential of the Debye screening layer.

7.5 Silicon Nanowire Field-effect Transistors for Affinity Sensing

The diameters of nanowires are comparable to the size scale of many biological species, such as proteins and nucleic acids. The nanowires (NWs) configured as FETs have been used as sensitive real-time electricity-based detectors of biological species.[17,39] As a "gold standard semiconductor", silicon presents high mobility carriers (100–1500 cm^2 V^{-1} s^{-1}). Silicon nanowires can be fabricated through a top-down approach based on the CMOS fabrication process or through a bottom-up approach based on a vapor–liquid–solid (VLS) growth mechanism where a metal nanocluster catalyst is used to direct the material growth along one direction.

The typical structure of a silicon nanowire FET in a liquid gated configuration is presented in Figure 7.8. As shown in this figure, at a constant source–drain voltage, the conductance of the SiNW changes in response to the binding of biological species, depending on their own charge and on the doping of the nanowire. The sensitivity of the nanowire is critically dependent on its diameter; the higher sensitivity for smaller diameters is due to the larger surface–volume ratio. In addition, lower NW doping densities result in higher sensitivity.[39] For the SiNW FET, the native oxide on the NW surface serves as the gate oxide (Figure 7.8), and the thinner the oxide, the higher the sensitivity.

Since 2015, fourteen papers have been published on SiNW FETs for the detection of biomarkers of different diseases: bladder cancer,[94] oral cancer,[95] lung cancer,[96,97] COVID-19,[98] prostate cancer,[99] vascular endothelial growth factor,[100] neurological disorder,[101] heart disease,[102,103] parasitic infections,[104,105] Alzheimer's disease,[106] and influenza.[107] Except for one

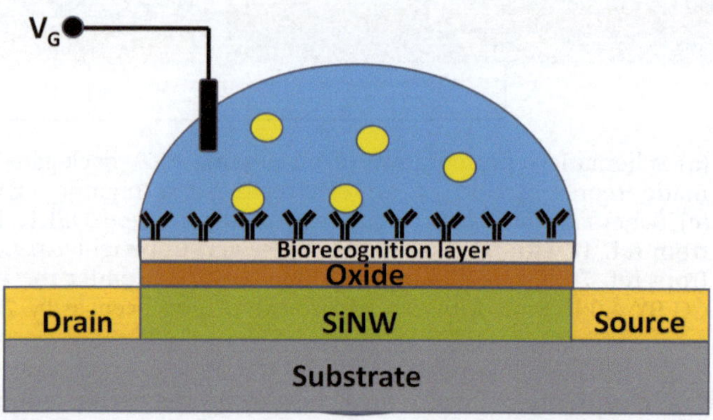

Figure 7.8 Scheme of a silicon nanowire field-effect transistor. Reproduced from ref. 39 with permission from Elsevier, Copyright 2021.

paper that presents a bottom-up approach for the VLS-CVD preparation of Si NWs,[103] in the other papers, Si NWs were prepared through a top-down approach in CMOS technology. All these biomarkers were detected in diluted ionic media, except in two articles where the detection was carried out in a high-ionic strength environment. In ref. 103, the chemically-controlled antigen-dissociation detection approach was applied. A suitable "chemical environment" (addition of ethylene glycol), which enabled the clear-up splitting of the dissociation regime window into two sub-regimes, was created, thus allowing the complete washing of the nonspecifically adsorbed salts and biomolecules while significantly delaying the dissociation of the specific antigen–antibody pairs on the SiNW surface. It was then demonstrated the direct and quantitative detection of protein biomarkers (cTnT and cTnI), down to concentrations in the fM range, from unprocessed whole minuscule samples of only a few microliters.

A modification of a SiNW surface with a porous and biomolecule-permeable polymer layer (PEG) was used to increase the effective Debye screening length and then the detection sensitivity.[105] In the first step, the replacement of the whole anti-IgG antibody (Wab) by a reduced anti-IgG antibody (Fab) improves the sensitivity of detection after amplification by the R18 aptamer. The Fab/PEG-SiNWFET immunosensor exhibited a lower detection limit of 1 pg mL^{-1}, not achievable with the other configurations of the SiNW FET interface.

7.6 Field Effect Transistors Based on Other Inorganic Nanomaterials for Affinity Sensing

Layers of the 2D material molybdenum disulfide (MoS_2) are held together by weak van der Waals forces, and a pristine MoS_2 monolayer is only ~0.65 nm thick. Monolayer MoS_2 exhibits a direct energy bandgap of ~1.9 eV, which lowers leakage current and makes it an emerging material for designing highly sensitive FET biosensors. In all the works published since 2015, multilayers of MoS_2 were grown or deposited as exfoliated flakes on Si/SiO_2. An Al_2O_3 nanofilm was deposited on the MoS_2 layer as a dielectric layer to protect the device from liquid erosion and enhance the electrostatic charge effect of the target molecules. Back-gated FET structures were prepared. Biomarkers of different diseases were detected with these MoS_2-based FETs, such as mental stress,[108] cardiovascular disease,[109] colonic schistosomiasis,[110] and prostate cancer.[111] Detection limits in the fM range were obtained.

Nanostructured zinc oxide (ZnO) is a semiconductor with a wide bandgap energy of 3.37 eV. ZnO is a suitable material for biosensing applications due to its biocompatibility, low toxicity to humans, and high isoelectric point. The tuning of the surface defects in an electrolyte-gated ZnO FET biosensor by modifying the experimental conditions of electron beam evaporation allowed to ensure a low detection limit of PSA, the biomarker of prostate cancer, and to assure a reliable long-term performance.[112]

An amorphous indium gallium zinc oxide (a-IGZO) bilayer was fabricated by changing the partial pressure of O_2 in the chamber to fabricate bilayers with different amounts of O vacancies. The low-O vacancy layer, on top of the active layer, provides a self-passivation effect, without extra passivation layers.[113] After the immobilization of biotin, the electrolyte-gated FET, through APTES grafting on the a-IGZO surface, allowed streptavidin detection with a detection limit of 1 pg mL^{-1}, with good selectivity.

A microfluidic field-effect transistor biosensor made of a rolled-up indium nitride microtube was designed in order to reduce the electrostatic screening of the biosensing surface and then prove the device's sensitivity for charge detection in electrolytic environments.[114] Multiple rolled-up InN microtube FET biosensors fabricated on the same substrate were integrated with a microfluidic channel for convenient fluid handling and shared the same external electrode (inserted into the microchannel outlet) for gating voltage modulation.

The detection limit for the HIV-1 pg41 antibody, as a model disease marker, in 0.01 × PBS solution is 0.1 ng mL^{-1}, which is 20 times lower than that obtained with the standard ELISA kit.

Nanostructured indium oxide (In_2O_3) is a semiconductor with a bandgap energy of 2.8 eV. The affinity of an aptamer for dopamine was determined in artificial cerebrospinal fluid, showing that the identified aptamer adopts a conformation optimized for indium oxide-based FET detection of dopamine when Mg^{2+} and Ca^{2+} are present. Moreover, this aptamer was found to show a higher sensitivity for dopamine *versus* norepinephrine, as well as other metabolites. The detection limit for dopamine was found to be 10 fM.[115]

7.7 Carbon Nanotube Field-effect Transistors for Affinity Sensing

Carbon nanotubes can be obtained with metallic or semi-conducting properties. For the fabrication of CNT-FETs, CNTs are deposited from a dispersion of CNTs[116–122] or CNTs are directly grown on the Si/SiO_2 substrate.[118,123–127]

Horizontally aligned p-type CNTs are obtained by a controlled chemical vapor deposition (CVD) growth,[123] and suspended CNTs between platinum electrodes were fabricated by utilizing the surface tension of liquid silver.[124] The resistance of the CNTs increased, leading to a decrease in the drain current of the liquid-gated CNT-FET when the hybridization of DNA increased.[124] The same effect was obtained with the immunodetection of interleukin 6.[123]

When a dispersion of CNTs in dichlorobenzene is deposited on a Kapton surface, a network of CNTs is formed. The network consists predominantly of CNT bundles with some individual CNTs. The small number of metallic (m) tubes and semiconducting tubes of different bandgaps (s′) than the principal s-CNT lead to the formation of m–s junctions and s–s′ junctions in the network in addition to m–m and s–s connections. The detection of

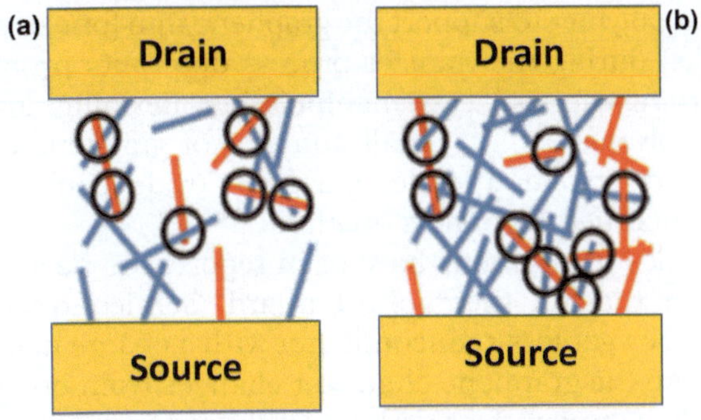

Figure 7.9 Schematic of the formation of metal–semiconducting CNT bundle junctions on a (a) sparse (10 min) and (b) dense CNT network (80 min). Bundles with only semiconducting tubes are in blue and those with metallic tubes are in red. The metal–semiconducting junctions are circled. Reproduced from ref. 128 with permission from Elsevier, Copyright 2019.

potassium ions through a liquid-gated CNT-FET functionalized with a specific aptamer presents a sensitivity of response depending on the time of deposition of the CNT dispersion. The sensitivity of detection has been related to the increased dominance of key m–s junctions on the conductance of the network close to percolation, as shown in the scheme presented in Figure 7.9.[128]

The CNT-FETs for affinity sensing described in the papers published since 2015 allow the detection of biomarkers of different diseases: acute respiratory syndrome,[123] cancer,[118] oral cancer,[129] and acute myocardial infarction.[120] Other CNT-FETs allow the detection of microorganisms such as bacteria[121] and COVID-19.[122] Hazardous priority substances were detected by CNT-FETs, including alkylphenol[116] and atrazine.[117]

7.8 Graphene Field-effect Transistors for Affinity Sensing

Due to the extraordinary electrical properties of graphene, since 2015, more than fifty papers have reported affinity sensors based on graphene field-effect transistors (G-FETs). Graphene is made up of carbon atoms bound to three others with a 120° bond angle, resulting in a hexagonal lattice arrangement of sp^2-hybridized carbon with a high specific surface area (2630 $m^2\,g^{-1}$). For the CVD fabrication of graphene films, the carbon atoms decomposed from diverse hydrocarbon sources at high temperatures diffuse on the metal substrate to form initial nuclei, which further expand and coalesce to form graphene layers. The growth procedure will end once the substrates are fully covered with a graphene film, which is mainly referred to as the so-called "self-limiting surface reaction mechanism" that has been observed in Cu and on Ni. After one decade of research efforts, PMMA (poly(methyl methacrylate)) is still the most widely used polymer to support the graphene film (one or two layers, with minimum cracks) during the transfer process due to its prominent features like transparency, cheapness, easy manipulation, flexibility and solubility in several organic solvents.[3,130–143] Small domains of graphene can also be obtained as reduced graphene oxide, graphene oxide being obtained from graphite by the modified Hummers' method.[144–154]

Carrier mobilities in graphene have been reported to be about 2 orders of magnitude larger than in the "gold-standard" semiconductor silicon. Intrinsic graphene is a gapless semiconductor with a charge neutrality point or Dirac point (where the graphene character changes from being electron-like to hole-like) at the Fermi surface with no free carriers at $T = 0$ K. This system has an empty conduction band and a filled valence band; infinitesimal doping using an external gate voltage (back-gate (Figure 7.10(A)) or liquid-gate (Figure 7.10(B)) in a field-effect transistor generates the electron transition from the valence to the conductance band, thereby changing the intrinsic state to the extrinsic one. By varying the external gate voltage, the system is tuned from being electron-like to being hole-like through its intrinsic nature at the charge neutrality point (CNP), then transducing graphene's ambipolar behavior (Figure 7.10(C)).

A graphene biosensor-based FET works mainly through the electrostatic gating effect, which is defined as a consequence of an additional effective gate voltage formed by the collection of the adsorbed biomolecules within

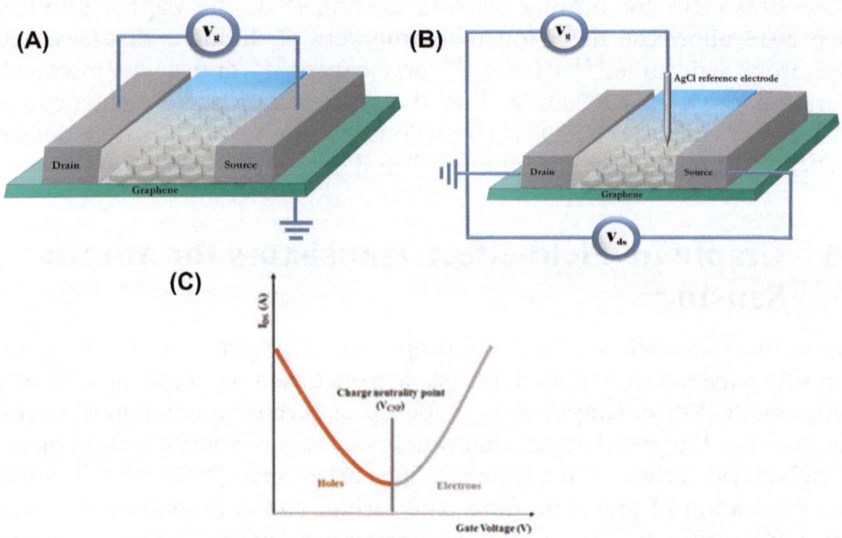

Figure 7.10 (A) Back-gated graphene field effect transistor. (B) Liquid-gated graphene field effect transistor. (C) Typical ambipolar transfer characteristics of a G-FET showing that two types of carriers in graphene can continuously be modulated from holes (red color) to electrons (grey color) using the field effect. The V_{CNP} is located at the transition between the hole and electron regime.

the Debye screening length. Such drain current variation can be used as a sensing metric for biomolecule adsorption. Another type of sensing metric is to follow the transconductance variation through the slope of the transfer curve's linear part of the p or n branch. Additionally, the ambipolar behavior of the G-FET devices offers a supplementary sensing mechanism through the measurement of the Dirac point shift (ΔV_D) or the charge neutrality point (ΔV_{CNP}) as a function of the surface charge-induced concentration.

For the immobilization of the receptor molecules, covalent functionalization of the graphene surface causes considerable variation in the graphene structure, including the changes of electronic properties caused by the disruption of π electrons and transformation of the carbon atom hybridization from the sp^2- to sp^3-hybridized state. The non-covalent functionalization using specific aromatic linker molecules like PBASE (1-pyrenebutanoic acid succinimidyl ester) (Figure 7.11), which are possibly anchored onto the graphene surface *via* π–π interaction, does not alter the electrical properties of graphene or its structural integrity.

A positive shift of the Dirac point is observed (Figure 7.12), explained by a p-doping effect *via* charge transfer between the pyrene unit and graphene. A positive shift of 0.20 V was observed by Seo *et al.*,[3] and a more positive shift is observed after the immobilization of the SARS-CoV-2 spike antibody.

In addition to diluting the buffer salt concentration, co-immobilization using polymers such as polyethylene glycol (PEG) could locally reduce the Debye screening effect in the vicinity of the biorecognition molecules. Following this strategy, Andoy *et al.*[134] used graphene modified with an aptamer and PEG for detecting thyroid-stimulating hormone (TSH). The developed G-FETs exhibited a very low LOD ($\simeq 0.2 \times 10^{-15}$ M) and a high sensitivity toward THS protein in high ionic strength and whole serum.

Nonetheless, designing short receptors is still preferable for several research groups. It is an effective and straightforward method to reduce the distance separating the transistor surface and biomolecule analytes. This point was demonstrated by Kanai *et al.*,[155] who used a variable antibody fragment to detect a small antigen peptide through an open-sandwich immunoassay based on a liquid-gated G-FET. The sensitivity was increased by a factor of 10 compared to an ELSA test.

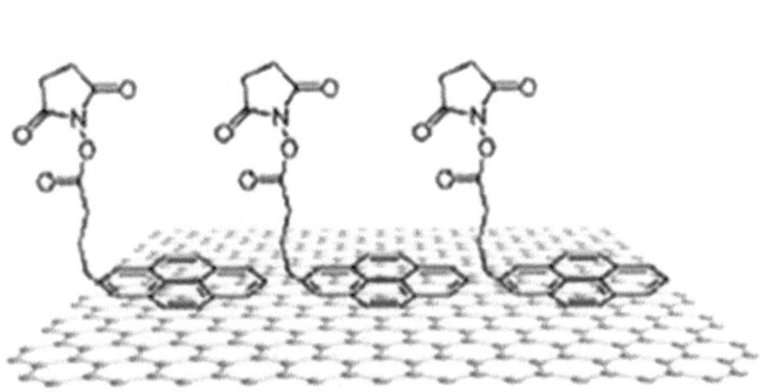

Figure 7.11 The schematic diagram of graphene functionalization using the specific aromatic linker molecule PBASE.

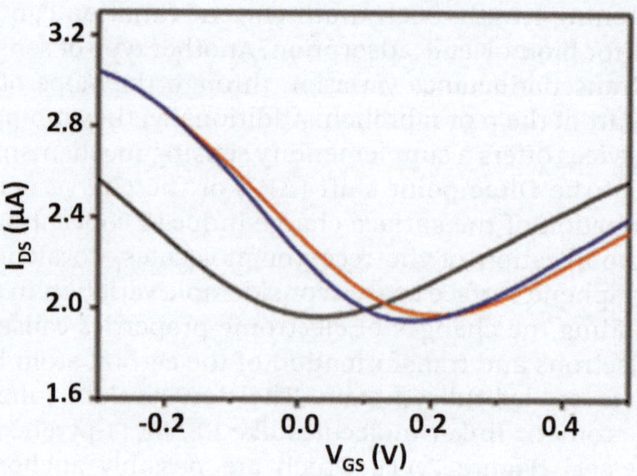

Figure 7.12 Electrical characterization of pristine, PBASE-modified and SARS-CoV-2 antibody-immobilized graphene. Measurement of the transfer curve of the COVID-19 FET sensor in steps of the antibody immobilization ($V_{DS} = 0.01$ V). Black curve: pristine graphene, red curve: PBAS-modified graphene, blue curve: SARS-CoV-2 antibody-PBAS-modified graphene. Reproduced from ref. 3 with permission from American Chemical Society, Copyright 2020.

An extremely low LOD of 600 zM in buffer solution was recorded by Hwang *et al.*,[139] who developed a deformed (crumpled) graphene monolayer channel for miRNA detection based on FET biosensors. The critical factor in reaching such lower LOD values depends on the following strategy to overcome the Debye length screening effect. Hwang *et al.* combined the benefits of PNA's uncharged and short structure with the advantage of a crumpled graphene surface, which allows a larger shift of the Dirac point, compared to a flat graphene surface.

The numerous graphene FETs developed since 2015 allow the detection of biomarkers of different diseases, such as tuberculosis,[156,157] cancer,[137,141,147,148,150,158,159] trophoblastic tumors,[131,132,160] mental stress,[161–163] hyperthyroidism,[134] cardiovascular disorder,[136] rheumatoid arthritis,[136] HIV,[136] bone disease,[155] Alzheimer's disease,[126,164,165] bacterial metabolic volatile indole,[166] and heart failure.[153] Some of them allow the detection of viruses, such as hepatitis B,[146,167] COVID-19,[3,151,168] Japanese encephalitis,[169] avian influenza,[169] and human papillomavirus;[152] the detection of toxins,[170] such as aflatoxin B1,[145] okadaic acid,[171] and geosmin;[172] the detection of drugs[173] and the detection of pesticides, such as chlorpyrifos.[174] Many G-FETs were used for the detection of DNA sequences,[139,140,149,175–184] miRNA,[144] and RNA.[133]

When comparing the analytical performance of G-FETs for the detection of cortisol, it is observed that an electrolyte-gated G-FET based on graphene nanoplatelets functionalized with a monoclonal cortisol antibody[161] gave a detection limit of 10 pg mL^{-1} and a dynamic range of 10 pg mL^{-1}–10 μg mL^{-1}; another electrolyte-gated G-FET based on graphene nanoplatelets functionalized with a mouse monoclonal cortisol antibody[162] gave a detection limit of 0.85 ng mL^{-1} and a dynamic range of 1 pg mL^{-1}–10 ng mL^{-1} and an extended gate G-FET based on a platinum extended gate electrode covered with single-layer graphene decorated with a 61-basepair aptamer[163] gave a detection limit of 0.2 nM (72.4 ng mL^{-1}) and a dynamic range of 1 nM (362 pg mL^{-1})–10 μM (3.62 μg mL^{-1}). In the third G-FET, the use of a short aptamer allows detection in the Debye layer; nevertheless, the same amplitude (3 decades) of dynamic range was also obtained with the antibody-functionalized G-FET,[171] which can be explained by the Donnan effect.[2]

For the detection of the hepatitis B virus, three G-FETs were functionalized with a monoclonal antibody; two of them were based on graphene nanogrids (graphene deposited on nanoporous silicon oxide through the electrophoretic deposition method) and led to a limit of detection of 0.05 fM[167] and 0.1 fM[146] Hep-B virus, and one is based on the deposition of ERGO on FTO covered glass and led to a detection limit of 1 fM[185] due to a lower quality of graphene.

7.9 Comparison of the Detection Limits Obtained with Different Types of FET-based Affinity Sensors

Cortisol, a biomarker of mental stress, was detected with different types of FETs: with a silicon-based FET, a detection limit of 50 μM was obtained;[186] with an AlGaN/GaN FET, a detection limit of 1 pM was obtained;[77] with an OFET detection limits of 2.76 pM[81] and 27.3 pM[187] were obtained; with a MoS_2 based FET, a detection limit of 2.76 aM was obtained;[108] with a graphene-based FET, detection limits of 27.6 pM,[161] 2.34 pM[162] and 0.2 nM[163] were obtained. This comparison shows that the silicon-based FET gives a detection limit in the micromolar range, while the other FET gives a detection limit in the picomolar range. A very low detection limit was obtained with the MoS_2-based FET in the attomolar range.

Prostate-specific antigen (PSA) is a biomarker of prostate cancer; its molecular weight is 30 000 Daltons. It was detected with different types of FETs: with an AlGaN/GaN FET detection limits of 100 fg mL^{-1} [74] and 1 ng mL^{-1} [188] were obtained; with a SiNW FET, a detection limit of 23 fg mL^{-1} [199] was obtained; with a MoS_2 based FET, a detection limit of 1 fg mL^{-1} was obtained;[111] with a ZnO FET, a detection limit of 0.06 fM (2.4 ag mL^{-1}) was obtained;[112] with a graphene FET, a detection limit of 1 pg mL^{-1} was obtained.[150] This comparison shows that lower detection limits are obtained with the nanomaterial-based FET, in the range of femtogram per milliliter, the lower one being obtained with the ZnO FET.

The detection of DNA was performed with different types of FETs: with a silicon-based FET, a detection limit of 1 μM was obtained;[71] with an AlGaN/GaN FET, a detection limit of 10 fM was obtained;[78] with an OFET, a detection limit of 9.77 pM was obtained;[92] with a SWCNT FET a detection limit of 10 aM was obtained;[124] with graphene FETs, ten DNA sensors were reported with different detection limits of 10 nM,[180,183] 1 nM,[149] 1 pM,[179] 100 fM,[175] 2.3 fM,[189] 1 fM,[176,190] 10 aM,[140] and 600 zM.[139] This comparison shows that lower detection limits are obtained with SWCNT FETs and graphene FETs. Both carbonaceous nanomaterial-based FETs present a higher field effect for detecting DNA hybridization than FETs based on other nanomaterials. The lower detection limit was obtained with crumpled graphene,[146] with this morphology modifying the Debye length, which increases in the concave regions.[165]

7.10 Problems and Commercialization Perspectives

From a number of recently published works on affinity sensors based on field-effect transistors, graphene-FETs appear to be more popular. Due to the large-scale production of CVD-grown graphene on CU catalyst, the window towards industrial-scale production of high-quality graphene is opened. Graphene FET chips and back-gated OFET interdigitated substrates are commercialized for research by Sigma-Aldrich (Saint Louis, Missouri, USA), as well as an OFET miniprober. Graphene field-effect transistors (G-FETs) for sensing applications are also commercialized by MSE Supplies (Tucson, Arizona, USA).

The high sensitivity of FET-based biosensors makes them a promising platform for electronic biosensors. However, their wide application is limited by the accuracy and reliability of their sensing results. In order for biosensors to be valuable in both research and commercial settings, they must be able to identify target molecules, have suitable biometric components, and have the potential to be more widely used compared to current detection technologies (Figure 7.13).

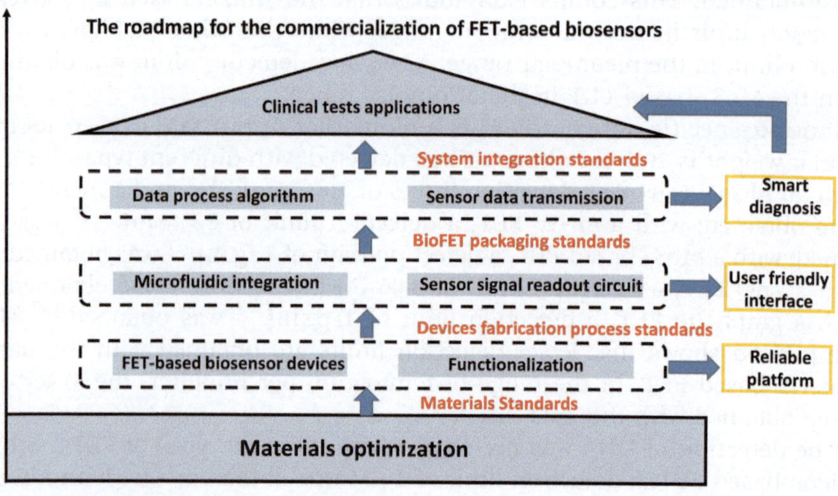

Figure 7.13 Schematic representation of the FET-based biosensor industrialization process.

The following engineering issues need to be addressed in order to commercialize FET-based biosensors:

1. The optimization of the device structure to reduce fluctuations and improve its signal-to-noise ratio (SNR), stability (caused by baseline drift in complex storage environments), and sustainability.
2. The development of a controlled surface biofunctionalization process for multiple target detection. Multiplex detection is particularly attractive in medical diagnostic scenarios, where more than one analyte can be used as a biomarker for a single disease state. For example, nanomaterials should be integrated with micro-biochips (labs-on-a-chip) for sample processing and analysis for multiple clinical diagnosis.
3. The improvement of the reproducibility and affordability of large-scale manufacturing.
4. The development of reliable, low-cost packing technology.
5. The addressal of the challenge of sensor calibration for various applications.
6. The simplification of user interventions.
7. The establishment of standard performance indicators of product quality.
8. The performance of extensive clinical testing to prove the reliability and safety of the product.
9. The addressal of the lifetime and baseline drift of the devices in complex storage environments, as well as the challenge of detecting multiple biomarkers for a single disease.

Among the many challenges, the consistency and reproducibility of the performance of FET-based biosensor devices is one of the most important challenges, as this plays a decisive role in the reliability and accuracy of the results. Therefore, every process—including materials synthesis, device fabrication, and biomodification—needs to be handled carefully (Figure 7.13). Another challenge is system integration (Figure 7.13), as current measurements have mainly been performed under ideal conditions such as pure buffer solutions, whereas real physiological samples are much more complex and can introduce interference and contamination effects.

In addition, the development of parallel and integrated biosensor arrays with specific functions through spatially resolved surface functionalization for multiplexing is still a challenge. However, FET-based biosensors offer several significant advantages over traditional methods such as PCR, mass spectrometry, and enzyme-linked immunosorbent assays (ELISAs), including simplicity, low cost, portability, ultra-high sensitivity, excellent selectivity, and label-free, real-time electrical detection. As these technical issues are addressed, it is expected that academic research will translate into commercially viable prototypes in the near future.

The future of FET-based biosensors is promising, with opportunities for the development of multi-targets and multi-functional arrays and the integration of microfluidic and signal processing circuits for self-adaptive disease diagnosis. The low-temperature fabrication process of graphene enables low-effort 3D integration and high-throughput processing of sensing data. These smart microsystems have the potential to revolutionize the way we diagnose and manage diseases.

7.11 Conclusion

This chapter shows that nanomaterial-based field effect transistors for the affinity sensing of biomarkers and DNA sequences present growing interest in the scientific community. Very low detection limits and very short response times, compared to classical ELISA tests, were obtained. Graphene FET chips and back-gated OFET interdigitated substrates are already commercialized for research. Nevertheless, many key points should be addressed before a large-scale commercialization of smart microsystems integrating multi-target arrays and microfluidic and signal processing circuits for disease diagnosis.

Acknowledgements

The authors acknowledge the financial support of the EU H2020 WIDESPREAD Program entitled Bionanosens grant agreement #951887.

References

1. P. Bergveld, *Sens. Actuators, B*, 2003, **88**, 1–20.
2. M. Kaisti, *Biosens. Bioelectron.*, 2017, **98**, 437–448.
3. G. Seo, G. Lee, M. J. Kim, S.-H. Baek, M. Choi, K. B. Ku, C.-S. Lee, S. Jun, D. Park, H. G. Kim, S.-J. Kim, J.-O. Lee, B. T. Kim, E. C. Park and S. I. Kim, *ACS Nano*, 2020, **14**, 5135–5142.
4. D. Barceló, *Case Stud. Chem. Environ. Eng.*, 2020, **2**, 100042.
5. R. Antiochia, *Microchim. Acta*, 2020, **187**, 639.
6. K. S. V. Santhanam, *Med. Devices Sens.*, 2021, **4**, e1014.
7. A. Poghossian, M. Jablonski, D. Molinnus, C. Wege and M. J. Schöning, *Front. Plant Sci.*, 2020, **11**, 598103.
8. A. Demeke Teklemariam, M. Samaddar, M. G. Alharbi, R. R. Al-Hindi and A. K. Bhunia, *Mol. Cell. Probes*, 2020, **54**, 101662.
9. S. Forouhi and E. Ghafar-Zadeh, *Micromachines*, 2020, **11**, 1003.
10. M. Srivastava, N. Srivastava, P. K. Mishra and B. D. Malhotra, *Sci. Total Environ.*, 2021, **754**, 142363.
11. S. Suleman, S. K. Shukla, N. Malhotra, S. D. Bukkitgar, N. P. Shetti, R. Pilloton, J. Narang, Y. Nee Tan and T. M. Aminabhavi, *Chem. Eng. J.*, 2021, **414**, 128759.
12. N. P. Shetti, A. Mishra, S. D. Bukkitgar, S. Basu, J. Narang, K. Raghava Reddy and T. M. Aminabhavi, *ACS Appl. Bio Mater.*, 2021, **4**, 1178–1190.

13. K. Y. Goud, K. K. Reddy, A. Khorshed, V. S. Kumar, R. K. Mishra, M. Oraby, A. H. Ibrahim, H. Kim and K. V. Gobi, *Biosens. Bioelectron.*, 2021, **180**, 113112.
14. A. Panahi, D. Sadighbayan, S. Forouhi and E. Ghafar-Zadeh, *Biosensors*, 2021, **11**, 103.
15. E. E. Etienne, B. B. Nunna, N. Talukder, Y. Wang and E. S. Lee, *Bioengineering*, 2021, **8**, 98.
16. G. Thriveni and K. Ghosh, *Biosensors*, 2022, **12**, 647.
17. D. Tran, T. Pham, B. Wolfrum, A. Offenhäusser and B. Thierry, *Materials*, 2018, **11**, 785.
18. M. Eswaran, B. Chokkiah, S. Pandit, S. Rahimi, R. Dhanusuraman, M. Aleem and I. Mijakovic, *Small Methods*, 2022, **6**, 2200809.
19. S. Wang, X. Qi, D. Hao, R. Moro, Y. Ma and L. Ma, *J. Electrochem. Soc.*, 2022, **169**, 027509.
20. A. Rhouati, A. Teniou, M. Badea and J. L. Marty, *Sensors*, 2021, **21**, 1485.
21. X. Chen, T. Dong, X. Wei, Z. Yang, N. M. Matos Pires, J. Ren and Z. Jiang, *Biosens. Bioelectron.*, 2019, **142**, 111453.
22. P. Salvo, F. M. Vivaldi, A. Bonini, D. Biagini, F. G. Bellagambi, F. M. Miliani, F. Di Francesco and T. Lomonaco, *Biosensors*, 2020, **10**, 155.
23. P. Sharma, A. Panchal, N. Yadav and J. Narang, *Int. J. Biol. Macromol.*, 2020, **155**, 685–696.
24. B. Li, H. Tan, D. Jenkins, V. Srinivasa Raghavan, B. G. Rosa, F. Güder, G. Pan, E. Yeatman and D. J. Sharp, *Carbon*, 2020, **168**, 144–162.
25. P. Mohankumar, J. Ajayan, T. Mohanraj and R. Yasodharan, *Measurement*, 2021, **167**, 108293.
26. S. Samota, R. Rani, S. Chakraverty and A. Kaushik, *Mater. Sci. Semicond. Process.*, 2022, **141**, 106404.
27. S. Dolati, J. Soleymani, S. Kazem Shakouri and A. Mobed, *Clin. Chim. Acta*, 2021, **514**, 107–121.
28. F. Farshchi and M. Hasanzadeh, *Biomed. Pharmacother.*, 2021, **134**, 111153.
29. K. Malecka, E. Mikuła and E. E. Ferapontova, *Sensors*, 2021, **21**, 736.
30. C. Ibau, M. K. Md Arshad and S. C. B. Gopinath, *Biosens. Bioelectron.*, 2017, **98**, 267–284.
31. D. Sadighbayan, M. Hasanzadeh and E. Ghafar-Zadeh, *TrAC, Trends Anal. Chem.*, 2020, **133**, 116067.
32. V. Kumar, K. Vaid, S. A. Bansal and K.-H. Kim, *Biosens. Bioelectron.*, 2020, **165**, 112382.
33. I. Ben Chabchoubi, N. Belkhamssa, M. Ksibi and O. Hentati, *Trends Environ. Anal. Chem.*, 2021, **30**, e00127.
34. S. Szunerits and R. Boukherroub, *Curr. Opin. Electrochem.*, 2018, **12**, 141–147.
35. I. Prattis, E. Hui, P. Gubeljak, G. S. Kaminski Schierle, A. Lombardo and L. G. Occhipinti, *Trends Biotechnol.*, 2021, **39**, 1065–1077.
36. M. Singh, N. Kaur and E. Comini, *J. Mater. Chem. C*, 2020, **8**, 3938–3955.
37. S. Kim and H. Yoo, *Micromachines*, 2021, **12**, 565.

38. K. T. Upadhyay and M. K. Chattopadhyay, *Mater. Sci. Eng., B*, 2021, **263**, 114849.
39. A. Zhang, J.-H. Lee and C. M. Lieber, *Nano Today*, 2021, **38**, 101135.
40. T. Sakata, S. Nishitani and T. Kajisa, *RSC Adv.*, 2020, **10**, 16999–17013.
41. I. M. Bhattacharyya, S. Cohen, A. Shalabny, M. Bashouti, B. Akabayov and G. Shalev, *Biosens. Bioelectron.*, 2019, **132**, 143–161.
42. C. Kotlowski, P. Aspermair, H. U. Khan, C. Reiner-Rozman, J. Breu, S. Szunerits, J.-J. Kim, Z. Bao, C. Kleber, P. Pelosi and W. Knoll, *Flexible Printed Electron.*, 2018, **3**, 034003.
43. T. Minamiki, Y. Sasaki, S. Su and T. Minami, *Polym. J.*, 2019, **51**, 1–9.
44. M. L. Coluccio, S. A. Pullano, M. F. M. Vismara, N. Coppedè, G. Perozziello, P. Candeloro, F. Gentile and N. Malara, *Micromachines*, 2020, **11**, 123.
45. R. A. Picca, K. Manoli, E. Macchia, L. Sarcina, C. Di Franco, N. Cioffi, D. Blasi, R. Österbacka, F. Torricelli, G. Scamarcio and L. Torsi, *Adv. Funct. Mater.*, 2020, **30**, 1904513.
46. X. Yao, Y. Zhang, W. Jin, Y. Hu and Y. Cui, *Sensors*, 2021, **21**, 995.
47. R. Forsyth, A. Devadoss and O. Guy, *Diagnostics*, 2017, 7, 45.
48. Laxmi, B. Mahapatra, R. V. Krishna and P. K. Patel, Erode, India, 2021, p. 020011.
49. A. Tiwari, *Graphene bioelectronics*, Elsevier, Amsterdam, Netherlands, 2018.
50. R. Ahmad, T. Mahmoudi, M.-S. Ahn and Y.-B. Hahn, *Biosens. Bioelectron.*, 2018, **100**, 312–325.
51. A. de Moraes and L. Kubota, *Chemosensors*, 2016, **4**, 20.
52. Vu and Chen, *Sensors*, 2019, **19**, 4214.
53. D. Sung and J. Koo, *Biomed. Eng. Lett.*, 2021, **11**, 85–96.
54. S. Cheng, S. Hideshima, S. Kuroiwa, T. Nakanishi and T. Osaka, *Sens. Actuators, B*, 2015, **212**, 329–334.
55. H.-J. Jang, J. Ahn, M.-G. Kim, Y.-B. Shin, M. Jeun, W.-J. Cho and K. H. Lee, *Biosens. Bioelectron.*, 2015, **64**, 318–323.
56. H. Hayashi, R. Toyama, R. Takibuchi, S. Hideshima, S. Kuroiwa, N. Kaneko, K. Horii, K. Ohashi, T. Momma and T. Osaka, *Electrochemistry*, 2021, **89**(2), 134–137.
57. K. K. Han, S. Mangmool, K. Sathirakul, K. Chansiri, W. Jeamsaksiri and M. Yasawong, *Thai J. Pharm. Sci.*, 2020, **44**(3), 159–167.
58. D. Vozgirdaite, H. Ben Halima, F. G. Bellagambi, A. Alcacer, F. Palacio, N. Jaffrezic-Renault, N. Zine, J. Bausells, A. Elaissari and A. Errachid, *Chemosensors*, 2021, **9**, 26.
59. H. Ben Halima, F. G. Bellagambi, A. Alcacer, N. Pfeiffer, A. Heuberger, M. Hangouët, N. Zine, J. Bausells, A. Elaissari and A. Errachid, *Anal. Chim. Acta*, 2021, **1161**, 338468.
60. H. Ben Halima, F. G. Bellagambi, F. Brunon, A. Alcacer, N. Pfeiffer, A. Heuberger, M. Hangouët, N. Zine, J. Bausells and A. Errachid, *Talanta*, 2022, 123802.
61. H. Ben Halima, F. G. Bellagambi, M. Hangouët, A. Alcacer, N. Pfeiffer, A. Heuberger, N. Zine, J. Bausells, A. Elaissari and A. Errachid, *Talanta*, 2023, **251**, 123759.

62. H. Ben Halima, F. G. Bellagambi, M. Hangouët, A. Alcacer, N. Pfeiffer, A. Heuberger, N. Zine, J. Bausells and A. Errachid, *Electroanalysis*, 2022, elan.202200141.
63. H. Ben Halima, N. Zine, J. Bausells, N. Jaffrezic-Renault and A. Errachid, *Micromachines*, 2022, **13**, 2235.
64. H. B. Halima, N. Zine, J. Gallardo-Gonzalez, A. E. Aissari, M. Sigaud, A. Alcacer, J. Bausells and A. Errachid, in *2019 20th International Conference on Solid-State Sensors, Actuators and Microsystems & Eurosensors XXXIII (TRANSDUCERS & EUROSENSORS XXXIII)*, IEEE, Berlin, Germany, 2019, pp. 1067–1070.
65. S. Hideshima, H. Hayashi, H. Hinou, S. Nambuya, S. Kuroiwa, T. Nakanishi, T. Momma, S.-I. Nishimura, Y. Sakoda and T. Osaka, *Sci. Rep.*, 2019, **9**, 11616.
66. Ó. Gutiérrez-Sanz, N. Haustein, M. Schroeter, T. Oelschlaegel, M. S. Filipiak and A. Tarasov, *Sens. Actuators, B*, 2019, **295**, 153–158.
67. C. Baldacchini, A. F. Montanarella, L. Francioso, M. A. Signore, S. Cannistraro and A. R. Bizzarri, *Sensors*, 2020, **20**, 6364.
68. L. Könemund, L. Neumann, F. Hirschberg, R. Biedendieck, D. Jahn, H.-H. Johannes and W. Kowalsky, *Sci. Rep.*, 2022, **12**, 4397.
69. M.-Á. García-Chamé, Ó. Gutiérrez-Sanz, E. Ercan-Herbst, N. Haustein, M. S. Filipiak, D. E. Ehrnhöfer and A. Tarasov, *Biosens. Bioelectron.*, 2020, **159**, 112129.
70. C.-R. Wu, S.-L. Wang, P.-H. Chen, Y.-L. Wang, Y.-R. Wang and J.-C. Chen, *Sens. Actuators, B*, 2021, **334**, 129567.
71. M. Kaisti, A. Kerko, E. Aarikka, P. Saviranta, Z. Boeva, T. Soukka and A. Lehmusvuori, *Sci. Rep.*, 2017, 7, 15734.
72. H. Lee, M. Bae, S.-H. Jo, J.-K. Shin, D. Son, C.-H. Won, H.-M. Jeong, J.-H. Lee and S.-W. Kang, *Sensors*, 2015, **15**, 18416–18426.
73. J. Yang, P. Carey, F. Ren, M. A. Mastro, K. Beers, S. J. Pearton and I. I. Kravchenko, *Appl. Phys. Lett.*, 2018, **113**, 032101.
74. S. Yang, L. Gu, X. Ding, B. Miao, Z. Gu, L. Yang, J. Li and D. Wu, *IEEE Electron Device Lett.*, 2018, **39**, 1592–1595.
75. J. Wang, Z. Gu, X. Liu, L. Zhao, H. Peng and J. Li, *Analyst*, 2020, **145**, 2725–2730.
76. T.-Y. Tai, A. Sinha, I. Sarangadharan, A. K. Pulikkathodi, S.-L. Wang, G.-Y. Lee, J.-I. Chyi, S.-C. Shiesh, G.-B. Lee and Y.-L. Wang, *Anal. Chem.*, 2019, **91**, 5953–5960.
77. K. Woo, W. Kang, K. Lee, P. Lee, Y. Kim, T.-S. Yoon, C.-Y. Cho, K.-H. Park, M.-W. Ha and H. H. Lee, *Biosens. Bioelectron.*, 2020, **159**, 112186.
78. N. Espinosa, S. U. Schwarz, V. Cimalla and O. Ambacher, *Sens. Actuators, B*, 2015, **210**, 633–639.
79. T. Minamiki, T. Minami, Y. Sasaki, S. Wakida, R. Kurita, O. Niwa and S. Tokito, *Sensors*, 2016, **16**, 2033.
80. J. Song, J. Dailey, H. Li, H.-J. Jang, P. Zhang, J. T.-H. Wang, A. D. Everett and H. E. Katz, *Adv. Funct. Mater.*, 2017, **27**, 1606506.

81. H.-J. Jang, T. Lee, J. Song, L. Russell, H. Li, J. Dailey, P. C. Searson and H. E. Katz, *ACS Appl. Mater. Interfaces*, 2018, **10**, 16233–16237.
82. H. J. N. P. D. Mello, B. B. Junior and M. Mulato, *Sens. Actuators, A*, 2021, **318**, 112481.
83. E. Yu. Poimanova, P. A. Shaposhnik, D. S. Anisimov, E. G. Zavyalova, A. A. Trul, M. S. Skorotetcky, O. V. Borshchev, D. Z. Vinnitskiy, M. S. Polinskaya, V. B. Krylov, N. E. Nifantiev, E. V. Agina and S. A. Ponomarenko, *ACS Appl. Mater. Interfaces*, 2022, **14**, 16462–16476.
84. G. Palazzo, D. De Tullio, M. Magliulo, A. Mallardi, F. Intranuovo, M. Y. Mulla, P. Favia, I. Vikholm-Lundin and L. Torsi, *Adv. Mater.*, 2015, **27**, 911–916.
85. M. Magliulo, D. De Tullio, I. Vikholm-Lundin, W. M. Albers, T. Munter, K. Manoli, G. Palazzo and L. Torsi, *Anal. Bioanal. Chem.*, 2016, **408**, 3943–3952.
86. P. Seshadri, K. Manoli, N. Schneiderhan-Marra, U. Anthes, P. Wierzchowiec, K. Bonrad, C. Di Franco and L. Torsi, *Biosens. Bioelectron.*, 2018, **104**, 113–119.
87. R. Massey, S. Bebe and R. Prakash, *IEEE Sens. Lett.*, 2020, **4**, 1–4.
88. M. Sensi, M. Berto, S. Gentile, M. Pinti, A. Conti, G. Pellacani, C. Salvarani, A. Cossarizza, C. A. Bortolotti and F. Biscarini, *Chem. Commun.*, 2021, **57**, 367–370.
89. J. L. Gall, S. Vasilijević, N. Battaglini, G. Mattana, V. Noël, R. Brayner and B. Piro, *Electrochim. Acta*, 2021, **372**, 137881.
90. C. Sun, R. Li, Y. Song, X. Jiang, C. Zhang, S. Cheng and W. Hu, *Anal. Chem.*, 2021, **93**, 6188–6194.
91. T. Minamiki, T. Minami, Y.-P. Chen, T. Mano, Y. Takeda, K. Fukuda and S. Tokito, *Commun. Mater.*, 2021, **2**, 8.
92. H. J. N. P. D. Mello, B. Bachour Junior and M. Mulato, *Sens. Actuators, A*, 2021, **318**, 112481.
93. T. Minamiki, T. Minami, R. Kurita, O. Niwa, S. Wakida, K. Fukuda, D. Kumaki and S. Tokito, *Appl. Phys. Lett.*, 2014, **104**, 243703.
94. H.-C. Chen, Y.-T. Chen, R.-Y. Tsai, M.-C. Chen, S.-L. Chen, M.-C. Xiao, C.-L. Chen and M.-Y. Hua, *Biosens. Bioelectron.*, 2015, **66**, 198–207.
95. Y. Zhang, R. Chen, L. Xu, Y. Ning, S. Xie and G.-J. Zhang, *Anal. Sci.*, 2015, **31**, 73–78.
96. A. Gao, N. Lu, Y. Wang and T. Li, *Sci. Rep.*, 2016, **6**, 22554.
97. A. Gao, X. Yang, J. Tong, L. Zhou, Y. Wang, J. Zhao, H. Mao and T. Li, *Biosens. Bioelectron.*, 2017, **91**, 482–488.
98. A. Wasfi, F. Awwad, J. G. Gelovani, N. Qamhieh and A. I. Ayesh, *Nanomaterials*, 2022, **12**, 2638.
99. G. Presnova, D. Presnov, V. Krupenin, V. Grigorenko, A. Trifonov, I. Andreeva, O. Ignatenko, A. Egorov and M. Rubtsova, *Biosens. Bioelectron.*, 2017, **88**, 283–289.
100. S. I. Zida, C.-C. Yang, Y. L. Khung and Y.-D. Lin, *J. Med. Biol. Eng.*, 2020, **40**, 601–609.
101. J.-H. Lee, E.-J. Chae, S. Park and J.-W. Choi, *Nano Convergence*, 2019, **6**, 13.

102. S.-M. Chang, S. Palanisamy, T.-H. Wu, C.-Y. Chen, K.-H. Cheng, C.-Y. Lee, S.-S. F. Yuan and Y.-M. Wang, *Sci. Rep.*, 2020, **10**, 22027.
103. M. Zverzhinetsky, V. Krivitsky and F. Patolsky, *Biosens. Bioelectron.*, 2020, **170**, 112658.
104. D. Rani, Y. Singh, M. Salker, X. T. Vu, S. Ingebrandt and V. Pachauri, *Anal. Bioanal. Chem.*, 2020, **412**, 6777–6788.
105. C.-A. Vu, P.-H. Pan, Y.-S. Yang, H. W.-H. Chan, Y. Kumada and W.-Y. Chen, *Sensors*, 2021, **21**, 650.
106. C.-A. Vu, W.-Y. Chen, Y.-S. Yang and H. W.-H. Chan, *Sens. Actuators, B*, 2021, **329**, 129150.
107. K. A. Malsagova, T. O. Pleshakova, A. F. Kozlov, R. A. Galiullin, V. P. Popov, F. V. Tikhonenko, A. V. Glukhov, V. S. Ziborov, I. D. Shumov, O. F. Petrov, V. M. Generalov, A. A. Cheremiskina, A. G. Durumanov, A. P. Agafonov, E. V. Gavrilova, R. A. Maksyutov, A. S. Safatov, V. G. Nikitaev, A. N. Pronichev, V. A. Konev, A. I. Archakov and Y. D. Ivanov, *Biosensors*, 2021, **11**, 119.
108. H. Park, S. Baek, B. Jeong, A. Sen, S. Kim, Y. C. Park, S. Kim and Y. J. Kim, Research Square, 2020, preprint, DOI: 10.21203/rs.3.rs-64919/v1.
109. N. R. Dalila, M. K. M. Arshad, S. C. B. Gopinath, M. N. M. Nuzaihan and M. F. M. Fathil, *Microchim. Acta*, 2020, **187**, 588.
110. M. Ma, L. Chao, Y. Zhao, J. Ding, Z. Huang, M. Long, F. Wang, J. Jiang and Z. Liu, *J. Phys. D: Appl. Phys.*, 2021, **54**, 245401.
111. Y. Zhang, D. Feng, Y. Xu, Z. Yin, W. Dou, U. E. Habiba, C. Pan, Z. Zhang, H. Mou, H. Deng, X. Mi and N. Dai, *Appl. Surf. Sci.*, 2021, **548**, 149169.
112. B. Chakraborty, R. Saha, S. Chattopadhyay, D. De, R. D. Das, M. K. Mukhopadhyay, M. Palit and C. RoyChaudhuri, *Appl. Surf. Sci.*, 2021, **537**, 147895.
113. H. Woo Son, J. H. Park, M.-S. Chae, B.-H. Kim and T. G. Kim, *Sens. Actuators, B*, 2020, **312**, 127955.
114. P. Song, H. Fu, Y. Wang, C. Chen, P. Ou, R. T. Rashid, S. Duan, J. Song, Z. Mi and X. Liu, *Biosens. Bioelectron.*, 2021, **190**, 113264.
115. N. Nakatsuka, J. M. Abendroth, K.-A. Yang and A. M. Andrews, *ACS Appl. Mater. Interfaces*, 2021, **13**, 9425–9435.
116. N. Belkhamssa, J. P. da Costa, C. I. L. Justino, P. S. M. Santos, S. Cardoso, A. C. Duarte, T. Rocha-Santos and M. Ksibi, *Talanta*, 2016, **158**, 30–34.
117. N. Belkhamssa, C. I. L. Justino, P. S. M. Santos, S. Cardoso, I. Lopes, A. C. Duarte, T. Rocha-Santos and M. Ksibi, *Talanta*, 2016, **146**, 430–434.
118. N. T. Tung, P. T. Tue, T. Thi Ngoc Lien, Y. Ohno, K. Maehashi, K. Matsumoto, K. Nishigaki, M. Biyani and Y. Takamura, *Sci. Rep.*, 2017, 7, 17881.
119. M. S. Filipiak, M. Rother, N. M. Andoy, A. C. Knudsen, S. Grimm, C. Bachran, L. K. Swee, J. Zaumseil and A. Tarasov, *Sens. Actuators, B*, 2018, **255**, 1507–1516.
120. S. G. Surya, S. M. Majhi, D. K. Agarwal, A. A. Lahcen, S. Yuvaraja, K. N. Chappanda and K. N. Salama, *J. Mater. Chem. B*, 2020, **8**, 18–26.

121. C.-H. Han and J. Jang, *Lab Chip*, 2021, **21**, 184–195.
122. M. Thanihaichelvan, S. N. Surendran, T. Kumanan, U. Sutharsini, P. Ravirajan, R. Valluvan and T. Tharsika, *Mater. Today: Proc.*, 2022, **49**, 2546–2549.
123. H. Chen, T. K. Choo, J. Huang, Y. Wang, Y. Liu, M. Platt, A. Palaniappan, B. Liedberg and A. I. Y. Tok, *Mater. Des.*, 2016, **90**, 852–857.
124. Y. Sun, Z. Peng, H. Li, Z. Wang, Y. Mu, G. Zhang, S. Chen, S. Liu, G. Wang, C. Liu, L. Sun, B. Man and C. Yang, *Biosens. Bioelectron.*, 2019, **137**, 255–262.
125. M. A. Zamzami, G. Rabbani, A. Ahmad, A. A. Basalah, W. H. Al-Sabban, S. Nate Ahn and H. Choudhry, *Bioelectrochemistry*, 2022, **143**, 107982.
126. H. Chen, M. Xiao, J. He, Y. Zhang, Y. Liang, H. Liu and Z. Zhang, *ACS Sens.*, 2022, 7, 2075–2083.
127. Y. Deng, L. Liu, J. Li and L. Gao, *Biosensors*, 2022, **12**, 776.
128. M. Thanihaichelvan, L. A. Browning, M. P. Dierkes, R. M. Reyes, A. V. Kralicek, C. Carraher, C. A. Marlow and N. O. V. Plank, *Biosens. Bioelectron.*, 2019, **130**, 408–413.
129. V. A. Pham Ba, Y. M. Han, Y. Cho, T. Kim, B. Y. Lee, J. S. Kim and S. Hong, *ACS Appl. Mater. Interfaces*, 2018, **10**, 17100–17106.
130. C. Zheng, L. Huang, H. Zhang, Z. Sun, Z. Zhang and G.-J. Zhang, *ACS Appl. Mater. Interfaces*, 2015, 7, 16953–16959.
131. K. Islam, A. Suhail and G. Pan, *Biosensors*, 2017, 7, 27.
132. B. Li, G. Pan, A. Suhail, K. Islam, N. Avent and P. Davey, *Carbon*, 2017, **118**, 43–49.
133. M. Tian, S. Xu, J. Zhang, X. Wang, Z. Li, H. Liu, R. Song, Z. Yu and J. Wang, *Adv. Condens. Matter Phys.*, 2018, **2018**, 1–6.
134. N. M. Andoy, M. S. Filipiak, D. Vetter, Ó. Gutiérrez-Sanz and A. Tarasov, *Adv. Mater. Technol.*, 2018, **3**, 1800186.
135. R. Campos, J. Borme, J. R. Guerreiro, G. Machado, M. F. Cerqueira, D. Y. Petrovykh and P. Alpuim, *ACS Sens.*, 2019, **4**, 286–293.
136. S. Islam, S. Shukla, V. K. Bajpai, Y.-K. Han, Y. S. Huh, A. Kumar, A. Ghosh and S. Gandhi, *Biosens. Bioelectron.*, 2019, **126**, 792–799.
137. D. Kwong Hong Tsang, T. J. Lieberthal, C. Watts, I. E. Dunlop, S. Ramadan, A. E. del Rio Hernandez and N. Klein, *Sci. Rep.*, 2019, **9**, 13946.
138. S. Wang, M. Z. Hossain, K. Shinozuka, N. Shimizu, S. Kitada, T. Suzuki, R. Ichige, A. Kuwana and H. Kobayashi, *Biosens. Bioelectron.*, 2020, **165**, 112363.
139. M. T. Hwang, M. Heiranian, Y. Kim, S. You, J. Leem, A. Taqieddin, V. Faramarzi, Y. Jing, I. Park, A. M. van der Zande, S. Nam, N. R. Aluru and R. Bashir, *Nat. Commun.*, 2020, **11**, 1543.
140. Y. Xia, Y. Sun, H. Li, S. Chen, T. Zhu, G. Wang, B. Man, J. Pan and C. Yang, *Talanta*, 2021, **223**, 121766.
141. S. Ramadan, R. Lobo, Y. Zhang, L. Xu, O. Shaforost, D. Kwong Hong Tsang, J. Feng, T. Yin, M. Qiao, A. Rajeshirke, L. R. Jiao, P. K. Petrov,

I. E. Dunlop, M.-M. Titirici and N. Klein, *ACS Appl. Mater. Interfaces*, 2021, **13**, 7854–7864.
142. S. Wang, M. Sun, Y. Zhang, H. Ji, J. Gao, S. Song, J. Sun, H. Liu, Y. Zhang and L. Han, *Anal. Chem.*, 2022, **94**, 14785–14793.
143. H.-J. Jang, X. Sui, W. Zhuang, X. Huang, M. Chen, X. Cai, Y. Wang, B. Ryu, H. Pu, N. Ankenbruck, K. Beavis, J. Huang and J. Chen, *ACS Appl. Mater. Interfaces*, 2022, **14**, 24187–24196.
144. B. Cai, L. Huang, H. Zhang, Z. Sun, Z. Zhang and G.-J. Zhang, *Biosens. Bioelectron.*, 2015, **74**, 329–334.
145. J. Basu, S. Datta and C. RoyChaudhuri, *Biosens. Bioelectron.*, 2015, **68**, 544–549.
146. J. Basu, N. Samanta, S. Jana and C. RoyChaudhuri, *Microelectron. Reliab.*, 2018, **91**, 154–159.
147. S. Mansouri Majd and A. Salimi, *Anal. Chim. Acta*, 2018, **1000**, 273–282.
148. C. Zhang, S. Cheng, K. Si, N. Wang, Y. Wang, P. Chen, H. Dong and W. Hu, *J. Mater. Chem. C*, 2020, **8**, 7436–7446.
149. S. Papamatthaiou, P. Estrela and D. Moschou, *Sci. Rep.*, 2021, **11**, 9815.
150. N. Mandal, V. Pakira, N. Samanta, N. Das, S. Chakraborty, B. Pramanick and C. RoyChaudhuri, *Talanta*, 2021, **222**, 121581.
151. J. Li, D. Wu, Y. Yu, T. Li, K. Li, M.-M. Xiao, Y. Li, Z.-Y. Zhang and G.-J. Zhang, *Biosens. Bioelectron.*, 2021, **183**, 113206.
152. P. Aspermair, V. Mishyn, J. Bintinger, H. Happy, K. Bagga, P. Subramanian, W. Knoll, R. Boukherroub and S. Szunerits, *Anal. Bioanal. Chem.*, 2021, **413**, 779–787.
153. I. Novodchuk, M. Kayaharman, I. R. Ausri, R. Karimi, X. S. Tang, I. A. Goldthorpe, E. Abdel-Rahman, J. Sanderson, M. Bajcsy and M. Yavuz, *Biosens. Bioelectron.*, 2021, **180**, 113114.
154. J. Sengupta and C. M. Hussain, *Carbon Trends*, 2021, **2**, 100011.
155. Y. Kanai, Y. Ohmuro-Matsuyama, M. Tanioku, S. Ushiba, T. Ono, K. Inoue, T. Kitaguchi, M. Kimura, H. Ueda and K. Matsumoto, *ACS Sens.*, 2020, **5**, 24–28.
156. S. Farid, X. Meshik, M. Choi, S. Mukherjee, Y. Lan, D. Parikh, S. Poduri, U. Baterdene, C.-E. Huang, Y. Y. Wang, P. Burke, M. Dutta and M. A. Stroscio, *Biosens. Bioelectron.*, 2015, **71**, 294–299.
157. Z. Wang, Z. Hao, X. Wang, C. Huang, Q. Lin, X. Zhao and Y. Pan, *Adv. Funct. Mater.*, 2021, **31**, 2005958.
158. L. Zhou, H. Mao, C. Wu, L. Tang, Z. Wu, H. Sun, H. Zhang, H. Zhou, C. Jia, Q. Jin, X. Chen and J. Zhao, *Biosens. Bioelectron.*, 2017, **87**, 701–707.
159. D. H. Kim, H. G. Oh, W. H. Park, D. C. Jeon, K. M. Lim, H. J. Kim, B. K. Jang and K. S. Song, *Sensors*, 2018, **18**, 4032.
160. C. Haslam, S. Damiati, T. Whitley, P. Davey, E. Ifeachor and S. Awan, *Diagnostics*, 2018, **8**, 5.
161. N. N. M. Maidin, R. A. Rahim, N. H. A. Halim, A. S. Z. Abidin, N. A. Ahmad and Z. Lockman, *AIP Conf. Proc.*, 2018, **2045**(1), 020022.

162. M. M. Nur Nasyifa, A. R. Ruslinda, N. H. Abdul Halim, A. S. Zainol Abidin, F. N. Mohd Faudzi, N. A. Ahmad, Z. Lockman, B. Rezek, A. Kromka and S. C. B. Gopinath, *J. Taiwan Inst. Chem. Eng.*, 2020, **117**, 10–18.
163. S. Sheibani, L. Capua, S. Kamaei, S. S. A. Akbari, J. Zhang, H. Guerin and A. M. Ionescu, *Commun. Mater.*, 2021, **2**, 10.
164. T. Bungon, C. Haslam, S. Damiati, B. O'Driscoll, T. Whitley, P. Davey, G. Siligardi, J. Charmet and S. A. Awan, in The 1st International Electronic Conference on Biosensors, MDPI, 2020, p. 14.
165. T. Bungon, C. Haslam, S. Damiati, B. O'Driscoll, T. Whitley, P. Davey, G. Siligardi, J. Charmet and S. A. Awan, *Front. Mol. Biosci.*, 2021, **8**, 651232.
166. Z. Lin, G. Wu, L. Zhao and K. W. C. Lai, *Nanomaterials*, 2021, **11**, 1155.
167. J. Basu and C. RoyChaudhuri, *Sensors*, 2016, **16**, 1481.
168. T.-R. Cui, Y.-C. Qiao, J.-W. Gao, C.-H. Wang, Y. Zhang, L. Han, Y. Yang and T.-L. Ren, *Molecules*, 2021, **26**, 6947.
169. A. Roberts, N. Chauhan, S. Islam, S. Mahari, B. Ghawri, R. K. Gandham, S. S. Majumdar, A. Ghosh and S. Gandhi, *Sci. Rep.*, 2020, **10**, 14546.
170. H. Gong, F. Chen, Z. Huang, Y. Gu, Q. Zhang, Y. Chen, Y. Zhang, J. Zhuang, Y.-K. Cho, R. H. Fang, W. Gao, S. Xu and L. Zhang, *ACS Nano*, 2019, **13**, 3714–3722.
171. J. Antunes, C. Justino, J. P. da Costa, S. Cardoso, A. C. Duarte and T. Rocha-Santos, *Microchem. J.*, 2018, **138**, 465–471.
172. S. J. Park, S. E. Seo, K. H. Kim, S. H. Lee, J. Kim, S. Ha, H. S. Song, S. H. Lee and O. S. Kwon, *Biosens. Bioelectron.*, 2021, **174**, 112804.
173. S. Xu, T. Wang, G. Liu, Z. Cao, L. A. Frank, S. Jiang, C. Zhang, Z. Li, V. V. Krasitskaya, Q. Li, Y. Sha, X. Zhang, H. Liu and J. Wang, *Sens. Actuators, B*, 2021, **326**, 128991.
174. S. Islam, S. Shukla, V. K. Bajpai, Y.-K. Han, Y. S. Huh, A. Ghosh and S. Gandhi, *Sci. Rep.*, 2019, **9**, 276.
175. G. Xu, J. Abbott, L. Qin, K. Y. M. Yeung, Y. Song, H. Yoon, J. Kong and D. Ham, *Nat. Commun.*, 2014, **5**, 4866.
176. C. Zheng, L. Huang, H. Zhang, Z. Sun, Z. Zhang and G.-J. Zhang, *ACS Appl. Mater. Interfaces*, 2015, **7**, 16953–16959.
177. M. T. Hwang, P. B. Landon, J. Lee, D. Choi, A. H. Mo, G. Glinsky and R. Lal, *Proc. Natl. Acad. Sci. U. S. A.*, 2016, **113**, 7088–7093.
178. J. Ping, R. Vishnubhotla, A. Vrudhula and A. T. C. Johnson, *ACS Nano*, 2016, **10**, 8700–8704.
179. S. Xu, J. Zhan, B. Man, S. Jiang, W. Yue, S. Gao, C. Guo, H. Liu, Z. Li, J. Wang and Y. Zhou, *Nat. Commun.*, 2017, **8**, 14902.
180. A. K. Manoharan, S. Chinnathambi, R. Jayavel and N. Hanagata, *Sci. Technol. Adv. Mater.*, 2017, **18**, 43–50.
181. J. Sun, X. Xie, K. Xie, S. Xu, S. Jiang, J. Ren, Y. Zhao, H. Xu, J. Wang and W. Yue, *Nanoscale Res. Lett.*, 2019, **14**, 248.
182. R. Hajian, S. Balderston, T. Tran, T. deBoer, J. Etienne, M. Sandhu, N. A. Wauford, J.-Y. Chung, J. Nokes, M. Athaiya, J. Paredes, R. Peytavi,

B. Goldsmith, N. Murthy, I. M. Conboy and K. Aran, *Nat. Biomed. Eng.*, 2019, **3**, 427–437.

183. Y. Zhang, Y. Ding, C. Li, H. Xu, C. Liu, J. Wang, Y. Ma, J. Ren, Y. Zhao and W. Yue, *Anal. Methods*, 2021, **13**, 1839–1846.
184. S. Balderston, J. J. Taulbee, E. Celaya, K. Fung, A. Jiao, K. Smith, R. Hajian, G. Gasiunas, S. Kutanovas, D. Kim, J. Parkinson, K. Dickerson, J.-J. Ripoll, R. Peytavi, H.-W. Lu, F. Barron, B. R. Goldsmith, P. G. Collins, I. M. Conboy, V. Siksnys and K. Aran, *Nat. Biomed. Eng.*, 2021, 1–13.
185. J. B. Pal, Detection of Biomolecules in High Ionic Strength physiological Solution Using Electrochemically Reduced Graphene oxide Field-Effect Transistor Immunosensor, 2020, https://wvvw.easychair.org/publications/preprint/p7mK (accessed June 17, 2021).
186. H. Hayashi, R. Toyama, R. Takibuchi, S. Hideshima, S. Kuroiwa, N. Kaneko, K. Horii, K. Ohashi, T. Momma and T. Osaka, *Electrochemistry*, 2021, **89**, 134–137.
187. R. Massey, S. Bebe and R. Prakash, *IEEE Sens. Lett.*, 2020, **4**, 1–4.
188. A. Varghese, C. Periasamy and L. Bhargava, *IEEE Trans. Nanotechnol.*, 2019, **18**, 747–755.
189. R. Hajian, S. Balderston, T. Tran, T. deBoer, J. Etienne, M. Sandhu, N. A. Wauford, J.-Y. Chung, J. Nokes, M. Athaiya, J. Paredes, R. Peytavi, B. Goldsmith, N. Murthy, I. M. Conboy and K. Aran, *Nat. Biomed. Eng.*, 2019, **3**, 427–437.
190. J. Ping, R. Vishnubhotla, A. Vrudhula and A. T. C. Johnson, *ACS Nano*, 2016, **10**, 8700–8704.

Subject Index